Theory of Probability

Sixth Edition

Boris V. Gnedenko (1912–1995)

Theory of Probability

Sixth Edition

Boris V. Gnedenko†
Moscow State University
Russia

Translated from the Russian by
Igor A. Ushakov
Qualcomm, Inc., San Diego, California

CRC Press
Taylor & Francis Group
Boca Raton London New York

CRC Press is an imprint of the
Taylor & Francis Group, an **informa** business

First Edition originally published in Russian about 1950, before the USSR was a member of the Universal Copyright Convention.

Sixth Edition originally published in Russian in 1988 as Курс теории вероятностей/*Course of Probability Theory* by Nauka, Moscow, USSR.

Sixth Edition expanded and updated.

First published in 1997 by Overseas Publisher Association

Published in 2020 by CRC Press
Taylor & Francis Group
6000 Broken Sound Parkway NW, Suite 300
Boca Raton, FL 3487-2742

First issued in paperback 2020

ISBN-13: 978-0-367-57931-9 (pbk)
ISBN-13: 978-9-05-699585-0 (hbk)

British Library Cataloguing in Publication Data

Gnedenko, B. V. (Boris Vladimirovich)
Theory of probability. — 6th ed.
1. Probabilities
I. Title
519.2

Contents

Translator's Preface ix

Author's Preface to the English Translation of the Russian Sixth Edition xi

Author's Preface to the Russian Sixth Edition xiii

From the Author's Preface to the Russian First Edition xv

Foreword xvii

Biography of Boris V. Gnedenko xix

1 INTRODUCTION 1

2 RANDOM EVENTS AND THEIR PROBABILITIES 5

- 1 Intuitive Understanding of Random Events 5
- 2 Sample Space. Classical Definition of Probability 8
- 3 Examples 16
- 4 Geometrical Probability 26
- 5 On Statistical Estimation of Unknown Probability 34
- 6 Axiomatic Construction of the Theory of Probability 37
- 7 Conditional Probability and the Simplest Basic Formulae 43
- 8 Examples 52
- 9 Exercises 59

3 SEQUENCES OF INDEPENDENT TRIALS 63

- 10 Introduction 63
- 11 The Local Limit Theorem 68
- 12 The Integral Limit Theorem 75
- 13 Applications of the Integral Theorem of DeMoivre–Laplace 84
- 14 Poisson's Theorem 90
- 15 Illustration of the Scheme of Independent Trials 95
- 16 Exercises 99

4 MARKOV CHAINS 105

17 Definition of a Markov Chain 105
18 Transition Matrix 106
19 Theorem on Limiting Probabilities 108
20 Exercises 112

5 RANDOM VARIABLES AND DISTRIBUTION FUNCTIONS 113

21 Fundamental Properties of Distribution Functions 113
22 Continuous and Discrete Distributions 120
23 Multidimensional Distribution Functions 125
24 Functions of Random Variables 134
25 The Stieltjes Integral 147
26 Exercises 153

6 NUMERICAL CHARACTERISTICS OF RANDOM VARIABLES 159

27 Mathematical Expectation 159
28 Variance 166
29 Theorems on Mathematical Expectation and Variance 172
30 Moments 179
31 Exercises 186

7 THE LAW OF LARGE NUMBERS 191

32 Mass Phenomenon and the Law of Large Numbers 191
33 Chebyshev's Form of the Law of Large Numbers 194
34 A Necessary and Sufficient Condition for the Law of Large Numbers 199
35 The Strong Law of Large Numbers 203
36 Glivenko's Theorem 210
37 Exercises 217

8 CHARACTERISTIC FUNCTIONS 219

38 The Definition and Simplest Properties of Characteristic Functions 219
39 The Inversion Formula and the Uniqueness Theorem 225
40 Helly's Theorem 231

41 Limit Theorems for Characteristic Functions 237
42 Positive–Semidefinite Functions 241
43 Characteristic Functions of Multi-Dimensional Random Variables 248
44 Laplace–Stieltjes Transform 253
45 Exercises 259

9 THE CLASSICAL LIMIT THEOREM 263
46 Statement of the Problem 263
47 Lindeberg's Theorem 266
48 The Local Limit Theorem 272
49 Exercises 279

10 THE THEORY OF INFINITELY DIVISIBLE DISTRIBUTIONS 281
50 Infinitely Divisible Distributions and Their Fundamental Properties 282
51 Canonical Representation of Infinitely Divisible Distributions 285
52 A Limit Theorem for Infinitely Divisible Distributions 291
53 Limit Theorems for Sums: Formulation of the Problem 294
54 Limit Theorems for Sums 296
55 Conditions for Convergence to the Normal and Poisson Distributions 300
56 Sum of a Random Number of Random Variables 303
57 Exercises 308

11 THE THEORY OF STOCHASTIC PROCESSES 311
58 Introduction 311
59 The Poisson Process 315
60 Death and Birth Processes 322
61 Conditional Distribution Functions and Bayes' Formula 333
62 The Generalized Markov Equation 337
63 Continuous Stochastic Processes. Kolmogorov's Equations 339
64 Purely Discontinuous Stochastic Process. The Kolmogorov–Feller Equations 348

65 Homogeneous Stochastic Processes with Independent Increments 357
66 The Concept of a Stationary Stochastic Process. Khinchine's Theorem on the Correlation Coefficient 363
67 The Notion of a Stochastic Integral. Spectral Decomposition of Stationary Processes 369
68 The Birkhoff–Khinchine Ergodic Theorem 373

12 ELEMENTS OF STATISTICS 379

69 Some Problems of Mathematical Statistics 379
70 The Classical Procedure for Estimating the Distribution Parameters 383
71 Exhaustive Statistics 394
72 Confidence Limits and Confidence Probabilities 396
73 Test of Statistical Hypothesis 403

APPENDIX: THE HISTORY OF PROBABILITY THEORY 413

Part 1 Main Concepts of Probability and Random Event 413

Part 2 Formation of the Foundation of Probability Theory 435

Part 3 Formation of the Concept of Random Variable 454

Part 4 History of Stochastic Processes 478

Tables of Function Values 485

References 489

Index 491

Translator's Preface

This book has been a traditional textbook on probability theory in Russia since it was published about 1950. Several generations of mathematical students all over the Soviet Union were taught by this book. Probably no other Russian textbook has been translated into so many languages: Bulgarian, Chinese, Czechoslovakian, English, French, German, Hungarian, Italian, Japanese, Polish, Portuguese, Romanian, Spanish, Swedish and Vietnamese.

Systematic, strong, deep and, at the same time, understandable presentation of material are distinctive features of the book. This is a rare example of a successful mathematical textbook written by an outstanding mathematician. Usually such books are overloaded by too-specific material or the author's personal achievements. The excellence of this book lies in the fact that Boris V. Gnedenko was not only a superb teacher but also an applied scientist and consultant to industry. He understood real problems and was able to explain their solutions on a very clear level.

I was honored to be a pupil and close friend of Dr. Gnedenko for more than thirty-five years. I remember well his ability to explain very difficult concepts with simple arguments and clear examples, and his capability to find the best explanations for a particular audience.

Boris Gnedenko also taught me appreciation for classical music, poetry and fine art; he was an extremely well-educated man in many areas. He taught by his own example, how to be a real human being: honest, principled and open to anyone who needed professional help. For all his pupils, Boris Gnedenko was a teacher with a capital "T."

During the last years of his life, Professor Gnedenko visited the United States twice. He lectured at George Washington University in Washington, DC; Harvard University in Boston, Massachusetts; and the University of North Carolina in Chapel Hill. He consulted at MCI Telecommunications in Richardson, Texas, and SOTAS, Inc. in Rockville, Maryland. These visits resulted in two books: *Probabilistic Reliability Engineering,* Boris V. Gnedenko and Igor A. Ushakov (1995, Wiley, New York) and *Statistical Reliability Engineering*, Boris V. Gnedenko, Igor V. Pavlov and Igor A. Ushakov (forthcoming, Wiley, New York).

During one visit he was encouraged to present an updated Russian edition of his famous textbook for translation in the United States. He helped me enormously during the translation by verifying some parts of the text by phone and correcting the spelling of names using his historical archives. Unfortunately, he died in late 1995 before completion of the translation.

This translation is my farewell and tribute to Boris Gnedenko.

In conclusion, I wish to express my thanks to Susanne Willson, who effectively helped by editing my translation of the historical appendices to the book, and Tatyana Ushakov, who prepared an English manuscript of the translation.

Author's Preface to the English Translation of the Russian Sixth Edition

The first edition of this book was published almost 50 years ago in Russia and has been through six editions in its original language. Over the years, the book has been translated into many languages. Translation of the last Russian edition, published in 1988, is now presented to English-speaking readers.

The newest edition differs essentially from previous versions. During recent years, books on reliability theory and queuing theory were published and then translated into English. So I decided to omit a chapter dedicated to these applied problems. I included a historical review on the development of probability theory and mathematical statistics. My deep belief is that knowledge of the history of any subject helps one to understand the subject better and more comprehensively. In addition, I made some structural changes and minor editorial corrections.

This book was written as a textbook for university students. In Russia it was a standard text for those who graduated from schools of mathematics, physics and mechanics. I know that the book was used for the same purposes in other countries.

Probability theory exemplifies a magnificent pure mathematical theory and at the same time a powerful applied mathematical tool.

Concerning theory, I remember a joke told by well-known American statistician George Doob. Attending a meeting of the Moscow Mathematical Society in the early 1960s, he remarked: "Some people think that probability theory is a part of mathematics. This is wrong. Indeed, mathematics is a part of probability theory."

This exaggeration reflects some real facts. Probability theory adopted and developed many other branches of mathematics: theory of measure, algebra of logic, theory of integral and differential equations, limit theorems, theory of complex variables, Fourie and Laplace–Stieltjes *transformations*, theory of optimization and many others. All of these methods are netted into an elegant and composite body.

Regarding use as an applied tool, we could enumerate countless areas of effective application of the theory of probability: demography and insurance (the first serious application historically), biology and medicine, telecommunications and transportation, reliability and safety of various systems and many, many others.

It seems to me that in the third millennium, the complexity of human activity in constructing global systems and performing complex technological processes (first of all, in telecommunications), and the prognosis of economic and social development of different regions will lead to the further evolution of probability theory and its application.

I believe this textbook will be useful for English-speaking readers living in some of the world's most technologically developed countries. I hope that reader and book will find each other.

I am grateful for the efforts in the translation of this book by Professor Igor A. Ushakov, my pupil and colleague. I appreciate the careful editing of the translation of the historical review by Susanne Willson. Also, I would like to thank Tatyana Ushakov, who spent enormous time and effort to prepare the final version of the manuscript of the English translation.

Two of my visits to the United States in 1991 and 1993 essentially helped to make the publication of this book a reality. In this connection I appreciate the support and sponsorship of Professor James Falk of George Washington University; Dr. William Hardy of MCI Telecommunications; and Dr. Peter Willson of SOTAS, Inc.

Moscow, *1995*

Author's Preface to the Russian Sixth Edition

More than a third of a century has gone by since the first edition of this book was published. Since then a number of excellent textbooks on probability theory have been published in my country and abroad. Significantly, most of these books give a strict theoretical presentation and show the power of mathematical *abstraction*. This book fulfills a quite different purpose; moving from intuitive conceptions and considering a large number of examples, it aims to present at least some of the applied research actively developing today.

This edition differs from earlier versions. Several sections with new results on an understandable level have been added; a small chapter addressed to mathematical statistics problems has been restored; and a brief history of the appearance and development of probability theory was written specifically for this edition. The latter material is based on some research by the author and his pupils. We must emphasize that many sides of the history of probability theory are still waiting to be investigated — in particular, the theory of stochastic processes. Everything concerning the classical probability theory is far from being known.

Everyone knows that abstract presentation of certain material is able to open the subject to the reader faster; beyond this the author might minimize the volume of the book. I believe that for the first meeting with mathematical subjects, especially with probability theory, presentation of the material should be accompanied by a number of examples that develop a probabilistic intuition and capability to tie abstract ideas and methods to practical problems. This ability is necessary for every mathematician and especially for those who intend to dedicate themselves to research and solution of practical problems. Moreover, a number of specialists in varied areas are faced with probabilistic problems in their everyday activities. For them, to enter into probabilistic concepts with the help of abstract methodology is a hard task. Abstract textbooks cannot help to build a bridge between practical needs and mathematical theory. However, this category of readers probably needs special types of textbooks written in different methodological and psychological styles.

After a book has been written, often the author sees all its imperfections and finds a number of passages that ought to be changed. I am such an author. I would be very grateful if readers would send me their comments.

I am happy to thank Yu. V. Prokhorov, B. A. Sevastyanov and D. M. Chibisov for the corrections and suggestions they made after reading the manuscript. Unfortunately, I could not incorporate all their suggestions into the book.

Moscow, *1988*

From the Author's Preface to the Russian First Edition

This textbook is divided into two parts — elementary and special. The latter part can be used as a basis for special courses dedicated to the summation of random variables, theory of stochastic processes, and mathematical statistics.

Probability theory is considered as a mathematical discipline, so obtaining of specific applied results is not a main issue. All examples serve as explanations of general concepts of the theory and point to their connection with applications. Simultaneously, these examples develop an ability to apply the probabilistic methods for solution of practical tasks. The reader should keep in mind some real problem that fills the mathematical model with content. This could help develop a probabilistic intuition allowing one to guess the answer to solving the problem in general terms even before a strict solution is arrived at with the help of analytical tools. I emphasize that it is impossible to study probability theory without systematic solution of the problem.

The first four sections of chapter 1 represent minor revisions to an unpublished manuscript by A. N. Kolmogorov.

I thank my dear teachers — A. N. Kolmogorov and A. Ya. Khinchine — who generously helped me with their advice and friendly discussions dedicated to the principal concepts of probability theory.

Moscow, *1950*

Foreword

In the over 40 years since the first Russian edition was published, Boris Gnedenko's ***Theory of Probability*** has become a classic. It has served as a popular text in probability for several generations of students — not only in Russia but throughout the world due to its translation into many languages. In German alone, there have been 11 editions; this translation of the sixth Russian edition is the fifth edition in English.

Boris Vladimirovich Gnedenko was one of the most productive and influential probabilists in Russia. His early research on limit distributions of sums of independent random variables culminated in a well-known monograph co-authored with A. N. Kolmogorov. Subsequently, his interests turned to various topics in applied probability including statistics, queueing, reliability, cybernetics and quality control; in several of these areas he wrote or co-authored authoritative books. Professor Gnedenko's interest in, and knowledge of, both pure and applied probability and his mastery of mathematical methods characterize this book. The presentation is lucid and rigorous; the reader's appetite is whetted with many examples drawn from a variety of fields.

Gnedenko distinguishes three definitions of probability: 1. as a quantitative measure of degree of certainty of the observer; 2. the "classical" definition of reducing the concept to the notion of equal likelihood; and 3. as relative frequency of occurrence of an event in a large number of trials. He rejects the first definition as subjective and psychological, not suitable for science. The classical and relative frequency definitions are synthesized and form the basis of the exposition. Many concepts are introduced in terms of the classical definition and then generalized to sets of events that cannot be partitioned into equally likely cases. This style makes it easy for the beginner to become familiar with the ideas of probability and understand how to relate them to models and problems in various areas.

The first few chapters cover rather standard material including characteristic functions and the central limit theorem. More advanced topics, such as infinitely divisible distributions and stochastic processes, are explored after the groundwork has been laid. A chapter on statistics introduces the main concepts and applies some of the results in probability developed earlier.

A substantial new chapter on the history of probability theory has been added. Such a history is particularly appropriate to this text because Russian mathematicians figure so prominently in the record; the names of Tchebycheff and Markov of the nineteenth century and Kolmogorov and Khinchine of the twentieth century are recognizable to every student of probability. Over the years, Boris Gnedenko wrote extensively on historical subjects, one of the first being his popular 1946 *Essays on the History of Mathematics in Russia.*

This book on probability differs from other texts in its mix of theory and application. While the mathematics is rigorous and complete, the theory is developed with applications in mind. Many examples demonstrate the usefulness of the results proved and motivate further study of the subject. On completion of this text, the reader is prepared to go on to more advanced theory and apply probabilistic methods to problems in various disciplines.

T. W. Anderson
Department of Statistics
Stanford University

Biography

Boris Gnedenko was born in Simbirsk, Russia, in 1912 into the family of a land surveyor. He was admitted to Saratov University, by special permission, when he was only fifteen years old and completed his studies in three years. After graduation, he lectured at Ivanov Textile Institute; he was hardly older than his students.

In 1934 he was invited by Alexander Khinchine to be a doctoral student at Moscow State University. There Gnedenko met one of Russia's greatest mathematicians, Andrei Kolmogorov, who became his lifelong friend. Both Khinchin and Kolmogorov significantly influenced Gnedenko's development as a scientist and a teacher.

At that time, one of the main directions in probability theory was analysis of limit distributions of sums of independent random variables. Boris Gnedenko continued work begun by Kolmogorov, Khinchine and P. Levy and in three years completed his Candidate dissertation[1] dedicated to the properties of infinitely divisible distributions. Extension of this research helped form the basis of his Doctor of Science thesis defended in 1939. Results of this work were included in a book by Gnedenko and Kolmogorov, *Limit Distribution for Sums of Independent Random Variables* (1949). This monograph, resolving the classical problem begun by Chebyshev, Lyapounov and Markov, was translated into many languages and became a classic.

Another important direction of Boris Gnedenko's work was solution of the problem of limit distribution of maximum order statistics of independent identically distributed random variables. In 1962 Kolmogorov emphasized that this problem, previously investigated by Frechet (1927), Fisher and Tippet (1928), and Gumbel (1935), received its complete solution only in Gnedenko's 1940 work. This work seriously influenced further development of asymptotic methods in probability theory and application. These two fundamental mathematical results became classics in modern probability theory.

In 1938 Gnedenko became a Professor at Moscow State University in the Department of Probability Theory headed by Kolmogorov. In 1945 he was elected Correspondent Member and in 1948 Academician of the Ukrainian Academy of Sciences. He moved to Lvov and became Chair of the department at Lvov University. Later Gnedenko became Chair of the Probability Theory Department at Kiev State University and Director of the Institute of Mathematics of the Ukrainian Academy of Sciences.

During this period he turned to mathematical statistics. This cycle of his work is recognized worldwide. Kolmogorov and N. Smirnov proved the first theorems on limit distributions of the maximal deviation of empirical distribution function from theoretical function, and of maximal deviation of two empirical distribution functions. Gnedenko found exact distributions for the case of finite samples. This simple and effective result allowed the reception of elementary proofs of known asymptotical results as well as the derivation of new proofs.

In 1950 Gnedenko published his textbook, *A Course on Probability Theory*. Republished several times, it became one of the main course texts on probability and statistics for many generations of mathematicians both in the former Soviet Union and abroad. Translated into many languages, the last Russian edition is now presented to the reader in English.

Gnedenko returned to Moscow in 1960 and again became a Professor in the Department of Probability Theory still headed by Kolmogorov. In 1966 he became the department Chair. This time was characterized by his intense activity in varied applications — queuing theory, reliability theory and applied statistics.

Gnedenko always dedicated his efforts to developing applied probabilistic methods. After graduation from Saratov University, he published three papers about multi-machine operation by a single operator. He wrote a paper, "Geiger-Müller Counters" (1941), which became a landmark of modern reliability theory. In 1964 and 1965 he published the first papers on limit distributions of survival time of renewable redundant systems; these papers stimulated development of work in probability applications.

As a first-class mathematician, Boris Gnedenko was never a "cabinet scientist." His social activity was outstanding. In Moscow he gathered a group of young mathematicians and engineers and involved them in the solution of important practical problems. In 1960 he organized the Moscow Seminar on Reliability and Quality Control at Moscow University. Within a year, this seminar was transformed into the All-Union Center on Reliability and Quality Control at the Moscow Polytechnic Museum. This Center, headed by Gnedenko, Ya. Sorin and myself, united Candidates and Doctors of Science who consulted with practical engineers and ran courses of continuing education programs. There were also two seminars on applied statistics, organized by Yu. Belyaev, and applied probabilistic models, organized by myself.

This Center played a very significant role in developing reliability theory and solution of practical problems in the area. During the almost 30 years of its existence, the Center has performed some 5,000 consultations, with from 5 to 10 participants, and presented 200 lectures, each with from 200 to 500 attendees. Engineers attended the lectures and consultations came from many locations in even the Far East and Central Asia Republics.

During the same period, Gnedenko organized and headed the journal *Reliability and Quality Control* and led a section on reliability and queuing theory for the journal *Engineering Cybernetics* (translated in the United States as *Soviet Journal of Computer and System Sciences*).

In 1965 he published *Mathematical Methods in Reliability Theory*, with Yu. Belyaev and A. Solovyev and in 1966 *Elements of Queuing Theory*, with I. Kovalenko; both books were translated into many languages. In 1983, with a group of his pupils and colleagues — E. Barzilovich, Yu. Belyaev, V. Kashtanov, I. Kovalenko, A. Solovyev, and myself — Gnedenko published *Aspects of Mathematical Theory of Reliability*, which accumulated new — and mostly original — results developed in this area during the previous 10–15 years.

Gnedenko always joined his theoretical research and teaching with intense consulting activity. Returning to Moscow from the Ukraine, he began to consult at the State Committee on Standards and several R&D institutes of the electronics and telecommunications industries. In 1992 he began consulting at SOTAS, Inc. in Rockville, Maryland.

During preparation of this edition Gnedenko visited the United States in 1991 and 1993, lecturing and consulting at various locations discussed in my preface.

For many years Gnedenko was interested in the history of mathematics, in particular, the development of probability and statistics. His 1946 *Essays on the History of Mathematics in Russia* was the only book of its kind at the time. He wrote several essays about great mathematicians and their work: Gauss, Euler, Bernoulli, Ostrogradsky, Chebyshev, Markov, Lobachevsky, Lyapounov, Khinchine and Kolmogorov. This edition includes an excellent analytical review on the history of probability and statistics.

In conclusion, I quote from Kolmogorov's 1961 paper written for Gnedenko's fiftieth birthday: "Academician Boris Vladimirovich Gnedenko by common international recognition is one of the most outstanding mathematicians working in the area of probability and statistics. He joins exclusively masterful possession of the methods of classical calculus with a deep understanding of modern aspects of probability theory and constant interest to its applications."

Igor Ushakov

Note

1. In Russia there are two levels of scientific degrees. The lower degree is called Candidate of Science and corresponds to the American PhD. The Russian higher scientific degree is Doctor of Science and has no counterpart in the United States.

Chapter 1

INTRODUCTION

The goal of this book is to present the basis of probability theory – a mathematical discipline studying the principle of random events.

The beginning of probability theory can be traced to the middle of the seventeenth century and connected with the names of Hyugens (1629–1695), Pascal (1623–1662), Fermat (1601–1665), and Jacob Bernoulli (1654–1705). In correspondence between Pascal and Fermat, one can find the first notions which step-by-step crystallized into a new branch of mathematics. The problems of interest were stimulated by tasks connected with hazard games; these problems lay beyond mathematics of that time. We should understand that the outstanding scientists analyzing hazard games foresaw a fundamental philosophical role of the science of studying random events. They were convinced that on the basis of mass random events there could be built strict mathematical principles. Only a current state of natural sciences led to the situation that hazard games were a main experimental basis for probability theory. Of course, it put its seal on the formal mathematical methods restricted by simple arithmetical calculations and combinatorial technique. Development of probability theory under the influence of natural sciences, in first order, physics, showed that the classical methods are still interesting even now.

Serious demands from the practice of natural and social sciences (first of all, theory of errors, gunfire tasks, and demography) led to the further development of probability theory and involved advanced analytical tools. An extraordinary role in developing probabilistic methods belongs to DeMoivre (1667–1754), Laplace (1749–1827), Gauss (1777–1855), and Poisson (1781–1840). The work by Lobachevsky (1792–1856), the creator of Non-Euclidean geometry, dedicated to the theory of errors of measurements on a sphere, also has a close relation to this direction.

From the middle of the nineteenth century up to the twenties of the current century, the development of probability theory is connected mainly with names of Russian scientists – Chebyshev (1821–1894), Markov (1856–1922), and Lyapounov (1857–1918). The success of the Russian science was prepared by the activity of Bounyakovsky (1804–1889) who widely propagated research in applied probability, especially in insurance and demography. He wrote the first textbook on probability in Russian which influenced the development of works in applied and theoretical probability. The main influence on the development of probability theory was due to works by Chebyshev, Markov, and Lyapounov who introduced and widely used the concept of random variable. Chebyshev's results connected with the Law of Large Numbers, Markov chains, and Lyapounov's central limit theorem will be examined later in this book.

The modern state of probability theory is characterized by an increasing interest in it and its wide penetration into practical applications. The Russian probabilistic school continues to dominate in the area. Among the first generation of Soviet scientists, the names of Bernshtein (1880–1968), Kolmogorov (1903–1987), and Khinchine (1894–1959) are prominent. During the presentation of the material in this book, we will inform the reader about main results and their influence on the development of probability theory. Thus, even in Chapter 1 we will discuss the fundamental works by Bernshtein, Mises (1883–1953), and Kolmogorov which are the basis of modern probability theory. In the 1920's, Khinchine, Kolmogorov, Slutsky (1880–1948), and Levy (1886–1971) established a connection between probability theory and the metric theory of functions. This allowed the discovery of a final solution of the problems which were formulated by Chebyshev and widened the content of probability theory. In the '30's Kolmogorov and Khinchine created the theory of stochastic processes which contains now the main direction of analysis in probability theory. This theory can serve as an excellent example of organic synthesis of mathematics and physics where a mathematician with an understanding of a physical problem is able to create for its solution an adequate mathematical tool. This part of probability theory will be considered in Chapter 10.

After molecular theory of substance became widely accepted, the use of probability theory was inevitable in chemistry and molecular physics. Notice that from the molecular physics viewpoint, any substance consists of a huge number of particles which are in continuous movement and interacting. The nature of these particles, the principles of their interaction, the character of their movement, etc., are not well known to us. The only reliable information is that there are many such particles and, in a homogeneous substance, their characteristics are close. Naturally, in such conditions traditional mathematical methods became powerless. For instance, differential equations methods cannot lead to success. Indeed, we usually know very little about these particles. Even if we knew almost everything about them, it would be impossible to make a solution with the help of standard methods of mechanics because of the enormous number of equations describing the physical object.

At the same time such an approach would be unacceptable from a methodological viewpoint. Really, we are interested in behavior of the mass of moving particles rather than in a collection of individual behaviors. The mass behavior cannot be obtained by a simple summation of behavior of individual particles. Moreover, in some limits the mass behavior becomes independent of the behavior of individual particles. Doubtless, an investigation of the new phenomena needs a new mathematical technique. What requirements should these methods satisfy? First of all, the huge number of particles must make the analysis of the phenomenon easier. Further, the lack of detailed knowledge about individual behavior of a particle must not be a principle obstacle for the analysis. These requirements are best of all satisfied by probability theory.

To avoid the erroneous impression that probability theory is used because of lack of knowledge, we emphasize the following. The philosophical basis for the use of probability theory lies in the fact that "mass" phenomena generate *new principles*. When a analyzing a complex natural phenomenon one needs to take into account only crucial features and avoid incorporating superfluous ones.

The measure of the fruitfulness of mathematical formalization is concordance between theory and practice. The development of the natural sciences, especially physics, confirm the extreme usefulness of probability theory for the purposes of mathematical modelling.

This connection of probability theory with practical needs explains why this branch of mathematics has flourished in recent decades. New results allow the solutions to new practical problems. New problems force the development of the theory. Of course, everything said before does not mean that probability theory is just an applied branch of mathematics. On the contrary, the experience of the last decades demonstrates that it became an elegant mathematical discipline with its own problems and methods of proving.

It was said in the beginning that probability theory deals with "random events." The strict definition will be given in Chapter 1. Here we restrict ourselves to several remarks. In everyday life we refer to random events as something rare and irregular, coming against usual perception. We will reject such an understanding. Random events in probability theory possess a set of properties, in particular, they occur as a mass phenomenon. Such phenomena are described by an enormously large number of equal or almost equal objects and do not significantly depend on the nature of individual objects. Such mass phenomena are usual in physics, econometrics, telecommunications, and military applications. Statistical quality control of mass production is exclusively based on probabilistic concepts. One of the most serious problems of contemporary engineering is a reliability problem which uses as its dominant tool varied probabilistic methods. It is time to mention that in its turn the needs generated by reliability theory have an influence on probability theory.

At this point it is appropriate to remember words said by the Founder of the Russian school of probability theory, Chebyshev: "Closeness of theory and practice gives the most fruitful results. Not only practice gains from this. Science is developing under the influence of practice: new objects for investigation appear, new sides of known objects become open.... Science profits from new applications of old methods, but even more so it gains from new methods by choosing a reliable guide among practical applications."

Chapter 2

RANDOM EVENTS AND THEIR PROBABILITIES

1 INTUITIVE UNDERSTANDING OF RANDOM EVENTS

For a very long time people have investigated and used for practical purposes only so-called deterministic laws. Most of the knowledge obtained in school courses on physics, chemistry, and mathematics relates to this field. Consider some examples.

If a pyramid has as its base a square with a side equal to a and its height equals h, then the volume of the pyramid equals $(1/3)a^2h$.

If a body falls to the Earth's surface, then the path coming in t seconds after the beginning of the fall equals $S=(gt^2)/2$.

If chemically pure water is heated up to 100°C under the atmospheric pressure 760 mm of mercury, it begins to vaporize.

The number of such example's can be increased with no limit. But not all situations which we meet in practical and scientific activity can be described by such forms of laws. For example, consider several questions. What is the number of tomorrow's traffic incidents in the New York area? What is the maximal level of water that will be observed at the Mississippi River in the St. Louis area at the next flood? How long will it take to repair the fiber trunk of telecommunication network cut by digging? All these and similar questions possesses one specific property: it is impossible to give a unique answer because the circumstances and nature of all of these events are full of uncertainty. In all similar cases we say that the event under consideration is random.

Random events intervene in our everyday life and – besides – the modern viewpoint is that all of nature's processes are random. So we need to learn how to analyze them with the help of special mathematical methods.

It is relevant to note that we deal with random phenomena not from time to time but regularly because they often play an essential role in the structure of an investigated real process. For instance, when designing a telephone network one needs to take into account a randomness of moments of calls and their duration. A seaport administration deals with random arrivals of ships which are sometimes far from pre-planned. The loading and unloading of a ship also have a random duration depending on a number of factors such as loading facilities, quantity and character of the cargo, its packing, etc. Thus in both cases, either telephone network design or seaport operations planning, one must take into account a number of random factors. Analogous problems arise in many other areas of human activity. This circumstance forced scientists to undertake intense research in probability and statistics during the last three centuries. A brief review of the history of the development of probability theory, mathematical statistics and theory of stochastic processes is given in the end of this book.

Before starting a presentation of the main results of probability theory, we need to formulate principal concepts of the subject. Up to now we gave only descriptive and intuitive formulations of "random" phenomena. But probability theory deals not with all of those phenomena which are referred to in our everyday life as "random". We must formulate some specific properties of phenomena which are considered as random in the probabilistic sense.

First of all, probability theory deals with phenomena which can be–in principle–realized an infinite number of times and in unchanged conditions. Let us consider some examples. Classical historical examples involve hazard games such as dice and cards. In both cases a researcher can repeat his experiments practically without restriction. A similar situation arises when a researcher investigates traffic in a modern telephone system. Thousands of calls can be observed in practically stable conditions. If there are different loadings of the system during night hours or on weekends, the researcher has the opportunity to select data by these attributes.

Further, probability theory deals with events which possess a so-called *statistical stability* or, in other words, which have a stable frequency of occurrence. This requirement should be considered in more detail.

Assume that we perform a sequence of trials in each of which some specified event A may or may not occur. These trials are performed under the same conditions and previous results of trials do not influence the forthcoming results (it is said that the trials are independent). Let μ denote the number of occurrences of event A in some predetermined number of trials, for instance, in n sequential trials. Then the frequency, i.e., the ratio μ/n, for large n is close to a constant for statistically stable events A and this frequency only changes slightly from one series of trials to another.

Checking the statistical stability is a difficult problem and we will consider it later.

Let us now emphasize that probability theory does not deal with unique events which can be never reproduced. For instance, it is impossible to say anything about the probability that a particular student will not have passed a current exam on probability theory because a single experiment is under consideration and there is no possibility to repeat this experiment under the same conditions. We can express only our subjective judgement based on our knowledge of this student. The task of probability theory is to investigate objective regular properties of a phenomenon which do not depend on subjective viewpoints of one or another researcher. So, while some events are very interesting and intriguing, probability theory cannot investigate them if there is no possibility to make many independent repetitions. For instance, all of the following events–

- in August 6, 1999, there will be an earthquake rated 8 on the Richter scale in Pasadena, California;
- before 2000 AIDS will be totally curable;
- in 2001 a poet of Longfellow's talent will be born in the USA – bear a feature of uniqueness and the answers to them (positive or negative) will be known at the appropriate time. Right now these hypotheses are uncertain: they may happen or may not but probability theory has no relation to them.

Probability theory studies such random events which can be objectively evaluated in terms of frequency of their occurrence. For instance, such evaluation can be formulated as:

The probability that under some specified complex of conditions $\mathfrak{S}$ *the event A will have occurred equals p.*[1]

Such kind of events are called stochastic or probabilistic. They appear in varied situations dealing with natural and social processes and different technical applications.

---

[1]Here we will still avoid to give a definition of the probability and rely on the intuition of the reader.

In practice we meet two types of trial repetitions:

(1) the same physical object is a subject of consecutive trials under some fixed conditions;
(2) different identical physical objects are tested at the same conditions.

It is basic to consider deterministic phenomena for which one can formulate the following statement:

Under some complex of conditions $\mathfrak{G}$ *an event A always occurs with necessity.*

In the next section it will be shown that the probability of a random event can be measured by a number between 0 and 1. A unit corresponds to the event which always occurs under the complex of conditions $\mathfrak{G}$. Such events are called true. If an event is impossible, its probability equals 0. So, a deterministic event can be considered as a particular case of random ones.

The fact that a probability of the event occurrence under some specified conditions exists is an essential statement even though the probability is unknown. This hypothesis needs to be confirmed and tested. The problem is in establishing of the dependence between the event occurrence and its characteristic conditions. This problem has philosophical aspects and still has not been solved. This problem is so difficult that even among scientists there are attempts to declare that the probabilistic properties are related only to the subjective opinion of the investigator about a possibility of an event occurrence.

Last time such "probabilistic" statements are often made to different events which cannot be characterized as random at all. For instance, sometimes one uses the so-called "subjective probability" for measurement of uncertain events.

2 SAMPLE SPACE. CLASSICAL DEFINITION OF PROBABILITY

In the previous section there were no formal definitions of random event or probability. Let us offer them now. We will simultaneously try to help the reader to understand probability on an intuitive level. This purpose forces us to introduce a definition of probability step by step, repeating the historical path of its development. Such an approach allows us to avoid a formal understanding and to go from simple statements to those more and more complex.

Let us begin with the so-called classical definition which is more of a method of computation in some specified situations than a definition. The classical definition of probability is based on the concept of equal likelihood of outcomes based on a real symmetry. The concept of equally likely (equiprobable) events is primary and is not a subject of determination. It can be explained through simple examples.

In the throwing at random of a perfectly cubical die made of completely homogeneous material, the equally likely events are the appearance of any of the specific number of points (from ace to six) because by symmetry no one face has objective preference over any other.

Assume that at each trial there is possible one of n mutually exclusive[2] and equally likely events $E_1, mE_2, \ldots, E_n$. Each such outcome is called *elementary*.

One often is interested in the probability of a random event which is composed from several elementary ones. For instance, throwing a die, one can find the probability that the points on its face exceed 3, i.e., there could be 4, 5 or 6. In general, if we are interested in occurrence of one of the specified elementary random events $E_{i_1}, E_{i_2}, \ldots, E_{i_m}$ then we say that we are interested in the occurrence of the event A consisting of the occurrence of one of n the elementary events above-mentioned.

The probability of a random event A is the ratio of the number of mutually exclusive events composing this event (i.e., the number m) to the entire number of all possible elementary events (i.e., n). The probability of the random event A denotes with $\mathbf{P}(A)$.

In accordance with the definition given above,

$$\mathbf{P}(A) = \frac{m}{n}.$$

For example, in a throwing a die, the complete group of mutually exclusive and equally probable events consists of the events

$$E_1, E_2, E_3, E_4, E_5, E_6,$$

[2] The term exclusive was introduced by T. Bayes. it means that if some outcome E_k has been realized then no other of $n-1$ outcomes can occur at the same trial.

consisting, respectively, in the occurrence of 1, 2, 3, 4, 5, and 6 points. The event

$$C = E_2 + E_4 + E_6,$$

consisting in the occurrence of an even number of points is divisible into three events which are mutually exclusive and equally probable events. Hence, the probability of the event C equals

$$P(C) = 3/6 = 1/2.$$

From the definition given-above, it is also evident that

$$P(E_i) = 1/6, \quad 1 \leqslant i \leqslant 6,$$
$$P(E_1 + E_2) = 2/6 = 1/3$$

and so forth.

Let us now consider the throwing of two dice. If the dice are fair, each of the 36 possible combinations of numbers on the two dice sides may be regarded as equally probable. Thus, the probability of occurrence, say, of 12 points equal 1/36. Occurrence of 11 points may turn up in two ways: a five on the first die and a six on the second one, or vice versa. Therefore, the probability of occurrence of 11 points in total equals 2/36 = 1/18. The reader can easily verify that the probability of occurrence of any specific number of points is given by the following table 1:

Table 1

Number of points	2	3	4	5	6	7	8	9	10	11	12
Probability	1/36	2/36	3/36	4/36	5/36	6/36	5/36	4/36	3/36	2/36	1/36

Let us compute a total number of random events which can be formed from n elementary ones. Obviously, one can arrange $\binom{n}{m}$ events each of which contains m some elementary events ($1 \leqslant m \leqslant n$). If $m = n$, any event happens always, i.e., it is a certain event. Thus in a whole there are

$$\sum_{1 \leqslant m \leqslant n} \binom{n}{m} = 2^n - 1$$

events. Add to these one more which does not correspond to either elementary event, i.e., it consists of an empty set. Obviously, this event never happens. Consequently, the total number of all of random events in this case equals 2^n.

Current considerations relate not only to the classical definition of probability but also to further generalizations. Let us consider a fixed set of conditions $\mathfrak{S}$ and some family S of events $A, B, C, \ldots$ each of which must either occur or not if the set of conditions $\mathfrak{S}$ is realized. Certain relations can exist between the events of the family S which we shall continually make use of and which, therefore, we need to study.

1) If for every realization of the set of conditions $\mathfrak{S}$ for which the event A occurs the event B also occurs, then one says that A ***implies*** B. This relation is denoted by $A \subset B$, or $B \supset A$.

2) If A implies B and at the same time B implies A, i.e., if for every realization of the set of conditions $\mathfrak{S}$ either A and B both take place or both do not take place, then one says that the events A and B are ***equivalent***. This relation is denoted by $A = B$.

3) The event consisting in the simultaneous occurrence of A and B is called the ***product*** (or ***intersection***) of the events A and B. This relation is denoted by AB (or $A \cap B$).

4) The event consisting in the occurrence of at least one of the events A or B is called the ***sum*** (or ***union***) of the events A and B. This relation is denoted by $A + B$ (or $A \cup B$).

5) The event consisting in the occurrence of A and the non-occurrence of B is be called the ***difference*** of the events A and B. This relation is denoted by $A - B$.

6) The events consisting in non-occurrence of A are called ***complimentary*** for event A. Such an event is denoted by $\bar{A}$. Suppose the set of conditions $\mathfrak{S}$ consists in the selection at random of a point inside the square shown in Figure 1. Introduce two events, A and B. Event A corresponds to the situation where the point selected lies inside the left-hand circle. Analogously, event B corresponds to the situation where the point selected lies inside the right-hand circle.

Then events

$$A, \bar{A}, B, \bar{B}, A + B, AB, A - B, B - A, \overline{A - B}$$

are the events consisting in the selected point falling inside the regions that have been shaded in the correspondingly labeled diagrams of Figure 1. (Incidentally, diagrams of this kind are known as ***Venn diagrams***.)

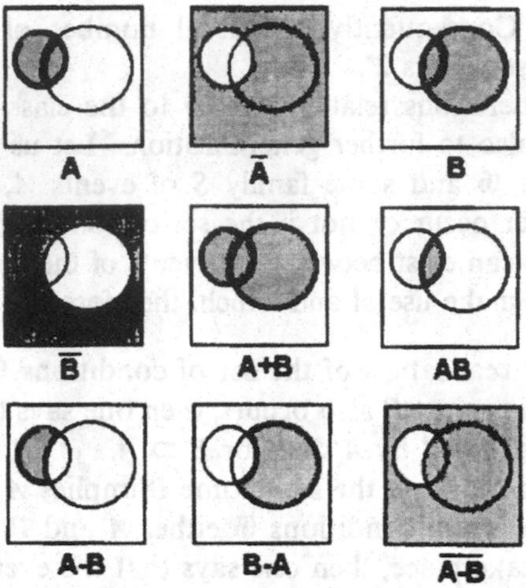

FIGURE 1. Venn diagram.

Let us consider another example. Assume that the set of conditions $\mathfrak{S}$ consists in a single throw of a die onto a table. Let A, B, C, and D denote, respectively, the event that a six, a three, an even number, and a multiple of three turn up. Then the events A, B, C, and D are characterized by the following relations:

$$A \subset C, \quad A \subset D, \quad B \subset D, \quad A + B = D, \quad CD = A.$$

The definition of the sum and product of two events is generalized to any number of events:

$$A + B + \cdots + N$$

denotes the event consisting of the occurrence of at least one of the events $A, B, \cdots, N$ and

$$AB \cdots N$$

denotes the event consisting of the simultaneous occurrence of all of the events $A, B, \ldots, N$.

7) An event is called *certain* if it must inevitably occur (whenever the set of conditions $\mathfrak{S}$ is realized). For example, if a pair of dice is thrown, then it is certain that the sum of the numbers thrown will not be smaller than two.

8) An event is called *impossible* if it never occurs (for any realization of the set of conditions 𝔖). For example, if a pair of dice is thrown, then it is impossible for 13 points to appear. It is clear that all certain events are equivalent to one another. It is therefore permissible to denote all certain events by a single letter. The letter Ω is used for this purpose. All impossible events are likewise equivalent. Any impossible event is denoted by the sign $\varnothing$.

For complimentary events A and A^- two following relations are true simultaneously:

$$A + \bar{A} = \Omega, \quad A\bar{A} = \varnothing.$$

For example, if for a single die throwing C denotes an even number of points then

$$\Omega - C = \bar{C}$$

is event consisting of occurrence of odd number of points.

9) Two events A and B are called *mutually exclusive* if their joint occurrence is impossible, i.e., if

$$AB = \varnothing.$$

If

$$A = B_1 + B_2 + \cdots + B_n$$

and the events B_i are mutually exclusive in pairs, i.e.,

$$B_i B_j = \varnothing \quad \text{for} \quad i \neq j,$$

then one says that the event A is decomposable into the mutually exclusive events $B_1, B_2, \ldots, B_n$. For example, in the throwing of a single die the event C consisting of the throw of an even number is decomposable into the mutually exclusive events E_2, E_4, and E_6 consisting respectively of the occurrence of a two, a four, and a six.

The events $B_1, B_2, \ldots, B_n$ compose a complete group of events if at least one of them must definitely occur (if the set 𝔖 is accomplished), i.e., if

$$B_1 + B_2 + \cdots + B_n = \Omega.$$

The complete groups of pairwise mutually exclusive events are especially important for our future consideration. For example, such a group is formed by a family of events

$$E_1, E_2, E_3, E_4, E_5, E_6$$

consisting, respectively, of the occurrence of a 1, 2, 3, 4, 5, and 6 points in a single throw of a die.

10) In every problem in probability theory, one has to deal with some specific set of conditions $\mathfrak{G}$ and some specific family of events S each of which occur or do not occur after each realization of the set of conditions $\mathfrak{G}$. It is proper to make the following assumptions concerning this family of events:

a) If the events A and B belong to the family S then so do the events AB, $A + B$, and $A - B$.

b) The family S contains the certain event and the impossible event.

A family of events satisfying these assumptions is called a *field of events.*

Note that equivalent events can replace each other in all probabilistic statements.

Consider some complete system G of pairwise exclusive equally likely events (elementary events) $E_1, E_2, \ldots, E_n$ and also consider a system of events S consisting of impossible event $\varnothing$, all of the events E_k of the system G and all of the events A composed of elementary events of G.

For example, if the system G consists of three events E_1, E_2, and E_3, then the system S consists of events $\varnothing$, E_1, E_2, E_3, $E_1 + E_2$, $E_2 + E_3$, $E_1 + E_3$, $\Omega = E_1 + E_2 + E_3$.

It easy to guess that the system S is the field of events. Indeed, it is easy to find that the sum, difference and product of events from S belongs to S. The impossible event is in S by construction and the true event Ω can be always determined as $\Omega = E_1 + E_2 + \cdots + E_n$.

According to the definition given, every event A belonging to the field of events G constructed above has a well-defined probability

$$P(A) = m/n$$

assigned to it, where m is the number of events E_i of the original group G which are particular cases of the event A. Thus, the probability $P(A)$ may be regarded as a function of the event A defined over the field of events S.

This function possesses the following properties:

1. For every event A of the field S

$$\mathsf{P}(A) \geqslant 0.$$

2. For the certain event Ω.

$$\mathsf{P}(\Omega) = 1.$$

3. If the event A is decomposable into the mutually exclusive events B and C and all three of the events A, B, and C belong to the field S then

$$\mathsf{P}(A) = \mathsf{P}(B) + \mathsf{P}(C).$$

This property is called the theorem on the addition of probabilities.

Property 1 is obvious, since the ratio m/n cannot be negative. The second property is equally obvious, since all of n possible outcomes of the trial are all favorable to the certain event Ω and hence

$$\mathsf{P}(\Omega) = m/n = 1.$$

Let us prove Property 3. Suppose that m' is the number of outcomes E_i of the group G favorable to the event B and m'' is the number favorable to the event C. Since by assumption the events B and C are mutually exclusive, the outcomes E_i that are favorable to one of them are distinct from those that are favorable to the other. Thus, there are altogether $m' + m''$ events E_i that are favorable to either event A or event B, i.e., favorable to the event $B + C = A$. Therefore

$$\mathsf{P}(A) = \frac{m' + m''}{n} = \frac{m'}{n} + \frac{m''}{n} = \mathsf{P}(B) + \mathsf{P}(C),$$

Q.E.D.

At this point let us indicate some additional properties of probability.

4. The probability of the event $\bar{A}$ complementary to the event A is given by

$$\mathsf{P}(\bar{A}) = 1 - \mathsf{P}(A).$$

In fact, since

$$A + \bar{A} = \Omega$$

it follows from Property 2 that

$$\mathsf{P}(A + \bar{A}) = 1,$$

and since the events A and A^- are mutually exclusive, Property 3 implies that

$$\mathsf{P}(A + \bar{A}) = \mathsf{P}(A) + \mathsf{P}(\bar{A}).$$

The last two equations prove the proposition above.

5. The probability of the impossible event is zero. In fact, the events Ω and $\varnothing$ are mutually exclusive, so

$$\mathsf{P}(\Omega) + \mathsf{P}(\phi) = \mathsf{P}(\Omega),$$

from which it follows that

$$\mathsf{P}(\phi) = 0.$$

6. If the event A implies the event B. then

$$\mathsf{P}(A) \leqslant \mathsf{P}(B).$$

Indeed, the event B can be represented as the sum of two events A and $A^- B$. Using Properties 3 and 1, one obtains from this that

$$\mathsf{P}(B) = \mathsf{P}(A + \bar{A}B) = \mathsf{P}(A) + \mathsf{P}(\bar{A}B) \geqslant \mathsf{P}(A).$$

7. The probability of any event lies between 0 and 1.

The relations

$$\varnothing \subset A + \varnothing = A = A\Omega \subset \Omega,$$

hold for any event A, and from this and the preceding property it follows that the inequalities

$$0 = \mathsf{P}(\varnothing) \leqslant \mathsf{P}(A) \leqslant \mathsf{P}(\Omega) = 1.$$

hold.

3 EXAMPLES

Let us now consider several examples of the calculation of the probabilities of events using the classical definition of probability. The examples cited are purely illustrative in character and do not pretend to deliver to the reader all of the basic methods for computing probabilities.

Example 1. One throws three dice. What is more probable: to obtain the sum of points equal to 11 or to 12?

This problem, as will be seen from the historical review at the end of the book, was one of the earliest forming the basis of the concepts and theory of probability theory. It is said that a legend is connected with this problem. Once a mercenary came to Galilei (some said that it was Huygens) and asked the following question: Why by logic should both sums appear with the same frequency on the average, but experiments show that 11 appears more frequently? The mercenary showed a convincing arguments: both 11 and 12 could be reduced to the sum of three positive summands only by six different ways, namely:

$$11=1+5+5=1+4+6=2+3+6=2+4+5=3+3+5=3+4+4$$

and

$$12=1+5+6=2+4+6=2+5+5=3+4+5=3+3+6=4+4+4$$

From this, the mercenary assumed that the equal probability of both outcomes followed.

Galilei explained that each of these presentations exists with its own "weight." He argued in the following way. Let us label the dice as "first," "second," and "third." The number of points of each die will be put on the appropriate position. It is clear that the decomposition of type $1+5+5$ might occur not in one but in three different ways:

$$(1,5,5), (5,1,5), (5,5,1).$$

In an analogous way the presentation $(1+4+6)$ might occur in six different ways:

$$(1,4,6) = (1,6,4) = (4,1,6) = (4,6,1) = (6,1,4) = (6,4,1).$$

It is easy to calculate that the sum equal to 11 appears not in six but in 27 different ways. The sum equal to 11 appears in 25 different ways. This is because of the fact that $4+4+4$ occurs in only a single way.

Now notice that the total number of all of possible equally probable outcomes with three dice equals $6^3 = 216$. (This fact was known even in the twelfth century.)

Let A and B denote the random events to the appearance of the sums equal to 11 and 12, respectively. Then in accordance with the classical definition of the probability

$$\mathsf{P}(A) = 27/216 \quad \text{and} \quad \mathsf{P}(B) = 25/216.$$

It is known from mathematical statistics that for reliable discrimination of two probabilities with a difference less than 0.01 one needs to perform many thousand experiments.

Example 2. From a deck of 36 cards, three cards are drawn at random. Find the probability that there will be exactly one ace among them.

SOLUTION. The complete group of equally probable and mutually exclusive events in this problem consisting of all the possible combinations of three cards is $\binom{36}{3}$, and there are (532) such combinations. The number of favorable outcomes can be computed as follows. One ace can be chosen in $\binom{4}{1}$ different ways and the two remaining cards (non-aces) in $\binom{32}{2}$ different ways. Since to each given ace there correspond $\binom{32}{2}$ ways in which the two remaining cards can be selected, the total number of favorable cases will be $\binom{4}{1}\binom{32}{2}$. Thus the required probability equals

$$p = \frac{\binom{4}{1}\cdot\binom{32}{2}}{\binom{36}{3}} = \frac{\frac{4}{1}\cdot\frac{32\cdot 31}{1\cdot 2}}{\frac{36\cdot 35\cdot 34}{1\cdot 2\cdot 3}} = \frac{31\cdot 16}{35\cdot 3\cdot 17} = \frac{496}{1785} \approx 0.2778.$$

Example 3. Three cards are drawn at random from a deck of 36 cards. Determine the probability that there will be at least one ace among them.

SOLUTION. Let A denote the event in question. This event can be decomposed into the sum of the following three mutually exclusive events: A_1, the occurrence of one ace; A_2, the occurrence of two aces; and A_3, the occurrence of three aces.

By arguments analogous to those carried out in the preceding example, one easily establishes that the number of cases favorable to the event A_1 is $\binom{4}{1}\cdot\binom{32}{2}$; the number favorable to A_2 is $\binom{4}{2}\cdot\binom{32}{1}$; and the number favorable to A_3 is $\binom{4}{3}\cdot\binom{32}{0}$.

Since the total number of possible cases is equal to $\binom{36}{3}$, one has

$$\mathbf{P}(A_1) = \frac{\binom{4}{1}\cdot\binom{32}{2}}{\binom{36}{3}} = \frac{16\cdot 31}{3\cdot 35\cdot 17} \approx 0.2778,$$

$$\mathbf{P}(A_2) = \frac{\binom{4}{2}\cdot\binom{32}{1}}{\binom{36}{3}} = \frac{3\cdot 16}{3\cdot 35\cdot 17} \approx 0.0269,$$

$$\mathbf{P}(A_3) = \frac{\binom{4}{3}\cdot\binom{32}{0}}{\binom{36}{3}} = \frac{1}{3\cdot 35\cdot 17} \approx 0.0006.$$

By virtue of the theorem of addition of probabilities,

$$\mathbf{P}(A) = \mathbf{P}(A_1) + \mathbf{P}(A_2) + \mathbf{P}(A_3) = \frac{109}{3\cdot 119} \approx 0.3053.$$

This problem may be solved by another method. The complementary event A^- is the event that no aces occur among the cards drawn. It is obvious that three non-aces may be drawn from a deck of cards in $\binom{32}{3}$ different ways, and therefore

$$\mathbf{P}(\bar{A}) = \frac{C_{32}^3}{C_{36}^3} = \frac{32\cdot 31\cdot 30}{36\cdot 35\cdot 34} = \frac{31\cdot 8}{3\cdot 17\cdot 7} \approx 0.6947.$$

The required probability equals

$$\mathbf{P}(A) = 1 - \mathbf{P}(\bar{A}) \approx 0.3053.$$

REMARK: In both examples above, the expression "at random" means that all possible combinations of three cards are equally probable.

Example 4. Consider now an example close by its mathematical contents to Example 2. Such a situation often appears in quality control.

A box contains N physically identical balls, M of them white and $N-M$ of them black. One picks up n balls (without replacement). What is the probability that among them there will be m white balls?

From the condition of the problem it is clear that $m \leqslant n$, $m \underline{M} M$, and $n-m \leqslant N-M$.

The total number of all of possible outcomes equals $\binom{N}{n}$. The number of favorable cases can be calculated in the following way: there are $\binom{M}{m}$ different ways to pick up m white balls, and there are $\binom{N-M}{n-m}$

ways to pick up $n-m$ black balls. Thus the total number of favorable outcomes equals $\binom{M}{m}\binom{N-M}{n-m}$. Thus the required probability equals

$$p = \frac{\binom{N}{M}\cdot\binom{N-M}{n-m}}{\binom{N}{M}}$$

Example 5. A deck of 36 cards is divided at random into two equal parts. What is the probability that an equal number of red and black cards turns up in both parts? (Again the expression "at random" means that all possible divisions of the deck are equally probable.)

SOLUTION. One has to determine the probability that among 18 cards drawn at random from the deck 9 will be red and 9 black.

The desired probability equals

$$p = \frac{\binom{18}{9}\cdot\binom{18}{9}}{\binom{36}{18}} = \frac{(8!)^4}{36!\,(9!)^4}.$$

To show the real value of that probability without boring calculation let us use the Stirling Formula which gives the following asymptotic equality

$$n! \sim \sqrt{2\pi n}\, n^n e^{-n}$$

Thus one has

$$18! \sim 18^{18}\cdot e^{-18}\sqrt{2\pi\cdot 18},$$

$$9! \sim 9^9\cdot e^{-9}\sqrt{2\pi\cdot 9},$$

$$36! \sim 36^{36}\cdot e^{-36}\sqrt{2\pi\cdot 36},$$

and, consequently,

$$p \approx \frac{\left(\sqrt{2\pi\cdot 18}\cdot 18^{18}\cdot e^{-18}\right)^4}{\sqrt{2\pi\cdot 36}\cdot 36^{36}\cdot e^{-36}\left(\sqrt{2\pi\cdot 9}\cdot 9^9\cdot e^{-9}\right)^4}.$$

after some simple transforms

$$p \approx \frac{2}{\sqrt{18\pi}} \approx \frac{4}{15} \approx 0.26.$$

Naturally, a question could arise: what are the relations of all these probabilities to the real phenomena? Once the following experiment was performed during a lecture. Students brought several decks of 36 cards and 100 times performed an experiment dividing the deck at random. The results of this experiment are presented in Figure 2. The graph shows how the frequency μ/n changes depending on the number of trials performed. At the beginning the graph has significant deviation from the line $y = p \approx 0.26$. With the increase of n the graph comes closer to the line.

Example 6. There are n particles each of which might be in one of N cells with the probability $1/N$, $N > n$. Problem: To find the probability that (1) each of n specified cells contains one particle, and (2) n arbitrary cells each contains one particle.

SOLUTION. This problem plays an important role in modern statistical physics and depending on how the complete group of equally probable events is formed, this problem leads to the physical statistics of Boltzman, of Bose-Einstein, or of Fermi-Dirac.

In the Boltzmann statistics, the equally likely events are all the conceivable distributions, taking all the particles to be distinguishable

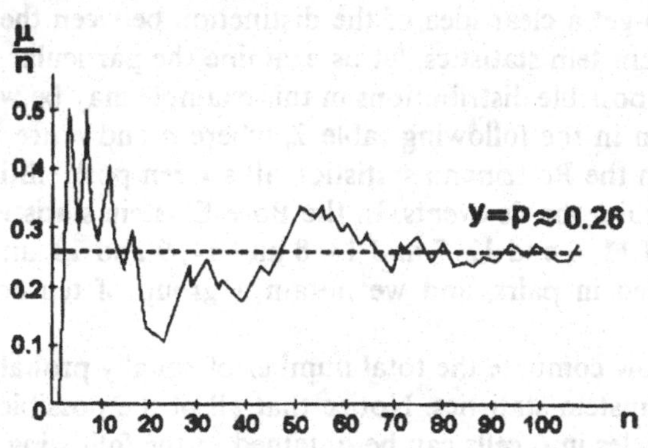

FIGURE 2. Sample of trajectories for Example 7.

and allowing any number of them from 0 up to n to be found in each of the cells. The total number of possible allocations can be computed in the following way. Each particle may be found in any one of N cells, so n particles might be allocated in N^n different ways. In the first of the above questions, the number of favorable cases will clearly be $n!$ and this implies that the probability of a single particle falling in each of n specified cells is

$$p_1 = \frac{n!}{N^n}.$$

In the second question, the number of favorable cases will be $\binom{N}{n}$ times larger, and this means that the probability of a single particle falling in each of n arbitrary cells equals

$$p_2 = \frac{\binom{N}{n}\cdot n!}{N^n} = \frac{N!}{N^n\cdot(N-n)!}.$$

In the Bose-Einstein statistics, those cases in which the various particles change places with each other are considered identical, i.e., it is important only how many particles are in a given cell, but not which ones. The complete group of equally probable events consists of all possible allocations of n particles in N cells. Each such allocation corresponds to some entire class of Boltzmann distributions differing among themselves not as to the number of particles in each of the given cells but only as regards the identity of the particles themselves. In order to get a clear idea of the distinction between the Boltzmann and Bose-Einstein statistics, let us examine the particular case $N = 4$, $n = 2$. The possible distributions in this example may be written in the form shown in the following table 2, where a and b are types of the particles. In the Boltzmann statistics, all sixteen possibilities represent different equiprobable events. In the Bose-Einstein statistics, however, cases 5 and 11, 6 and 12, 7 and 13, 8 and 14, 9 and 13, and 10 and 16 are identified in pairs, and we obtain a group of ten equally likely events.

Let us now compute the total number of equally probable events in the Bose-Einstein statistics. Notice that all of the possible allocations of the particles into cells can be obtained in the following way. Let us arrange the cells in sequence in a straight line and then allocate the

Table 2

Case	1	2	3	4	5	6	7	8	9	10	11	12	13	14	15	16
	ab				a	a	a				b	b	b			
Cells		ab			b			a	a		a			b	b	
			ab			b		b		a		a		a		b
				ab			b		b	b			a		a	a

particles into the cells side by side along the same line. Then consider all possible permutations of the particles and dividing walls of the cells. It is clear that this allows us to account for all possible ways in which the cells can be filled, taking into account both the order of the particles and the order of the dividing walls.

The number of such permutations is equal to $(N+n-1)!$. Among these permutations some are identical: every allocation among the cells is counted $(N-1)!$ times, because one has distinguished the dividing walls of the cells and, over and above this, every allocation has been counted $n!$ times, because we have taken into account not only the number of particles in a cell but also the order in which all the particles have been distributed amongst the cells. Thus, each allocation has been counted $n!(N-1)!$ times. So, from this it follows that the number of distinguishable allocations of the particles in the Bose-Einstein statistics equals

$$\frac{(n+N-1)!}{n!(N-1)!}.$$

Thus is found the number of equally probable events in the complete group of events. Now the question in the problem can be easily answered. In the Bose-Einstein statistics, the probabilities p_1 and p_2 are given by

$$p_1 = \frac{1}{\dfrac{(n+N-1)!}{n!(N-1)!}} = \frac{n!(N-1)!}{(n+N-1)!},$$

$$p_2 = \frac{\dbinom{N}{n}}{\dfrac{(n+N-1)!}{n!(N-1)!}} = \frac{N!(N-1)!}{(N-n)!(N+n-1)!}.$$

Finally, let us consider the Fermi-Dirac statistics. In this system of statistics, the particles are indistinguishable, and a cell may contain not more than one particle.

The total number of distinct allocations of the particles among the cells is easily computed: the first particle may occupy any one of N different places; the second, only one of $N-1$, the third, $N-2$; and finally, the n-th, $N-n+1$ different places. Thus, the number of different equally probable ways in which n particles can be allocated into N cells equals

$$N(N-1)\cdots(N-n+1)=\frac{N!}{(N-n)!}$$

It is easy to show that in the Fermi-Dirac statistics the desired probabilities equal

$$p_1=\frac{(N-n)!}{N!}.$$

$$p_2=1.$$

The example just considered shows how important it is to define precisely which events in a given problem are to be regarded as equally probable.

Example 7. There is a queue of $2n$ people at a cinema tiket booth. Among them n people have only ten-dollar bills and the remaining n have only five-dollar bills. The ticket price is five dollars. Each spectator buys only one ticket. There is no cash in the ticket booth when it opens. What is the probability that no customer will be required to wait for change?

Note that this problem is a reformulation of one of the practical problems met in the investigation of quality control of mass production.

All the possible arrangements of the customers on line are supposed to be equally probable. Let us use the following geometrical approach. Consider the plane xOy and suppose that the customers are arranged along the x-axis at the points with abscissas $1, 2, \ldots, 2n$ in the same order they occupy on line. The ticket booth is located at the origin. Let the value of $+1$ be assigned to each person having a ten-dollar bill, and the value of -1 to each one having a five-dollar bill.

Let us add these values moving from the origin to the right and mark the obtained sum as the corresponding ordinate. The graph of the trajectory of this process is depicted in Figure 3.

Let us compute the total number of trajectories that touch or intersect the line $y = 1$ at least once. These trajectories correspond to the case where at least one customer should wait for change. For this purpose, we construct a new fictitious trajectory, as follows: The new trajectory coincides with the initial one up to the first point of contact with the line $y = 1$, and from that point on it becomes a mirror reflection of the initial trajectory with respect to the line $y = 1$ (see Figure 3). Each new trajectory starts at the point (0, 0) and terminates at the point $(2n, 2)$. It follows that this trajectory has more one-step ascents then one-step descents (namely, $n + 1$ ascents and $n - 1$ descents). It follows that the total number of new trajectories equals $\binom{2n}{n-1}$. This means that the total number of favorable trajectories equals $\binom{2n}{n} - \binom{2n}{n-1}$. Thus the desired probability equals

$$p = 1 - \frac{\binom{2n}{n-1}}{\binom{2n}{n}} = 1 - \frac{n}{n+1} = \frac{1}{n+1}.$$

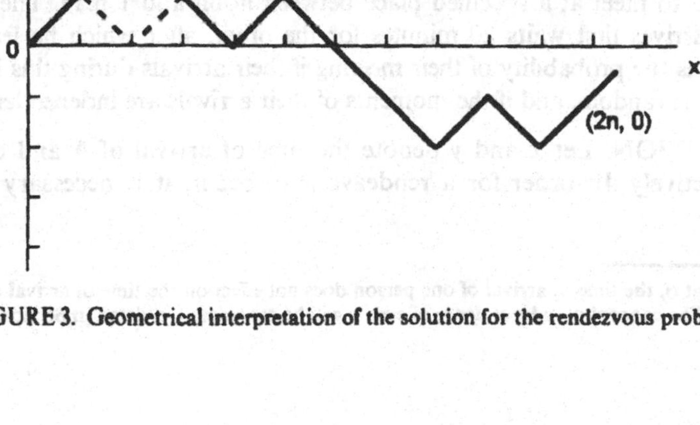

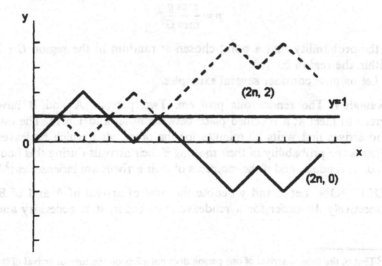

FIGURE 3. Geometrical interpretation of the solution for the rendezvous problem.

4 GEOMETRICAL PROBABILITY

The "classical" definition of probability based on the consideration of a finite group of equally probable events was considered as insufficient even at the very beginning of the development of the theory of probability. Even at that time, particular examples led to some modification of this definition and to a formulation of a concept of probability applicable to situations in which the set of conceivable outcomes was infinite. Here, nevertheless, the notion of "equal likelihood" of definite events played a central role.

The general problem which was posed, and which led to an extension of the concept of probability, can be formulated in the following way.

Suppose, for example, that there exists a region G in a plane and that it contains another region g. A point is thrown at random onto the region G, and the probability that the point falls in the region g is under investigation. The expression "thrown at random onto the region G" means that the probability of the point's falling in some part of the region G is proportional to its measure (i.e., its length, area, volume, etc.) and is independent of its location and its shape. Therefore, by definition,

$$p = \frac{\text{mes } g}{\text{mes } G}$$

is the probability that a point chosen at random in the region G fall within the region g.

Let us now consider several examples.

Example 1. The rendezvous problem. Two persons A and B have agreed to meet at a specified place between noon and 1 p. m. The one who arrives first waits 20 minutes for the other, after which he leaves. What is the probability of their meeting if their arrivals during this hour occur at random, and if the moments of their arrivals are independent[3] ?

SOLUTION. Let x and y denote the time of arrival of A and of B, respectively. In order for a rendezvous to occur, it is necessary and

[3] That is, the time of arrival of one person does not effect on the time of arrival of the other. The concept of independence of events will be considered in detail in Section 9.

sufficient that $|x-y| \leqslant 20$. Let x and y be the Cartesian coordinates of a point on a plane and one minute be taken as a unit of distance. All the possible outcomes will be depicted by the points inside a square of side 60, and the outcomes favorable to a meeting by the points of the shaded region (Figure 4).

The desired probability[4] equals the ratio of the area of the shaded figure to that of the entire square:

$$p=\frac{60^2-40^2}{60^2}=\frac{5}{9}.$$

This rendezvous problem has sometimes been applied by some practical engineers to the solution of the following maintenance problem. A service man takes care of a number of identical machines, each of which may require his attention at a random instant of time. It may

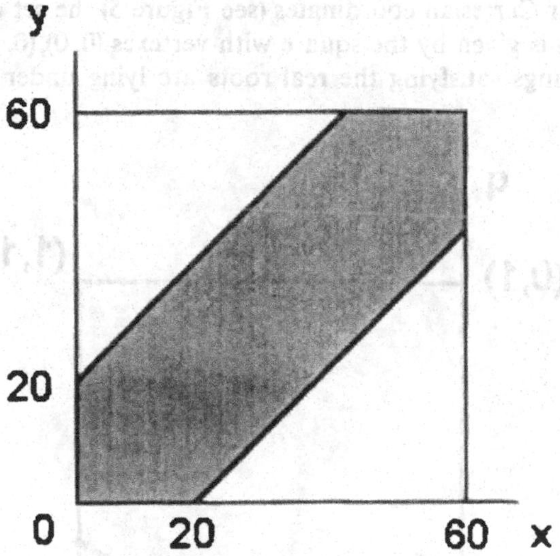

FIGURE 4. Graph of parabole for Example 2.

[4]In Section 9 it will be shown that since the arrival times of A and B are independent, the probability that A will arrive in the interval from x to $x+h$ and B in the interval from y to $y+s$ is $(h/60)$ $(s/60)$, i.e., proportional to the area of the rectangle of sides h and s.

happen that at some instant when the worker is occupied with one machine his attention is needed at another. The problem is to determine the average length of time that the machine is idle while waiting for service. Note that the scheme of the rendezvous problem is not adequate to this industrial problem (no predesignated moment of failure, the amount of time of service of the machines is not the same, etc.). In addition, one should mention the complexity of the computations entailed in the rendezvous problem for the case of a large number of participants.

Example 2. Coefficients p and q of the following equation

$$x^2 + px + q = 0$$

are chosen at random in interval (0, 1). The question is: what is the probability that the roots of the equation will be real numbers?

The roots are real numbers if the inequality $p^2 \geqslant 4q$ holds. In the rectangular Cartesian coordinates (see Figure 5) the set of all possible pairs (p, q) is given by the square with vertexes (0, 0), (0, 1), (1, 1), (1, 0). The meanings satisfying the real roots are lying under the parabola

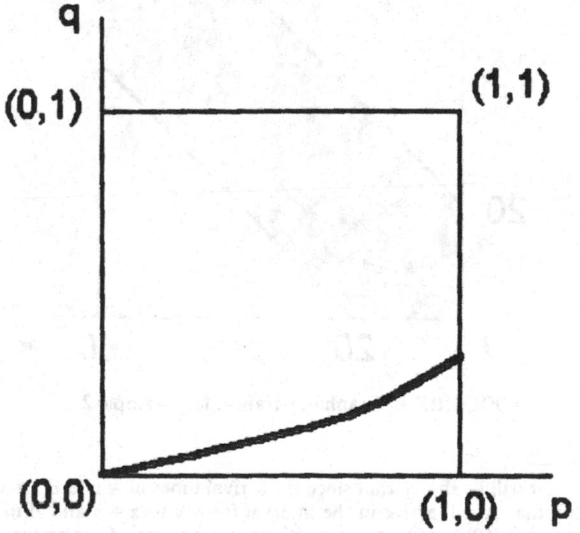

FIGURE 5. Geometrical interpretation for the Bertrand paradox.

$q = 0,25p^2$. Thus the required probability equals

$$\frac{\int_0^1 \frac{p^2}{4}\,dp}{1} = \frac{1}{12}$$

Problems of analogous types have interesting applications in the theory of numbers and in some scientific and engineering applications.

Example 3. Bertrand's paradox. The theory of geometrical probability has repeatedly been subjected to criticism because of the arbitrary definitions of the probability of events. Many authors have arrived at the conviction that in the case of an infinite number of outcomes, no definition of probability can be given that is objective and independent of the method of calculation. As a particularly bright representative of this skepticism, one may mention Joseph Bertrand, the French mathematician of the last century. In his textbook on the theory of probability, he cited a number of problems in geometrical probability in which the result depends on the method of solution. As an example of this, let us discuss one of the problems considered by Bertrand.

A chord of a circle is chosen at random. What is the probability that its length exceeds the length of a side of the inscribed equilateral triangle?

First Solution. By considerations of symmetry, one may fix the direction of the chord in advance. Let then the diameter perpendicular to this direction be drawn. Obviously, only the chords that intersect the diameter in the interval from one quarter to three quarters of its length will exceed the side of the equilateral triangle. Thus the required probability is 1/2.

Second Solution. By considerations of symmetry, one end-point of the chord may be fixed in advance. The tangent to the circle at this point and the two sides of the inscribed equilateral triangle with vertex at this point form three 60° angles. Only those chords falling within the middle angle are favorable cases. Thus, by this method of computation, the required probability turns out to be 1/3.

Third Solution. In order to fix the position of the chord, it suffices to give its midpoint. For the chord to satisfy the condition of the problem, it is necessary that its midpoint lie within the concentric circle with radius half that of the given circle. The area of this circle is one-fourth the area of the given circle. Thus, the required probability is 1/4.

Let us find out why the solution of this problem is not unique. Does it lie in the fact that it is fundamentally impossible to determine the probability when there are an infinite number of possible outcomes or does it lie in some inadmissible premises taken in the process of solving ? It is easy to see that solutions of three different problems were considered as the solution of the same problem because it was nowhere defined in what sense the chord is chosen at random. As a matter of fact, in the first solution, one considers that a circular cylindrical rod rolls along one of the diameters (Figure 6a). The set of all possible stopping points of the rod is the set of all points of the interval AB whose length is equal to the diameter. The equally possible events are taken to be the stopping of the rod in an interval of length h, regardless of where on the diameter the interval is located. In the second solution, a rod with one end hinged at a point on the circle is made to perform oscillations of 180° (Figure 6b). It is assumed here that the stopping of the rod on an arc of the circumference of length h depends only on the length of the arc and not on its location. Thus, the equally likely outcomes are taken to be the stopping of the rod upon any arcs of the circle of equal length. The inconsistency in the definition of the probability in the first and second solutions becomes quite clear after the following simple computation. According to the first setup, the probability of the rod stopping in the interval from A to x is x/D, where D is the length of the diameter. By elementary geometrical considerations, the probability that the projection of the

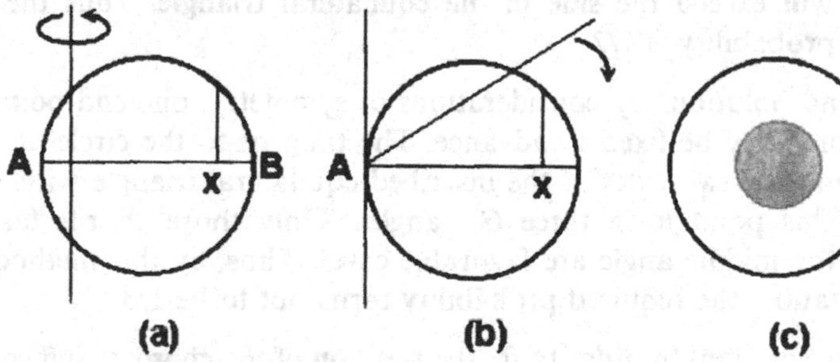

FIGURE 6. Geometrical interpretation for the Buffon problem.

point of intersection of the rod and the circle falls in the same interval equals

$$1-\frac{1}{\pi}\arccos\frac{2x-D}{D}\quad\text{for}\quad x\geqslant D/2$$

and

$$\frac{1}{\pi}\arccos\frac{D-2x}{D}\quad\text{for}\quad x\leqslant D/2.$$

Finally, in the third solution, one chooses a point at random in the given circle and asks what the probability is that the point falls within some smaller concentric circle (Figure 6c).

The difference in the formulation of the problem in all three cases is now quite clear.

Example 4. Buffon's needle problem. A plane is ruled with a series of parallel lines a distance of $2a$ apart. A needle of length $2l$ ($l<a$) is thrown at random[5] onto the plane. Determine the probability that the needle will intersect one of the lines.

SOLUTION. Denote the distance from the center of the needle to the closest line by x and the angle which the needle forms with this line by φ. The values of x and φ completely determine the position of the needle on the plane. All possible positions of the needle are then given by the coordinates of the points inside a rectangle with sides of length a and π. From Figure 7, one sees that a necessary and sufficient condition for the needle to intersect the line is that $x\leqslant l\sin\varphi$.

By virtue of the assumptions, the required probability is the quotient of the area of the shaded region in Figure 8 and the area of the rectangle:

$$p=\frac{1}{a\pi}\int_0^{\pi} l\sin\varphi\,d\varphi=\frac{2l}{a\pi}.$$

---

[5]The phrase "at random" in this example means the following. First, the center of the needle falls at random in a segment of lenth $2a$ perpendicular to the lines. Second, the probability that angle φ which the needle makes with the lines will lie between φ_1 and $\varphi_1+\Delta\varphi$ is proportional to $\Delta\varphi$. Third, the quantities x and φ are independent (see Section 7).

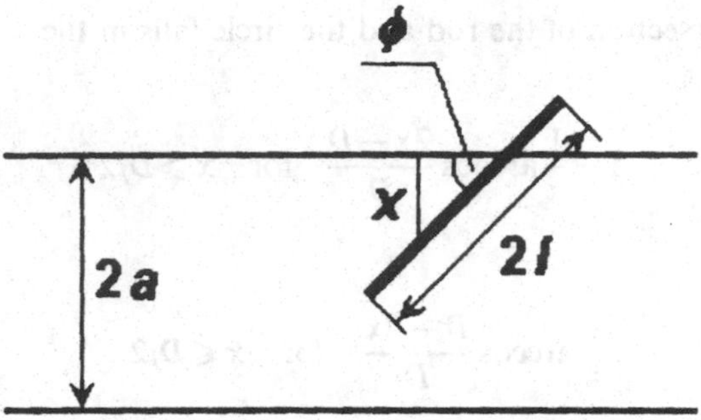

FIGURE 7. Graphical presentation of the probability for the Buffon problem.

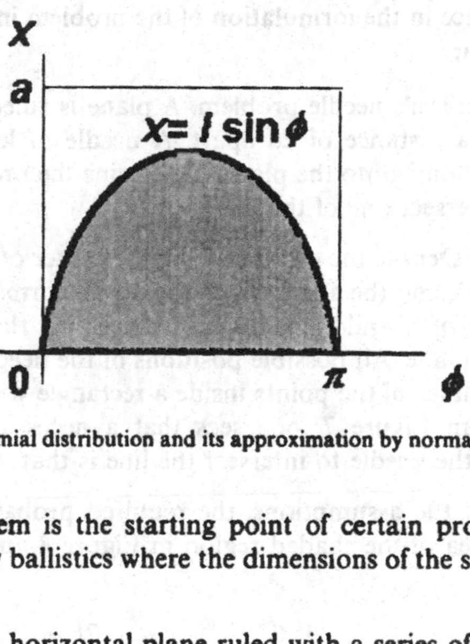

FIGURE 8. Binomial distribution and its approximation by normal distribution.

Buffon's problem is the starting point of certain problems in the theory of artillery ballistics where the dimensions of the shell are taken into account.

Example 5. On a horizontal plane ruled with a series of parallel lines a distance of $2a$ apart, a closed convex curve with a diameter less than $2a$ is thrown at random.[6] Determine the probability that this curve intersects one of the parallel lines.

[6]In this example, the phrase "at random" means that one takes some chord of the figure's boundary rigidly connected to the curve and throws it "at random" in the sense of the preceding example.

SOLUTION. To begin, suppose that the closed convex curve is an n-sided polygon. Let its sides be numbered from 1 to n. If the polygon intersects one of the lines, then this line crosses two of the sides. Denote the probability that the intersection is with the i-th and j-th sides by $p_{ij} = p_{ji}$. Obviously, the event A consisting in the intersection of one of the lines by the polygon can be expressed as the following sum of pairwise mutually exclusive events:

$$A = (A_{12} + A_{13} + \cdots + A_{1n}) + (A_{23} + A_{24} + \cdots + A_{2n})$$
$$+ \cdots + (A_{n-2,n-1} + A_{n-2,n}) + A_{n-1,n},$$

where A_{ij} ($i < j$, $i = 1, 2, \ldots$; $j = 2, \ldots$) denotes the event consisting in the intersection of the i-th and j-th sides with the line. By the theorem of addition of probabilities

$$p = \mathbf{P}(A) = [\mathbf{P}(A_{12}) + \mathbf{P}(A_{13}) + \cdots + \mathbf{P}(A_{1n})]$$
$$+ [\mathbf{P}(A_{23}) + \cdots + \mathbf{P}(A_{2n})] + \cdots + \mathbf{P}(A_{n-1,n})$$
$$= (p_{12} + p_{13} + \cdots + p_{1n}) + (p_{23} + p_{24} + \cdots + p_{2n}) + \cdots + p_{n-1,n}.$$

Using the equality $p_{ij} = p_{ji}$ one can rewrite the probability p in another way:

$$p = \frac{1}{2}[(p_{12} + p_{13} + \cdots + p_{1n}) + (p_{21} + p_{23} + \cdots + p_{2n})$$
$$+ \cdots + (p_{n1} + p_{n2} + \cdots + p_{n,n-1})].$$

But the sum

$$\sum_{j=1}^{n} p_{ij}$$

where $p_{ii} = 0$ is nothing other than the probability of intersection of the i-th side of the polygon with one of the parallel lines. If we denote the length of the i-th side of the polygon by $2l_i$, then from Buffon's problem it follows that

$$\sum_{j=1}^{n}{}' p_{ij} = \frac{2l_i}{\pi a},$$

It is not difficult to show that the concept "at random" defined in this way is independent of the choice of a chord.

and, consequently,

$$p = \frac{\sum_{i=1}^{n} 2l_i}{2\pi a}.$$

If the perimeter of the polygon is denoted by $2s$, then

$$p = \frac{s}{\pi a}.$$

Thus the probability p depends neither on the number of the polygon's sides nor on the length of the individual sides. From this one concludes that the formula also holds true for any convex closed curve, since the latter can be always regarded as the limit of convex polygons.

5 ON STATISTICAL ESTIMATION OF UNKNOWN PROBABILITY

The classical definition of probability meets insuperable difficulties of a fundamental nature if one passes from the simplest tasks to complex problems generated by the natural sciences and modern technology. First of all, the question arises how to find a reasonable way of selecting the "equally probable cases". For example, the arguments given above do not allow us to deduce the probability that an atom of some radioactive substance will disintegrate in a given interval of time. The same may be said about the gender of a newborn child.

The empirical methods of approximate estimation of the unknown probability of a random event appeared in the very beginning of the creation of probability theory.

Lengthy observations of a particular event A in a large number of repeated trials under the same set of conditions $\mathfrak{S}$ show that for a wide class of phenomena the number of occurrences or non-occurrences of the event A is subject to a stable law. Let us denote by μ the number of times the event A occurs in n independent trials. It turns out that for sufficiently large n the ratio μ/n in most of such series of observations becomes almost constant, with large deviations becoming less frequent as the number of trials increases. Moreover, if one considers a case well described by the classical definition of probability, this fluctuation of frequency is observed around the probability of the event, p. Later we will show that this fact has deep roots in Bernoulli's theorem.

Assume that an event A allows one to perform principally infinite number of independent trials under some fixed conditions $\mathfrak{G}$. If after a large number of observations it is found that the frequency of the event A fluctuated around some (in general, unknown) level, then we say that the event A has the probability of occurrence.

If the number of independent trials n is sufficiently large and the complex of condition $\mathfrak{G}$ is stable, the frequency can be taken as the desired value of the unknown probability.

At the same time trials allow conclusions of a different type. For example, let us assume that some prior information allows us to think that the probability of some event A equals p. But trials show that the frequency of the event is far from p. This circumstance can force us to change our prior hypotheses and undertake special investigations.

For example, if some particular face of a die appears essentially more frequently than 1/6, one can think that the die is not homogeneous.

As an illustration of the stability of the relative frequency in large numbers of trials, let us consider the distribution of newborn children according to sex and month of the year. The data are adopted from the book *Mathematical Methods of Statistics* by H. Cramer and constitutes part of the official Swedish vital statistics for the year 1935.

Figure 9 shows how the monthly relative frequency of birth of girls deviates from the corresponding relative frequency for the year.

We call attention to the fact that in the case of statistical definition the following properties of probability again take place:

1) The probability of the certain event equals 1;

2) The probability of the impossible event equals 0;

3) If a random event C is the sum of a finite number of mutually exclusive events $A_1, A_2, \ldots, A_n$, then its probability exists and equals the sum of the probabilities of each of the events

$$\mathbf{P}(C) = \mathbf{P}(A_1) + \mathbf{P}(A_2) + \cdots + \mathbf{P}(A_n).$$

In conclusion, it seems necessary to discuss the interpretation of probability given by R. von Mises, which is very widely used, especially among the natural scientists. Von Mises concludes that since the relative frequency deviates less and less from the probability p as the

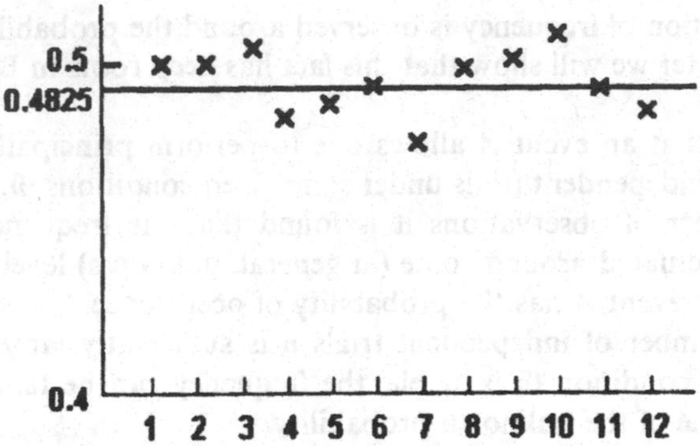

FIGURE 9. Illustration of possible outcomes of trials for n%3, k%3.

Table 3

Month	Total	Boys	Girls	Freq. (girls)
1	7280	3743	3537	0.486
2	6957	3550	3407	0.489
3	7883	4017	3866	0.490
4	7884	4173	3711	0.471
5	7892	4117	3775	0.478
6	7609	3944	3665	0.482
7	7585	3964	3621	0.462
8	7393	3797	3596	0.484
9	7203	3712	3491	0.485
10	6903	3512	3391	0.491
11	6552	3392	3160	0.482
12	7132	3761	3371	0.473
Year	88273	45682	42591	0.4825

number of experiments is continually increased, then one should have the limiting relation

$$p = \lim_{n \to \infty} \frac{\mu}{n}.$$

He proposed to regard this relation as the conceptual definition of probability. In his opinion, any prior definition is doomed to failure and only his empirical definition is capable of serving the interests of natural science, mathematics, and philosophy. Since the statistical definition is applicable in all practical and scientific situations whereas the classical definition based on equally probable events has only very limited applicability, von Mises proposed abandoning the latter entirely. Moreover he considered it altogether unnecessary to clarify the structure of the phenomena for which probability is an objective numerical property, since for him it suffices that the relative frequency is empirically stable.

Probability loses its objective character as a numerical characterization of some real phenomena in the frame of the concepts of von Mises. Discussion of the details of von Mises' theory is out of the scope of this book. The reader can find these details in his book *Probability, Statistics, and Truth.* A more extensive critique of this theory is contained in the articles by Khinchine.[7]

6 AXIOMATIC CONSTRUCTION OF THE THEORY OF PROBABILITY

Until recently, probability theory was not strictly formulated mathematical science with the fundamental concepts clearly defined. This lack of clarity frequently led to paradoxical conclusions (recall Bertrand's paradox). Naturally, the applications of the theory of probability to the study of natural phenomena were not based on a solid foundation, and at times met severe and reasonable criticism. It must be admitted that this situation did not bother scientists too much, and their naive probabilistic approach to various scientific problems led to important successes. The development of natural science at the beginning of the twentieth century put greater demands on probability theory. The systematic study of the fundamental concepts of probability theory became a necessity. That is particularly why a formal logical foundation for the theory of probability–its axiomatic construction–is of such great importance.

Probability theory as a mathematical science must have its basis in a giant experience accumulated by centuries of civilization. But the remaining part of the creation of the theory must be based on deduction. In other words, probability theory must be constructed in the same way as any other classical branch of mathematics – geometry, theoretical mechanics, abstract theory of groups, etc.

[7]The Frequency Theory of R. von Mises and contemporary ideas of probability theory. Problems of Philosophy, No. 1, pp. 91–102; No. 2, pp. 77–89, 1961 [in Russian].

The first such viewpoint was posed and developed, in 1917, by the renowned Russian mathematician Sergei N. Bernshtein. He based his work on the qualitative comparison of random events according to their larger or smaller probability.

There exists another approach proposed by Andrei N. Kolmogorov. This approach closely relates probability theory to the theory of sets and the modern measure-theoretical aspects of the theory of functions of a real variable. The present book follows the path taken by Kolmogorov.

The axiomatic construction of the foundations of probability theory proceeds from the fundamental properties of probability observed in the examples illustrating the classical and statistical definitions. Thus, the axiomatic definition of probability includes as a special case both the classical and the statistical definitions and overcomes the shortcomings of each. This basis allowed the development of a logically perfect modern probability theory which at the same time met the increased demands made upon the theory by modern science and technology.

In Kolmogorov's axiomatic construction of probability theory, the concept of a set (space) Ω consists of elementary events. Side by side with this set he considers a certain family $\mathfrak{J}$ of subsets of the set elementary events. A set $\mathfrak{J}$ is called an *algebra of sets* if the following conditions hold:

1) $\Omega \in \mathfrak{J}$, $\varnothing \in \mathfrak{J}$ ($\varnothing$ is an empty set);
2) $A \in \mathfrak{J}$ implies

$$\bar{A} \in \mathfrak{J}$$

3) $A \in \mathfrak{J}$ and $\bar{A} \in \mathfrak{J}$ together imply $A \cup B \in \mathfrak{J}$ and $A \cap B \in \mathfrak{J}$.

The set $\mathfrak{J}$ is called a σ-algebra if the following additional condition holds

4) $A_n \in \mathfrak{J}$ implies (for $n = 1, 2, \ldots$)

$$\bigcup_n A_n \in \mathfrak{J}, \quad \bigcap_n A_n \in \mathfrak{J}.$$

Operations over random events are understood as corresponding operations over sets. Table 4 represents the vocabulary for translation terms from set theory into the terms of probability theory.

Table 4

Notations	Terms	
	Set theory	Probability theory
Ω	Set, space	Space of elementary events, certain event
ω	Element of a set	Elementary event
A, B	Subsets	Random events
$A+B=A\cup B$	Union of subsets A and B	Sum of random events A and B
$AB=A\cap B$	Intersection of subsets A and B	Product of events A and B
A^-	Complimentary set to subset A	Complimentary event to event A
$A\backslash B$	Difference of sets A and B	Difference of events A and B
$\varnothing$	Empty set	Impossible event
$AB=A\cup B=\varnothing$	Sets A and B do not intersect each other	Exclusive events
$A=B$	Sets are equal	Events are equivalent
$A\subset B$	A is subset of B	Event A implies B

Let us now proceed to formulate the axioms that define probability.

AXIOM 1. With each random event A there is associated a non-negative number $\mathbf{P}(A)$ called its probability.

AXIOM 2. $\mathbf{P}(\Omega)=1$.

AXIOM 3. (Axiom of Addition of Probabilities). If the events $A_1, A_2, \ldots, A_n$ are pairwise mutually exclusive, then

$$\mathbf{P}(A_1+A_2+\cdots+A_n)=\mathbf{P}(A_1)+\mathbf{P}(A_2)+\cdots+\mathbf{P}(A_n).$$

For the classical definition of probability, it was not necessary to postulate the properties expressed by the second and third axioms, because one was able to prove them.

From these axioms one can deduce several important elementary consequences.

First of all, from the obvious equality

$$\Omega=\varnothing+\Omega$$

and Axiom 3 follows that

$$\mathbf{P}(\Omega)=\mathbf{P}(\varnothing)+\mathbf{P}(\Omega).$$

Thus :

1. The probability of the impossible event equals 0.
2. For any event A,

$$\mathsf{P}(\bar{A}) = 1 - \mathsf{P}(A).^{8}$$

3. For any random event A

$$0 \leqslant \mathsf{P}(A) \leqslant 1.$$

4. If event A implies event B then

$$\mathsf{P}(A) \leqslant \mathsf{P}(B).$$

5. Let A and B be two arbitrary events. Inasmuch as the summands in the sums $A+B=A+(B-AB)$ and $B=AB+(B-AB)$ are mutually exclusive events, it follows from Axiom 3 that

$$\mathsf{P}(A+B)=\mathsf{P}(A)+\mathsf{P}(B-AB) \quad \text{and} \quad \mathsf{P}(B)=\mathsf{P}(AB)+\mathsf{P}(B-AB).$$

From this follows the theorem of probabilities addition for two arbitrary events A and B:

$$\mathsf{P}(A+B)=\mathsf{P}(A)+\mathsf{P}(B)-\mathsf{P}(AB).$$

By virtue of the fact that $\mathsf{P}(AB)$ is non-negative, it follows that

$$\mathsf{P}(A+B) \leqslant \mathsf{P}(A)+\mathsf{P}(B)$$

For arbitrary events $A_1, A_2, \ldots, A_n$, the inequality

$$\mathsf{P}\{A_1+A_2+\cdots+A_n\} \leqslant \mathsf{P}(A_1)+\mathsf{P}(A_2)+\cdots+\mathsf{P}(A_n).$$

can be obtained by means of mathematical induction.

The system of Kolmogorov's axioms is consistent, since there exist real objects that satisfy these axioms. For example, let Ω be an arbitrary set containing a finite number of elements, i.e., $\Omega = \{a_1, a_2, \ldots, a_n\}$, and let $\mathfrak{Z}$ be the set of all the subsets

$$\{a_{i_1}, a_{i_2}, \ldots, a_{i_s}\}$$

$0 \leqslant i_1 \leqslant i_2 \leqslant \cdots \leqslant i_s \leqslant n$, $0 \leqslant s \leqslant n$, then setting $\mathsf{P}(a_1)=p_1$, $\mathsf{P}(a_2)=p_2, \ldots,$ $\mathsf{P}(a_n)=p_n$ where $p_1, p_2, \ldots, p_n$ are arbitrary non-negative numbers for

[8] This statement was formulated by Jacob Bernoulli in his Tractate.

which $p_1 + p_2 + \cdots + p_n = 1$ and

$$\mathbf{P}(a_{i_1}, a_{i_2}, \ldots, a_{i_r}) = p_{i_1} + p_{i_2} + \cdots + p_{i_r}$$

then all of Kolmogorov's axioms are satisfied.

However, this system of axioms is incomplete, since for a given set Ω one may select the probabilities in the set $\mathfrak{F}$ in different ways.

Thus, in the example considered above where one throws a die it was possible to set either

$$\mathbf{P}(E_1) = \mathbf{P}(E_2) = \cdots = \mathbf{P}(E_6) = 1/6 \tag{1}$$

or

$$\mathbf{P}(E_1) = \mathbf{P}(E_2) = \mathbf{P}(E_3) = 1/4,$$
$$\mathbf{P}(E_4) = \mathbf{P}(E_5) = \mathbf{P}(E_6) = 1/12 \tag{2}$$

and so on.

The incompleteness of the system of axioms of probability theory does not mean that the choice of the axioms is an unfortunate or insufficiently reasonable choice of them. It derives from the very nature of things: In various problems situations arise where it is necessary to study the same sets of random events, but with different probabilities. For example, we may have a pair of dice one of which is a true die (a perfect cube made of homogeneous material) and the other of which is not true. In the first case, the probability is given by the system of equations (1) and in the second case by, say, the system of equations (2).

For the further development of the theory, an additional assumption is required which is referred to as the extended axiom of probabilities addition. The necessity for introducing a new axiom is motivated by the fact that in probability theory we constantly have to consider events that decompose into an infinite number of sub-events.

EXTENDED AXIOM OF PROBABILITIES ADDITION. *If the event A is equivalent to the occurrence of at least one of the pairwise mutually exclusive events $A_1, A_2, \ldots, A_n, \ldots$ then*

$$\mathbf{P}(A) = \mathbf{P}(A_1) + \mathbf{P}(A_2) + \cdots + \mathbf{P}(A_n) + \cdots$$

Note that the extended axiom of probabilities addition can be replaced by the axiom of continuity, which is equivalent to it.

AXIOM OF CONTINUITY. *If the sequence of events $B_1, B_2, \ldots, B_n, \ldots$ is such that each succeeding event implies the preceding one and the product of all the events B_n is the impossible event, then*

$$\mathsf{P}(B_n) \to 0 \quad \text{for} \quad n \to \infty.$$

Let us prove the equivalence of the last two propositions.

1. The extended axiom of probabilities addition implies the axiom of continuity. In fact, suppose that $B_1, B_2, \ldots, B_n, \ldots$ are events such that

$$B_1 \supset B_2 \supset \cdots \supset B_n \supset \cdots$$

and for any $n \geqslant 1$,

$$\prod_{k \geqslant n} B_k = \phi. \tag{3}$$

Obviously,

$$B_n = \sum_{k=n}^{\infty} B_k \bar{B}_{k+1} + \prod_{k=n}^{\infty} B_k.$$

Since the events occurring in this sum are pairwise mutually exclusive, the extended axiom of probabilities addition gives

$$\mathsf{P}(B_n) = \sum_{k=n}^{\infty} \mathsf{P}(B_k \bar{B}_{k+1}) + \mathsf{P}\left(\prod_{k=n}^{\infty} B_k\right).$$

But the condition (3),

$$\mathsf{P}\left(\prod_{k=n}^{\infty} B_k\right) = 0,$$

therefore

$$\mathsf{P}(B_n) = \sum_{k=n}^{\infty} \mathsf{P}(B_k \bar{B}_{k+1}),$$

i.e., $\mathsf{P}(B_n)$ is the remainder of the convergent series

$$\sum_{k=1}^{\infty} \mathsf{P}(B_k \bar{B}_{k+1}) = \mathsf{P}(B_1).$$

Hence $\mathsf{P}(B_n)$ 0 as $n \to \infty$.

2. The axiom of continuity implies the extended axiom of probabilities addition.

Let the events $A_1, A_2, \ldots, A_n, \ldots$ be pairwise mutually exclusive and $A = A_1 + A_2 + \cdots + A_n + \cdots$

Let us set

$$B_n = \sum_{k=n} A_k.$$

It is evident that $B_{n+1} \subset B_n$. If the event B_n has occurred, then some event, say A_i $(i > n)$, has also occurred, and this implies, by virtue of the pairwise mutual exclusiveness of the events A_k, that the events $A_{i+1}, A_{i+2}, \ldots$ have not occurred. Thus, the events $B_{i+1}, B_{i+2}, \ldots$ are impossible and therefore the event

$$\prod_{k=n} B_k$$

is impossible. By the axiom of continuity, $\mathsf{P}(B_n) \to 0$ as $n \to \infty$. Since $A = A_1 + A_2 + \cdots + A_n + B_{n+1}$ we have by the ordinary axiom of probabilities addition

$$\mathsf{P}(A) = \mathsf{P}(A_1) + \mathsf{P}(A_2) + \cdots + \mathsf{P}(A_n) + \mathsf{P}(B_{n+1})$$

$$= \lim_{n \to} \sum_{k=1}^{n} \mathsf{P}(A_k) = \sum_{k=1} \mathsf{P}(A_k).$$

Thus, Kolmogorov's axiomatic approach allows us to construct probability theory as a part of the theory of measure and to consider a probability as a normed, completely additive, non-negative measure of a set.

The triplet $\{\Omega, \mathfrak{J}, \mathsf{P}(A)\}$ is called a probability space where Ω is a set of elementary events, $\mathfrak{J}$ is σ-algebra of subsets of Ω, and $\mathsf{P}(A)$ is the probability defined on σ-algebra $\mathfrak{J}$.

7 CONDITIONAL PROBABILITY AND THE SIMPLEST BASIC FORMULAS

It was already said that a certain set of conditions $\mathfrak{G}$ is the basis of the definition of the probability of an event. If there are no restrictions other than that the conditions $\mathfrak{G}$ are imposed on the probability $\mathsf{P}(A)$, then this probability is called unconditional.

However, in many cases, one has to determine the probability of an event under the condition that a certain event B has already occurred. Such a probability is called conditional and is denoted by the symbol $\mathbf{P}(A|B)$. This notation stands for the probability of the event A given that the event B has occurred. Strictly speaking, unconditional probabilities are also conditional, since the basic assumption of the theory is that a certain fixed set of conditions $\mathfrak{G}$ existed.

Example 1. A pair of dice is thrown. What is the probability of obtaining eight points (event A) if it is known that the number of points is even (event B)?

All the possible outcomes are indicated in Table 5. The entry in each cell corresponds to one of the possible events: the numbers in the parentheses are the points of the first and second dice, respectively.

The total number of possible outcomes is 36, of which 5 are favorable to the occurrence of the event A. Therefore, one has the unconditional probability $\mathbf{P}(A) = 5/36$. If the event B has occurred, then one out of 18 (and not 36) possibilities has been realized, and therefore the conditional probability is $\mathbf{P}(A|B) = 5/18$.

Example 2. Two cards are drawn in succession from a deck of playing cards. Determine (a) the unconditional probability that the second card is an ace (the first card drawn being unknown), and (b) the conditional probability that the second card is an ace if the first draw was an ace.

Let A denote the event that an ace appears in the second draw, and B the event that an ace appears in the first draw. It is clear that the equation

$$A = AB + A\bar{B}$$

holds.

Table 5

(1,1)	(2,1)	(3,1)	(4,1)	(5,1)	(6,1)
(1,2)	(2,2)	(3,2)	(4,2)	(5,2)	(6,2)
(1,3)	(2,3)	(3,3)	(4,3)	(5,3)	(6,3)
(1,4)	(2,4)	(3,4)	(4,4)	(5,4)	(6,4)
(1,5)	(2,5)	(3,5)	(4,6)	(5,5)	(6,5)
(1,6)	(2,6)	(3,6)	(4,6)	(5,6)	(6,6)

Since the events AB and $A\bar{B}$ are mutually exclusive, one has

$$\mathsf{P}(A)=\mathsf{P}(AB)+\mathsf{P}(A\bar{B}).$$

The drawing of two cards from a deck of 36 cards can be accomplished in $35\cdot 36$ different ways (taking into account the order in which they are drawn). Of these, $4\cdot 3$ cases are favorable to the event AB and $32\cdot 4$ cases are favorable to the event $A\bar{B}$. Therefore,

$$\mathsf{P}(A)=\frac{4\cdot 3}{36\cdot 35}+\frac{32\cdot 4}{36\cdot 35}=\frac{1}{9}.$$

If the first card is known to be an ace, then there are 35 cards remaining, of which three are aces. Hence,

$$\mathsf{P}(A|B)=3/35.$$

The general solution to the problem of finding a conditional probability can be given without difficulty in the case of the classical definition of probability. Indeed, suppose that of n exhaustive, mutually exclusive, and equally probable occurrences $A_1, A_2, \ldots, A_n$, m are favorable to the event A, k are favorable to the event B, and r are favorable to the event AB (clearly, $r\leqslant k$, $r\leqslant m$). If the event B has occurred, this implies that one of the events A_j favorable to B has occurred. Under this condition, r and only r of the events A_j are favorable to the occurrence of A. Thus,

$$\mathsf{P}(A|B)=\frac{r}{k}=\frac{r/n}{k/n}=\frac{\mathsf{P}(AB)}{\mathsf{P}(B)}. \tag{1}$$

The concept of independent events plays an important role in probability theory and its applications. In particular the main part of all results represented in this book are based on this concept.

The theorem of probabilities multiplication becomes very simple for independent events A and B

$$\mathsf{P}(AB)=\mathsf{P}(A)\cdot\mathsf{P}(B).$$

In exactly the same way, if $\mathsf{P}(A)>0$, then

$$\mathsf{P}(B|A)=\frac{\mathsf{P}(AB)}{\mathsf{P}(A)}. \tag{1'}$$

Of course, if B (or respectively, A) is the impossible event, then the equation (1) (or (1')) loses its meaning.

Note that examples 1 and 2 represent only motivations but not the proofs of (1) (and (1')).

If $\mathbf{P}(A)\,\mathbf{P}(B) > 0$ then each of equalities (1) and (1') is equivalent to the so-called theorem of probabilities multiplication by which

$$\mathbf{P}(AB) = \mathbf{P}(A)\mathbf{P}(B \mid A) = \mathbf{P}(B)\mathbf{P}(A \mid B), \tag{2}$$

i.e., the probability of the product of two events equals the product of the probability of one of them by the conditional probability of another under condition that the first one has occurred.

The theorem of probabilities multiplication is applicable when one of the events is impossible. So, if $\mathbf{P}(A) = 0$ then $\mathbf{P}(A|B) = 0$ and $\mathbf{P}(AB) = 0$.

It is said that an event A is independent of an event B if the relation

$$\mathbf{P}(A \mid B) = \mathbf{P}(A). \tag{3}$$

holds, i.e., if the occurrence of the event B does not affect the probability of the event A.[9]

If the event A is independent of the event B, then it follows from (2) that

$$\mathbf{P}(A)\,\mathbf{P}(B \mid A) = \mathbf{P}(B)\,\mathbf{P}(A).$$

From this, one finds

$$\mathbf{P}(B \mid A) = \mathbf{P}(B), \tag{4}$$

i.e., the event B is also independent of A. Thus, under the assumption, independence of events is mutual.

If independence of events A and B is defined via the equality

$$\mathbf{P}(AB) = \mathbf{P}(A)\,\mathbf{P}(B),$$

then this definition is always correct, including the situation where $\mathbf{P}(A) = 0$ or $\mathbf{P}(B) = 0$.

Let us next generalize the notion of the independence of two events to that of a collection of events.

---

[9] The concept of the conditional probability and independence of events as well as the formulation of the theorem of probabilities multiplication were created by A. DeMoivre in 1718.

The events $B_1, B_2, \ldots, B_n$ are called collectively independent if for any one of them, say, B_p, and arbitrary other events from them $B_{i_1}, B_{i_2}, \ldots, B_{i_r}$ ($i_n \neq p$), the events B_p and $B_{i_1}, B_{i_2}, \ldots, B_{i_r}$ are mutually independent.

By virtue of what has just been said, this definition is equivalent to the following: for any $1 \leqslant i_1 < i_2 < \cdots < i_r \leqslant s$ and r ($1 \leqslant r \leqslant s$)

$$\mathsf{P}(B_{i_1} B_{i_2} \cdots B_{i_r}) = \mathsf{P}(B_{i_1})\mathsf{P}(B_{i_2}) \cdots \mathsf{P}(B_{i_r}).$$

Note that for collective independence of several events it is not sufficient that they be pairwise independent. The following simple example shows this. Imagine that three faces of a tetrahedron are colored red, green, blue, and the fourth face has sectors with all of these three colors. Let A be the event that the face on which it lands contains red, B the event that it contains green, and C the event that it contains blue. Suppose the tetrahedron to be thrown once. It is easy to see that the probability that the tetrahedron lands on a face containing red is

$$\mathsf{P}(A) = 1/2.$$

Similarly, one can count

$$\begin{aligned} \mathsf{P}(B) = \mathsf{P}(C) = \mathsf{P}(A \mid B) = \mathsf{P}(B \mid C) = \mathsf{P}(C \mid A) \\ = \mathsf{P}(B \mid A) = \mathsf{P}(C \mid B) = \mathsf{P}(A \mid C) = 1/2, \end{aligned}$$

Thus the events A, B, and C are therefore pairwise independent. However, if we know that the events B and C have been realized, then the event A must also have been realized, i.e.,

$$\mathsf{P}(A \mid BC) = 1.$$

So, the events A, B and C are collectively dependent. This example is due to S. N. Bernshtein.

Formula (1′), which in the case of the classical definition was derived from the definition of conditional probability, will be taken as a definition in the case of the axiomatic definition of probability. Hence in the general case for $\mathsf{P}(A) > 0$ the relationship

$$\mathsf{P}(B \mid A) = \frac{\mathsf{P}(AB)}{\mathsf{P}(A)}.$$

is taken by definition. (In the case $\mathsf{P}(A) = 0$, the conditional probability $\mathsf{P}(B \mid A)$ remains undefined.) This enables one to carry over automatically to the general concept of probability all the definitions and results of the present section.

Suppose now that the event B can occur together with one and only one of the n mutually exclusive events $A_1, A_2, \ldots, A_n$. In other words, assume that

$$B = \sum_{i=1}^{n} BA_i, \tag{5}$$

where events BA_i and BA_j with distinct subscripts i and j are mutually exclusive. By the Theorem of probabilities addition, one has

$$\mathsf{P}(B) = \sum_{i=1}^{n} \mathsf{P}(BA_i).$$

Using the theorem of probabilities multiplication, one finds

$$\mathsf{P}(B) = \sum_{i=1}^{n} \mathsf{P}(A_i)\mathsf{P}(B \mid A_i).$$

This relation is known as the formula of total probability,[10] and plays a basic role in the theory to follow.

For illustrative purposes, let us consider two examples.

Example 1. There are five boxes: two boxes with two white balls and one black ball (call this box of type A_1); one box with ten black balls (type A_2); and two boxes with three white balls and one black ball (type A_3).

A box is selected at random, and a ball is then drawn at random from the selected box. What is the probability that the ball withdrawn is white (event B)?

SOLUTION. Since the ball can only be taken from a box of one of the three types A_1, A_2, A_3, then $B = A_1B + A_2B + A_3B$.

By the formula on total probability,

$$\mathsf{P}(B) = \mathsf{P}(A_1)\mathsf{P}(B \mid A_1) + \mathsf{P}(A_2)\mathsf{P}(B \mid A_2) + \mathsf{P}(A_3)\mathsf{P}(B \mid A_3).$$

[10] The formula of total probability was widely used by mathematicians in the beginning of the eighteenth century but it was formulated as the basic statement of probability theory only by P. Laplace at the end of the eighteenth century.

But

$$\mathbf{P}(A_1)=2/5,\ \mathbf{P}(A_2)=1/5,\ \mathbf{P}(A_3)=2/5,$$
$$\mathbf{P}(B\,|\,A_1)=2/3,\ \mathbf{P}(B\,|\,A_2)=0,\ \mathbf{P}(B\,|\,A_3)=3/4.$$

Therefore,

$$\mathbf{P}(B)=2/5\cdot 2/3+1/5\cdot 0+2/5\cdot 3/4=17/30.$$

Example 2. It is known that the probability of receiving k calls at a telephone exchange in an interval of time t is $P_t(k)$ $(k=0,1,2,\ldots)$. Assuming that the number of calls received during each of two consecutive intervals are independent, find the probability that s calls will be received during an interval of $2t$.

SOLUTION. Let $A^k_{a,a+\tau}$ denote the event consisting of k calls being received during an interval of time $[a, a\tau]$. Obviously, we can write following equality;

$$A^s_{0,2t}=A^0_{0,t}A^s_{t,2t}+\cdots+A^s_{0,t}A^0_{t,2t},$$

that means that the event $A^s_{0,2\tau}$ may be regarded as the sum of $s+1$ mutually exclusive events consisting in i calls during the first interval of time of length t and $s-i$ calls in the immediately following interval of time of the same length t $(i=0,1,2,\ldots,s)$. By the theorem on the addition of probabilities,

$$\mathbf{P}(A^s_{0,2t})=\sum_{i=0}^{s}\mathbf{P}(A^i_{0,t}A^{s-i}_{t,2t}).$$

By the theorem of multiplication of probabilities for independent events,

$$\mathbf{P}(A^i_{0,t}A^{s-i}_{t,2})=\mathbf{P}(A^i_{0,t})\,\mathbf{P}(A^{s-i}_{t,2t}).$$

Therefore, if we set

$$P_{2t}(s)=\mathbf{P}(A^s_{0,2t}),$$

then

$$P_{2t}(s)=\sum_{i=0}^{s}P_t(i)\cdot P_t(s-i). \tag{6}$$

We shall later see that under certain very general conditions,

$$P_t(k)=\frac{(at)^k}{k!}e^{-at}, \tag{7}$$

where a is some constant and $k=0,1,2,\ldots.$

From formula (6), one finds:

$$P_{2t}(s) = \sum_{i=0}^{s} \frac{(at)^s e^{-2at}}{i!(s-i)!} = (at)^s e^{-2at} \sum_{i=0}^{s} \frac{1}{i!(s-i)!}.$$

But

$$\sum_{i=0}^{s} \frac{1}{i!(s-i)!} = \frac{1}{s!} \sum_{i=0}^{s} \frac{s!}{i!(s-i)!} = \frac{1}{s!}(1+1)^s = \frac{2^s}{s!}.$$

Therefore

$$P_{2t}(s) = \frac{(2at)^s e^{-2at}}{s!} \qquad (s = 0, 1, 2, \ldots).$$

Thus, if formula (7) holds for an interval of time of length t, then it also holds for an interval twice as long and, as one can easily see, it continues to hold for an interval which is any arbitrary multiple of t.

One can now derive the important formula of Bayes, or as it is sometimes called, the formula for probabilities of hypotheses. Suppose, as before, that (5) holds. It is required to find the probability of the event A_i if it is known that event B has occurred. According to the theorem of multiplication of probabilities, one has

$$\mathbf{P}(A_i B) = \mathbf{P}(B)\mathbf{P}(A_i \mid B) = \mathbf{P}(A_i)\mathbf{P}(B \mid A_i).$$

Hence,

$$\mathbf{P}(A_i \mid B) = \frac{\mathbf{P}(A_i)\mathbf{P}(B \mid A_i)}{\mathbf{P}(B)},$$

and using the formula of total probability, one then finds:

$$\mathbf{P}(A_i \mid B) = \frac{\mathbf{P}(A_i)\mathbf{P}(B \mid A_i)}{\sum_{j=1}^{n} \mathbf{P}(A_j)\mathbf{P}(B \mid A_j)}.$$

These formulas are referred to as Bayes' formulas.[11] The general scheme for applying these formulas in the solution of practical problems is as follows. Suppose that an event B can occur under a number of

[11] T. Bayes did not derive these formulas. As a matter of fact, he obtained only formula (1) of this section. The formulas written here were obtained by P. Laplace at the end of the eighteenth century.

different conditions concerning the nature of which n hypotheses $A_1, A_2, \ldots, A_n$ can be made. Assume that the probabilities of these hypotheses $\mathsf{P}(A_i)$ are known beforehand. It is also known that the hypothesis A_i assigns a conditional probability $\mathsf{P}(B|A_i)$ to the event B. An experiment has been performed, and the event B has occurred as its result. This should lead to a reevaluation of the probabilities of the hypotheses A_i. Bayes' formula allows a quantitative solution of this problem to be obtained.

In artillery practice, one performs a so-called ranging fire for the purpose of making more precise the knowledge of firing conditions (for example, the accuracy of the gunfire). Bayes' formula finds wide use in this case. Let us merely content ourselves with a purely schematic example exclusively for illustrative purpose.

Example 1. There are five boxes as follows: two boxes, each containing 2 white balls and 3 black balls (type A_1); two boxes, each containing 1 white ball and 4 black balls (type A_2); and one box, containing 4 white balls and 1 black ball (type A_3).

From one of the boxes a ball is withdrawn at random. It turns out to be white (event B). What is the probability, after the experiment has been performed (the posteriori probability), that the ball was taken from a box of type A_3?

SOLUTION. By assumption

$$\mathsf{P}(A_1) = 2/5, \mathsf{P}(A_2) = 2/5, \mathsf{P}(A_3) = 1/5;$$

$$\mathsf{P}(B|A_1) = 2/5, \mathsf{P}(B|A_2) = 1/5, \mathsf{P}(B|A_3) = 4/5.$$

Bayes' formula then gives:

$$\mathsf{P}(A_3|B) = \frac{\mathsf{P}(A_3)\mathsf{P}(B|A_3)}{\mathsf{P}(A_1)\mathsf{P}(B|A_1) + \mathsf{P}(A_2)\mathsf{P}(B|A_2) + \mathsf{P}(A_3)\mathsf{P}(B|A_3)}$$

$$= \frac{\frac{1}{5}\cdot\frac{4}{5}}{\frac{2}{5}\cdot\frac{2}{5} + \frac{1}{5}\cdot\frac{2}{5} + \frac{1}{5}\cdot\frac{4}{5}} = \frac{4}{10} = \frac{2}{5}.$$

In exactly the same way, one finds:

$$\mathbf{P}(A_1 \mid B) = 2/5, \mathbf{P}(A_2 \mid B) = 1/5.$$

8 EXAMPLES

There are several somewhat more complicated examples of the use of the above theory.

Example 1.[12] Two players A and B continue a certain game until one of them is completely ruined. The capital of the first player is a dollars and that of the second, b dollars. The probability of winning at each play is p for player A and q for player B. Assume that there are no draws, i.e., $p + q = 1$. In each play one of the players wins the sum of one dollar. Find the probability of ruin of each player under the assumption that the results of a play do not depend on previous outcomes.

SOLUTION. Let p_n denote the probability that gambler A will ruin if he initially has n dollars. It is obvious that the probability of interest equals p_a and

$$p_{a+b} = 0 \quad \text{and} \quad p_0 = 1 \tag{1}$$

since in the first case gambler A accumulates all the possible money (he has won) and in the second case he has nothing (already ruined).

If gambler A has n dollars prior to a certain play, then he can be ruined in two distinct ways: (a) either he first wins the next play and afterwards will lose the entire game, or he loses both the next play and the entire game. Therefore, by the formula on total probability, one can write

$$p_n = p \cdot p_{n+1} + q \cdot p_{n-1}.$$

[12]This problem is formulated in a classical form of the "gambler ruin problem." Notice that there are other formulations of the same problem, for instance: a particle is on the axis in the point 0 and each second changes its position on one centimeter to the left with the probability p or one centimeter to the right with the Probability $q = 1 - p$. What is the Probability that the particle will appear from the right to a point with the coordinate b ($b > 0$) before it will appear from the left to a point with the coordinate a ($a < 0$)?

The gamber ruin problem was formulated and solved by Huygens. There is an assumption that the probability of the gambler ruin exists.

So, a finite difference equation for p_n is obtained. It is easy to see that it may be expressed in the following form:

$$q(p_n - p_{n-1}) = p(p_{n+1} - p_n). \tag{2}$$

Let us first solve this equation for $p = q = 1/2$. Under this assumption,

$$p_{n+1} - p_n = p_n - p_{n-1} = \cdots = p_1 - p_0 = c,$$

where c is a constant. Hence, one finds that

$$p_n = p_0 + nc.$$

Since $p_0 = 1$ and $P_{a+b} = 0$, one has

$$p_n = 1 - \frac{n}{a+b}.$$

Thus the probability of the ruin of A is

$$p_a = 1 - \frac{a}{a+b} = \frac{b}{a+b}.$$

In an analogous way, one finds that for $p = 1/2$ the probability of the ruin of B is

$$q_b = \frac{a}{a+b}.$$

In a general case, where $p \neq q$, one finds from (2) that

$$q^n \prod_{k=1}^{n} (p_k - p_{k-1}) = p^n \prod_{k=1}^{n} (p_{k+1} - p_k).$$

After simplifying with the use of (1), one obtains:

$$p_{n+1} - p_n = (q/p)^n (p_1 - 1).$$

Let us consider the difference $p_{a+b} - p_n$. Obviously,

$$p_{a+b} - p_n - \sum_{k=n}^{a+b-1} (p_{k+1} - p_k) = \sum_{k=n}^{a+b-1} (q/p)^k (p_1 - 1)$$

$$= (p_1 - 1)\frac{(q/p)^n - (q/p)^{a+b}}{1 - q/p}.$$

Since $p_{a+b}=0$, one has

$$p_n=(1-p_1)\frac{(q/p)^n-(q/p)^{a+b}}{1-q/p},$$

and in as much $p_0=1$,

$$1=(1-p_1)\frac{(q/p)^0-(q/p)^{a+b}}{1-q/p}.$$

Eliminating p_1 from the last two equations, one obtains

$$p_n=\frac{(q/p)^{a+b}-(q/p)^n}{(q/p)^{a+b}-1}.$$

Hence, the probability of the ruin of player A is

$$p_a=\frac{q^{a+b}-q^a q^b}{q^{a+b}-p^{a+b}}=\frac{1-(p/q)^b}{1-(p/q)^{a+b}}.$$

In an analogous way, one finds that the probability of the ruin of player B is

$$q_b=\frac{1-(q/p)^a}{1-(q/p)^{a+b}}.$$

From the above formulas one can draw the following conclusions: If the capital of one of the players, say B, exceeds by far the capital of player A, so that b may be regarded as infinite in comparison with a, the ruin of B is practically impossible if the players are equally skillful. This conclusion will be quite different if A plays better than B, and so $p>q$. Assuming $b\approx\infty$, one then finds that

$$q_b\sim 1-(q/p)^a$$

and

$$p_a\sim(q/p)^a.$$

From this one can draw the conclusion that a skillful gambler, even with small capital, stands less chance of being ruined than a gambler with a large amount of capital who is less skillful.

The solution of certain problems in physics and engineering can be reduced to the problem of the gambler's ruin.

Example 2. Find the probability that a machine that is operating at the time t_0 will not stop before the time $t_0 + t$, if it is known that: (a) this probability depends only on the length of the interval of time $(t_0, t_0 + t)$; (b) the probability that the machine will stop in the interval of time Δt is proportional to Δt except for infinitesimals of higher order[13] with respect to Δt.

SOLUTION. Let $p(t)$ denote the required probability. The probability that the machine will stop in the interval of time Δt equals

$$1 - p(\Delta t) = a\Delta t + o(\Delta t),$$

where α is some constant.

Determine the probability that the machine, which was operating at the time t_0 is still running at moment $t_0 + t + \Delta t$. For independent machines operating in distinct time intervals by the theorem of probabilities multiplication
one has

$$p(t + \Delta t) = p(t) \cdot p(\Delta t) = p(t)(1 - a\Delta t - o(\Delta t)).$$

Hence

$$\frac{p(t + \Delta t) - p(t)}{\Delta t} = -ap(t) - o(1). \tag{3}$$

Let us now pass to the limit, letting $\Delta t \to 0$. Since the limit of the right-hand side of equation (3) exists, it follows that the limit of the left-hand side also exists. As a result, one has:

$$\frac{dp(t)}{dt} = -ap(t).$$

The solution of this differential equation is the function

$$p(t) = Ce^{-at},$$

[13]Further, the fact that some α is infinitesimally smaller than a quantity β will be denoted as $\alpha = o(\beta)$. If the ratio α/β is restricted by an absolute value then the notation $\alpha = o(\beta)$ will be used.

where C is a constant. This constant is determined from the obvious condition that $p(0) = 1$. Therefore,

$$p(t) = e^{-at}.$$

The first condition of the problem imposes a strong restriction on the conditions under which the machine is operated; however, in some cases it is realized with a great degree of accuracy. As an example, one could consider the operation of an automatic loom. Many other practical problems can be reduced to the problem under consideration, for example, the question of the probability distribution of the mean free path of a molecule in the kinetic theory of gases.

Example 3. Compiling of mortality tables in demography is often based on the following assumptions.

(1) The probability that a person will die during the interval of time from t to $t + \Delta t$ equals

$$p(t, t + \Delta t) = a(t)\Delta t + o(\Delta t),$$

where $a(t)$ is a non-negative continuous function.

(2) The death (or survival) of a particular person in a given interval of time (t_1, t_2) does not depend on what happened prior to the time t_1.

(3) The probability of death at the moment of birth is zero.

Using the above assumptions, determine the probability that a person A will die before attaining age t.

SOLUTION. Let $\pi(t)$ denote the probability that A will survive to age t. Compute $\pi(t + \Delta t)$. It is obvious from the assumptions made in the problem that the equality

$$\pi(t + \Delta t) = \pi(t)\,\pi(t + \Delta t; t),$$

holds where $\pi(t + \Delta t; t)$ denotes the probability that A will survive to age $t + \Delta t$ if he already has lived to age t. In accordance with the first and second assumptions,

$$\pi(t + \Delta t; t) = 1 - p(t, t + \Delta t) = 1 - a(t)\Delta t - o(\Delta t);$$

therefore

$$\pi(t + \Delta t) = \pi(t)[1 - a(t)\Delta t - o(\Delta t)].$$

Hence, one finds that $\pi(t)$ satisfies the following differential equation:

$$\frac{d\pi(t)}{dt} = -a(t)\,\pi(t).$$

The solution of this equation, taking the third assumption in the problem into account, is the function

$$\pi(t) = e^{-\int_0^t a(z)dz}.$$

Thus, the probability of death before attaining age t is

$$1 - \pi(t) = 1 - e^{-\int_0^t a(z)dz}.$$

Compiling mortality tables for adult an population, one usually uses of Makeham's formula, which states that

$$a(t) = \alpha + \beta e^{\gamma t},$$

where α, β and γ are positive constants[14]. This formula was derived on the assumption that an adult may die from causes that do not depend upon age and from causes that do depend on age, where the probability of death from the second type of cause increases geometrically with increasing age. Under such an additional assumption, one obtains

$$\pi(t) = \exp\left\{-\alpha t - \frac{\beta}{\gamma}(e^{\gamma t} - 1)\right\}.$$

Example 4. In modern nuclear physics, a Geiger counter is used to measure the intensity of a source of radiation. A particle that hits the counter causes an electrical discharge of duration τ which blocks for this period of time the possible current registrations by the counter. Find the probability that the counter will register all the particles that hit it during the time t if the following conditions are fulfilled:

(1) The probability that k particles hit the counter in an interval of time t does not depend on how many particles hit it before this interval.

[14] Their values are determined by the conditions of the existence of a group of the population and, above all, by social factors.

(2) The probability that k particles hit the counter in the interval of time $[t_0, t_0+t]$ is given by the formula[15]

$$p_k(t_0, t_0+t) = \frac{(at)^k e^{-at}}{k!},$$

where a is a positive constant.
(3) τ is a constant.

SOLUTION. Let $A(t)$ denote the event that all particles hitting the counter during the time t have been registered, and $B_k(t)$ denote the event that k particles have hit the counter during the time interval t.

By virtue of the first condition of the problem, for $t \geqslant \tau$, one has

$$\mathbf{P}\{A(t+\Delta t)\} = \mathbf{P}\{A(t)\}\mathbf{P}\{B_0(\Delta t)\} + \mathbf{P}(A(t-\tau)\}\mathbf{P}\{B_0(\tau)\}\mathbf{P}\{B_1(\Delta t)\} + o(\Delta t),$$

and for $0 \leqslant t \leqslant \tau$

$$\mathbf{P}\{A(t+\Delta t)\} = \mathbf{P}\{A(t)\}\mathbf{P}\{B_0(\Delta t)\} + \mathbf{P}\{B_0(t)\} + \mathbf{P}\{B_1(\Delta t)\} + o(\Delta t).$$

For brevity, let us introduce $\pi(t) = \mathbf{P}\{A(t)\}$. Then by the second and third conditions, for $0 \leqslant t \leqslant \tau$

$$\pi(t+\Delta t) = \pi(t)e^{-a\Delta t} + e^{-a\Delta t}a\Delta t e^{-at} + o(\Delta t)$$

and for $t \geqslant \tau$

$$\pi(t+\Delta t) = \pi(t)e^{-a\Delta t} + \pi(t-\tau)e^{-a\Delta t}a\Delta t e^{-a\tau} + o(\Delta t).$$

Letting $\Delta t \to 0$, one obtains in the limit the differential equation for $0 \leqslant t \leqslant \tau$

$$\frac{d\pi(t)}{dt} = -a\pi(t) + ae^{-at}, \tag{4}$$

and for $t \geqslant \tau$

$$\frac{d\pi(t)}{dt} = -a[\pi(t) - \pi(t-\tau)e^{-a\tau}]. \tag{5}$$

[15]Later on it will be shown why here and in Example 2 it is assumed that the probability has such a form.

From equation (4), one finds that for $0 \leqslant t \leqslant \tau$

$$\pi(t) = e^{-at}(c + at).$$

Using the condition $\pi(0) = 1$, one can determine the constant c. Finally, for $0 \leqslant t \leqslant \tau$

$$\pi(t) = e^{-at}(1 + at). \tag{6}$$

For $\tau \leqslant t \leqslant 2\tau$, the probability $\pi(t)$ is determined from the equation

$$\begin{aligned}\frac{d\pi(t)}{dt} &= -a[\pi(t) - \pi(t-\tau)e^{-a\tau}] \\ &= -a[\pi(t) - e^{-a(t-\tau)}(1 + a(t-\tau))e^{-a\tau}] \\ &= -a[\pi(t) - e^{-at}(1 + a(t-\tau))].\end{aligned}$$

The solution of this equation gives

$$\pi(t) = e^{-at}\left(c_1 + at + \frac{a^2(t-\tau)^2}{2!}\right).$$

The constant c_1 can be found on the basis of (6)

$$\pi(\tau) = e^{-a\tau}(1 + a\tau).$$

Thus, $c_1 = 1$ and for $\tau \leqslant t \leqslant 2\tau$ one has

$$\pi(t) = e^{-at}\left[1 + at + \frac{a^2(t-\tau)^2}{2!}\right].$$

It can be shown by mathematical induction that for $(n-1)\tau \leqslant t \leqslant n\tau$ the equality

$$\pi(t) = e^{-at} \sum_{k=0}^{n} \frac{a^k[t-(k-1)\tau]^k}{k!}.$$

holds.

9 EXERCISES

1. *A*, *B* and *C* are random events.
Explain the meaning of the relations:
(a) $ABC = A$; (b) $A + B + C = A$?

2. Simplify the expressions:
(a) $(A+B)(B+C)$; (b) $(A+B)(A+\bar{B})$; (c) $(A+B)(A+\bar{B})(\bar{A}+B)$.

3. Prove the relations:
(a) $\overline{(A+B)} = \bar{A}\bar{B}$; (b) $\overline{AB} = \bar{A}+\bar{B}$;
(c) $\overline{A_1 + A_2 + \cdots + A_n} = \bar{A}_1 \bar{A}_2 \cdots \bar{A}_n$
(d) $\overline{A_1 A_2 \cdots A_n} = \bar{A}_1 + \bar{A}_2 + \cdots + \bar{A}_n$.

4. A four volume book collection is placed on a shelf in random order. What is the probability that the books are in proper order from right to left or left to right ?

5. The numbers 1, 2, 3, 4, and 5 are written down on five cards. Three cards are drawn at random in succession, and the digits thus obtained are written from left to right in the order in which they were drawn. What is the probability that the resulting three digit number turns out to be even?

6. In a lot consisting of N items, M are defective. What is the probability that in a random sample of size n there will be m ($m \leqslant M$) defective?

7. A quality control inspector examines the items in a lot consisting of m items of first grade and n items of second grade. A sample of the first b items chosen at random from the lot has shown that all of them are of second grade ($b \leqslant M$). What is the probability that of the next two items selected at random from those remaining at least one proves to be of second grade?

8. From a box containing m white balls and n black ones ($m > n$), one ball after another is drawn at random. What is the probability that at some point the number of white and black balls drawn will be the same?

9. A man writes letters to n addressees and puts each letter into a separate envelope. After this he addresses envelopes at random to all of his addressees. What is the probability that at least one letter will reach the addressee correctly?

10. In a hat there are n tickets numbered from 1 to n. The tickets are withdrawn one by one at random, without replacement. What is the probability that in at least one of the drawings the number on the ticket coincides with the number of the drawing?

11. An even number of balls is drawn at random from a box containing n white and n black balls (all of the different ways in which an even number of balls may be drawn are regarded as equally likely irrespective of their number). Find the probability that the same number of black balls and white balls are drawn.

12. *The problem* of *Chevalier de Mere*. Which is more probable: (a) throwing at least one ace with four dice or (b) at least one double ace in 24 throws of a pair of dice?

13. Three points a_1, a_2 and a_3 are selected at random on the interval $(0, a)$. Find the probability that three segments $(0, a_1)$, $(0, a_2)$ and $(0, a_3)$ are such that they allow the construction of a triangle.

14. N points are scattered at random and independently of one another inside a sphere of radius R.
(a) What is the probability that the distance from the center of the sphere to the nearest point will not be less than r?
(b) What is the limit of the probability found in part (a) if $R \to \infty$ and $(N/R^3) \to (4/3)\pi\lambda$?

REMARK: This problem is taken from astronomy: in the neighborhood of the sun, $\lambda \approx 0.0063$ if R is measured in parsecs.

15. The events $A_1, A_2, \ldots, A_n$ are independent, and $p(A_k) = p_k$. Find the probability of
(a) the occurrence of at least one of these events;
(b) the non-occurrence of all of them;
(c) the occurrence of exactly one (no matter which) of them.

16. Let $A_1, A_2, \ldots, A_n$ be any random events. Derive the formula

$$\mathsf{P}\left\{\sum_{k=1}^{n} A_k\right\} = \sum_{i=1}^{n} \mathsf{P}(A_i) - \sum_{1 \leqslant i < j \leqslant n} \mathsf{P}(A_i A_j) + \sum_{1 \leqslant i < j < k \leqslant n} \mathsf{P}(A_i A_j A_k) - \cdots (-1)^{n-1} \mathsf{P}(A_1 A_2 \cdots A_n).$$

Using this formula, solve Exercises 10 and 11.

17. The probability that a molecule which has collided with another at time $t = 0$ and undergone no further collisions with other molecules up to the time t will have a collision with another molecule in the interval of time $(t, t+t)$ is $\alpha t + o(\Delta t)$. Determine the probability that the time of free motion (i.e., the time between two successive collisions) is greater than t.

18. A bacterium reproduces by dividing into two bacteria. Assume that a bacterium divides in an interval of time Δt with a probability $\alpha t + o(\Delta t)$ which does not depend either on the number of previous divisions or on the number of existing bacteria. Determine the probability that if at time $t = 0$ there was one bacterium there will be n bacteria at time t.

Chapter 3

SEQUENCES OF INDEPENDENT TRIALS

10 INTRODUCTION

The performance of various tests and experiments is an inevitable condition of the development of science and applications. Before implementation of a new sort of corn or potato in modern agriculture, one must perform a huge volume of experimental confirmations to verify the advanced properties of the new type. Any technical equipment is tested for checking its reliability and maintainability. Analogous problems exist in pedagogy, medicine, economy, social studies. Everything new must be carefully tested before its wide implementation.

One also meets other situations where the subject of observation does not depend on the researcher. As an example, one may consider meteorology where a number of rainy days, air temperature and humidity, and other factors for different geographical areas are observed and collected for further statistical inference.

In scientific and applied activity one often must perform multiple trials under similar conditions. As a rule, outcomes of previous trials have no influence on the forthcoming outcomes. The simplest type of test is very important for practice: some event A might occur with the fixed probability p in each trial and this probability does not depend on the outcome of previous trials. This type of trial was introduced and investigated in detail by famous Swiss scientist Jacob Bernoulli (1654–1705) in his book *Ars Conjectandi* (The Art of Suggestions) published in 1713. After the author, this type of trial was called *Bernoulli trials*. This model attracts great attention, first, because of its exclusive role in the development of probability theory and its applications and, second, because of the great possibilities to expand and generalize this mathematical model. One should remember in connection with Bernoulli trials, of course, the Moivre-Laplace Theorems and the Law of Large Numbers.

Let us now consider the question: what does one understand as an elementary event? Obviously, this is a series of occurrences or non-occurrence of some event A in sequential trials. Let us connect an occurrence of the event A with the value of 1 and the opposite event with 0. Then the elementary event in this case for n trials consists of alternating series of ones and zeros. For instance, if $n=3$ all possible elementary events are written as the following triplets: (0, 0, 0), (0, 0, 1), (0, 1, 0), (1, 0, 0), (0, 1, 1), (1, 0, 1), (1, 1, 0), (1, 1, 1). The meaning of all of these events is obvious.

It is easy to see that for n trials one has 2^n different possible outcomes.

Let us now introduce a probabilistic measure on the set of possible events. There is the unique possibility. Let the probability that the event A occurs at trial k equal p, and $q=1-p$. Occurrence or non-occurrence of the event A for Bernoulli trials are independent. It follows from the theorem of multiplication that the probability that event A occurs in m specified trials (for instance, with the numbers s_1, $s_2, \ldots, s_m$) and does not occur in remaining $n-m$ trials equals $p^m q^{n-m}$. This probability does not depend on the order of numbers $s_1, s_2, \ldots, s_m$.

The simplest problem concerning the Bernoulli trials consists in determining the probability $P_n(m)$ that in n trials an event A will take place m times $(0 \leqslant m \leqslant n)$.

We just have found that the probability that the event A will have occurred in m given trials and will not have occurred in the remaining $n-m$ trials equals $p^m q^{n-m}$. By the addition theorem for probabilities, the desired probability is equal to the probability just calculated summed for all the different ways in which m occurrences of event A and $n-m$ non-occurrences of this event take place. One knows from combinatorial analysis that the number of such combinations equals

$$\binom{n}{m} = \frac{n!}{m!(n-m)!}$$

and, consequently,

$$P_n(m) = \binom{n}{m} p^m q^{n-m} \quad (m = 0, 1, 2, \ldots, n). \tag{1}$$

The equation (1) is called the *Bernoulli formula*. It is easy to see that the probability $P_n(m)$ is the coefficient of x^m in the binomial expansion of $(q+px)^n$ in powers of x. Because of this property, the set of probabilities $P_n(m)$ is called the binomial distribution law.

By a slight modification of the above reasoning, one can easily generalize the result obtained. If in each trial one of k mutually exclusive events $A_1, A_2, \ldots, A_k$ can occur, and the probability of occurrence of the event A_j in each trial is p_j, then the probability that in n trials the event A_1 will take place m_1 times, the event $A_2 - m_2$ times,..., and the event $A_k - m_k$ times $(m_1 + m_2 + \cdots + m_k = n)$ is given by

$$P_n(m_1, m_2, \ldots, m_k) = \frac{n!}{m_1!m_2!\cdots m_k!} p_1^{m_1} p_2^{m_2} \ldots p_k^{m_k}. \qquad (1')$$

One can also easily satisfy oneself that this probability is the coefficient of

$$x_1^{m_1} x_2^{m_2} \ldots x_k^{m_k}$$

in the multinomial expansion of $(p_1x_1 + p_2x_2 + \cdots + p_kx_k)^n$. Naturally, the collection of probabilities given by (1') is called the multinomial distribution law. This distribution has a wide application in natural sciences, econometrics and engineering.

Since all of possible outcomes consist in the occurrence of the event A 0,1,2,..., n times, then it is clear that

$$\sum_{m=0}^{n} P_n(m) = 1.$$

This statement can be derived without probabilistic arguing. Indeed, $p+q=1$ and by binomial formula

$$\sum_{m=0}^{n} P_n(m) = (p+q)^n = 1^n = 1.$$

Let us consider several numerical examples concerning the Bernoulli's trials. In these examples, the computations of the probabilities will not be carried to the final numerical completion because it is more reasonable to postpone this until convenient methods of computation developed below.

Example 1. The probability of a particular manufactured item being defective equals 0.005. What is the probability that, of 10,000 items chosen at random, a) exactly 40, b) no more than 70 will be defective?

In this example, $n=10{,}000$, $p=0.005$. Therefore, by formula (1), one finds:

$$\text{a) } P_{10000}(40)=\binom{10{,}000}{40}(0.995)^{9960}(0.005)^{40}.$$

The probability that there will be no more than seventy defective items is equal to the sum of the probabilities of the number of defective items being equal, respectively, to $0,1,2,\ldots$, and 70. Thus,

$$\text{b) } \mathsf{P}\{\mu\leqslant 70\}=\sum_{m=0}^{70} P_n(m)=\sum_{m=0}^{70}\binom{10{,}000}{m}0.995^{10000-m}\,0.005^m.$$

Example 2. There are two containers A and B each having a volume of 1,000 cc. Each of them contains $2.7\cdot 10^{22}$ molecules of a gas. The containers are connected so that a free exchange of molecules can occur between them but there is no exchange with the outside. Find the probability that after some specified time there will be at least 0.00000001% more molecules in one container than in another.

The probability that a given molecule is in one container or another is the same and is equal to 1/2. Thus the situation is as if $5.4\cdot 10^{22}$ trials were being performed in each of which the probability of a molecule being in container A is 1/2. Let μ be the number of molecules in container A, and $5.4\cdot 10^{22}-\mu$ molecules be in container B. One must determine the probability that

$$|\mu-(5.4\cdot 10^{22}-\mu)|\geqslant\frac{5.4\cdot 10^{22}}{10^{10}}=5.4\cdot 10^{12}.$$

In other words, one has to find the probability

$$p=\mathsf{P}\{|\mu-2.7\cdot 10^{22}|\geqslant 2.7\cdot 10^{12}\}.$$

According to the addition theorem, $p=\Sigma P_n(m)$ where the sum is taken over those values of m for which $|m-2.7\cdot 10^{22}|\geqslant 2.7\cdot 10^{12}$.

The examples considered show that the solution of real problems needs simple approximate formulas for the probabilities $P_n(m)$ and for sums

$$\sum_{m=s}^{t} P_n(m)$$

for fixed t and large values of n. In addition, the task of approximate calculation of the probabilities $P_n(m)$ for different m and large n arises. This problem will be solved later. Here we present some elementary facts concerning the behavior of the probabilities $P_n(m)$ depending on m. For $0 \leqslant m < n$, a simple calculation shows that

$$\frac{P_n(m+1)}{P_n(m)} = \frac{n-m}{m+1} \cdot \frac{p}{q}.$$

Hence it follows that

$$P_n(m+1) > P_n(m),$$

if $(n-m)p > (m+1)q$, i.e., if $np - q > m$;

$$P_n(m+1) = P_n(m),$$

if $m = np - q$ and, finally,

$$P_n(m+1) < P_n(m),$$

if $m > np - q$.

One sees that the probability $P_n(m)$ first increases with increasing m, then attains a maximum, and with the further increase of m, decreases. If $np - q$ is an integer, then the probability $P_n(m)$ takes on its maximum value for two values of m, namely, $m_0 = np - q$ and $m_0' = np - q + 1 = np + p$. If, however, $np - q$ is not an integer, then $P_n(m)$ attains a maximum for the smallest integer $m = m_0$ greater than m_0. The number m_0 is called the *most probable value* of μ. If $np - q$ is an integer, then μ has two most probable values: m_0 and $m_0' = m_0 + 1$.

Note that if $np - q < 0$, then

$$P_n(0) > P_n(1) > \cdots > P_n(n),$$

and if $np - q = 0$, then

$$P_n(0) = P_n(1) > P_n(2) > \cdots > P_n(n).$$

Later one will see that for large values of n all the probabilities $P_n(m)$ tend to be close to zero, and only for m near the most probable value of m_0 do they differ at all noticeably from zero. This fact will be proved later on. At this point consider an illustrative numerical example.

Example 3. Let $n = 50$, $p = 1/3$. There are two most probable values: $my = np - q = 16$ and $m_0 + 1 = 17$. The values of the probabilities $P_n(m)$ to four decimal places are given in Table 6.

Table 6

m	$P_n(m)$	m	$P_n(m)$	m	$P_n(m)$
<5	0.0000	13	0.0679	23	0.0202
5	0.0001	14	0.0879	24	0.0113
6	0.0004	16	0.1077	25	0.0059
7	0.0012	16	0.1178	26	0.0028
8	0.0033	17	0.1178	27	0.0012
9	0.0077	18	0.1080	28	0.0005
10	0.0157	19	0.0910	29	0.0002
11	0.0287	20	0.0704	30	0.0001
12	0.0470	21	0.0503	>30	0.0000
		22	0.0332		

11 THE LOCAL LIMIT THEOREM

Considering the numerical examples in the preceding section, one could come to the conclusion that for large values of n and m the calculation of the probabilities $P_n(m)$ involves considerable difficulties. This fact was mentioned by mathematicians in the beginning of the eighteenth century in works on demography. It became clear that there was a need for asymptotic formulas that would enable one to determine probabilities $P_n(m)$ and their sums. This problem was exhaustively solved by French mathematician Abraham DeMoivre (1667–1754) who lived all his life in England. Later on two of his excellent theorems expounded below were a subject for application and further generalization. The first of them is often called the Local Laplace Theorem but we shall refer to it as the Local DeMoivre-Laplace Theorem for the sake of doing historical justice.

THE LOCAL DEMOIVRE-LAPLACE THEOREM. *If the probability of occurrence of some event A in n independent trials is constant and equal to p* ($0 < p < 1$), *then the probability* $P_n(m)$ *that the event A occurs exactly m times in these trials satisfies the relation*

$$\sqrt{npq}\, P_n(m) : \frac{1}{\sqrt{2\pi}} e^{-(1/2)x^2} \to 1 \tag{1}$$

as $n \to \infty$ *uniformly in all m for which*

$$x = x_{mn} = \frac{m - np}{\sqrt{npq}} \tag{2}$$

are contained in some finite interval.

PROOF. The proof given is based on the well-known Stirling's formula (which, incidently, was simultaneously discovered by DeMoivre).

$$s! = \sqrt{2\pi s}\, s^s e^{-s} e^{\theta_s}$$

In this formula the power θ_s satisfies to the inequality

$$|\theta_s| \leqslant \frac{1}{12s}. \tag{2'}$$

Notice that inequality (2) can be written in the form

$$m = np + x\sqrt{npq}.$$

It follows that

$$n - m = nq - x\sqrt{npq}.$$

The latter equalities show that if x remains in some closed interval, then numbers m and $n-m$ infinitely increase with $n \to \infty$. After this remark one can use the Stirling formula. It allows one to obtain

$$P_n(m) = \frac{n!}{m!(n-m)!} p^m q^{n-m}$$

$$= \frac{1}{\sqrt{2\pi}} \sqrt{\frac{n}{m(n-m)}}\, e^{-\theta} \left(\frac{n}{m} p\right)^m \left(\frac{n}{n-m} q\right)^{n-m},$$

where

$$\theta = \theta_n - \theta_m - \theta_{n-m} < \frac{1}{12}\left(\frac{1}{n} + \frac{1}{m} + \frac{1}{n-m}\right).$$

From this expression one can see that for any interval $a \leqslant x \leqslant b$, the value of θ uniformly goes to 0 as $n \to \infty$. Consequently, the multiplier $e^{-\theta}$ under the same conditions uniformly goes to 1.

Consider now the value of

$$\ln A_n = \ln\left(\frac{n}{m} p\right)^m \left(\frac{n}{n-m} q\right)^{n-m} = -\left(np + x\sqrt{npq}\right)$$

$$\ln\left(1 + x\sqrt{\frac{q}{np}}\right) - \left(nq - x\sqrt{npq}\right) \ln\left(1 - x\sqrt{\frac{p}{np}}\right).$$

Under the assumptions of the theorem for large n the values of

$$\sqrt{\frac{q}{np}} \quad \text{and} \quad \sqrt{\frac{p}{nq}}$$

can be arbitrarily small, so one can use expansion of the logarithm. Taking the first two terms, one has

$$\ln\left(1+x\sqrt{\frac{q}{np}}\right)=x\sqrt{\frac{q}{np}}-\frac{1}{2}\frac{qx^2}{np}+O\left(\frac{1}{n^{3/2}}\right),$$

$$\ln\left(1-x\sqrt{\frac{p}{nq}}\right)=-x\sqrt{\frac{p}{nq}}-\frac{1}{2}\frac{px^2}{nq}+O\left(\frac{1}{n^{3/2}}\right).$$

Simple calculations show that

$$\ln A_n=-\frac{x^2}{2}+O\left(\frac{1}{\sqrt{n}}\right)$$

and uniformly by n in any finite interval x

$$A_n : e^{-x^2/2} \to 1 \qquad (n\to\infty).$$

Then

$$\sqrt{npq}\cdot\sqrt{\frac{n}{M(n-m)}}\to 1$$

uniformly in any finite interval x. This argument proves the theorem.

In a similar way one can prove an analogous theorem for the polynomial distribution.

THE LOCAL THEOREM FOR POLYNOMIAL DISTRIBUTION. *If the probabilities $p_1, p_2, \ldots, p_k$ that the respective events $A_1^{(s)}, A_2^{(s)}, \ldots, A_k^{(s)}$ occur in trials do not depend on the number of the trial and differ from* 0 *and* 1 *($0<p_i<1$, $i=1,2,\ldots,k$), then the probability $P_n(m_1, m_2, \ldots, m_k)$ that in n independent trials the events $A_i^{(s)}$ ($i=1,2,\ldots,k$) occur m_i times ($m_1+m_2+\cdots+m_k=n$) satisfies the relation*

$$\sqrt{n^{k-1}}\,P_n(m_1, m_2, \ldots, m_k) : \frac{e^{-(1/2)\Sigma_{i=1} q_i x_i^2}}{(2\pi)^{(k-1)/2}\sqrt{p_1 p_2 \ldots p_k}} \to 1$$

$(n\to\infty)$

uniformly in all the m_i for which

$$x_i = \frac{m_i - np_i}{\sqrt{np_i q_i}}$$

belongs to arbitrary finite intervals $a_i \leqslant x_i \leqslant b_i$.

PROOF: From the equality

$$\sum_{i=1}^{k} m_i = n$$

the relation

$$\sum_{i=1}^{k} x_i \sqrt{np_i q_i} = 0$$

follows. This permits to express any x_i via the remaining variables. In addition, notice that

$$\sum_{i=1}^{k} p_i = 1.$$

Obviously for $k = 2$ one can obtain the Local DeMoivre-Laplace theorem as a particular case.

Example 1. In Example 2 of the preceding section, it was necessary to determine $P_n(m)$ for $n = 10{,}000$, $m = 40$, and $p = 0.005$. By the Local DeMoivre-Laplace Theorem just proved, one has:

$$P_n(m) \sim \frac{1}{\sqrt{2\pi npq}} e^{-1/2\,((m-np)/\sqrt{npq})^2}$$

In this example,

$$\sqrt{npq} = \sqrt{10000 \cdot 0.005 \cdot 0.995} = \sqrt{49.75} \sim 7.05.$$

$$\frac{m - np}{\sqrt{npq}} \sim -1.42.$$

Consequently,

$$P_n(m) \sim \frac{1}{7.05\sqrt{2\pi}} e^{-1 \cdot 42^2/2}.$$

The values of the function

$$\varphi(x) = \frac{1}{\sqrt{2\pi}} e^{-x^2/2}$$

have been tabulated, and a short table is given at the end of the book. From this table one finds that:

$$P_n(m) \sim \frac{0.1456}{7.05} = 0.00206.$$

The exact computations without using the DeMoivre-Laplace Theorem, give:

$$P_n(m) \sim 0.00197.$$

In order to illustrate the kind of approximation given by the De Moivre-Laplace Theorem, and also to give a geometrical interpretation of the analytical transformations used in the proof, let us discuss a numerical example.

Let the probability p equal 0.2. In Tables 7–9 are given the values of m,

$$x = \frac{m - np}{\sqrt{npq}}$$

probabilities $P_n(m)$, quantities $(\sqrt{npq})\, P_n(m)$, and also the function

$$\varphi(x) = \frac{1}{\sqrt{2\pi}} e^{-x^2/2}$$

to four decimal places, for $n = 4, 25$, and 100 trials, respectively. In Figure 10, the ordinates depict the probabilities $P_n(m)$ for various

Table 7

$n = 4$					
m	0	1	2	3	4
$P_n(m)$	0.4096	0.4096	0.1536	0.0256	0.0016
x	−1.00	0.25	1.50	2.75	4.00
$(\sqrt{npq})\, P_n(m)$	0.3277	0.3277	0.1229	0.0205	0.0013
$\varphi(x)$	0.2420	0.3867	0.1295	0.0091	0.0001

Table 8

$n = 25$

m	x	$P_n(m)$	$\sqrt{npq}\,P_n(m)$	$\varphi(x)$
0	−2.5	0.0037	0.0075	0.0175
1	−2.0	0.0236	0.0472	0.0540
2	−1.5	0.0708	0.1417	0.1295
3	−1.0	0.1358	0.2715	0.2420
4	−0.5	0.1867	0.3734	0.3521
5	0.0	0.1960	0.3920	0.3989
6	0.5	0.1633	0.3267	0.3521
7	1.0	0.1108	0.2217	0.2420
8	1.6	0.0623	0.1247	0.1295
9	2.0	0.0294	0.0589	0.0540
10	2.5	0.0118	0.0236	0.0175
11	3.0	0.0040	0.0080	0.0044
12	3.5	0.0012	0.0023	0.0009
13	4.0	0.0003	0.0006	0.0001
14	4.5	0.0000	0.0000	0.0000
>14	>4.5	0.0000	0.0000	0.0000

integer values of the abscissa m. It is apparent from the diagram that with increasing n the quantities $P_n(m)$ decrease uniformly. In order to keep the points $[m, P_n(m)]$ in the figure from being so close to the m-axis as to be practically indistinguishable from it, we have chosen radically different scales for the coordinate axes.

If $x_n = (m - nq)/\sqrt{npq}$ and $\sqrt{npq}\,P_n(m)$ are introduced in place of m and $P_n(m)$, respectively, as abscissa and ordinate, we obtain

(1) a translation of the origin to the point $(np, 0)$, which lies near the abscissa corresponding to the maximum of the ordinate of $P_n(m)$;

(2) an increase in the unit of length along the abscissa axis by a factor $\sqrt{npq}$ (in other words, a compression of the diagram by a factor of $\sqrt{npq}$ in the direction of the abscissa); and

(3) a diminution in the unit of length along the ordinate axis by a factor of $\sqrt{npq}$ (in other words, an expansion of the diagram in the direction of the ordinate by a factor of $\sqrt{npq}$).

Figures 10 a, b, and c depict the curve $y = \varphi(x)$ and the points $[m, P_n(m)]$ after the transformation just described has been made, i.e., the points $[x_n, y_n(m)]$. One sees that for $n = 25$, the points $[x_n, y_n(m)]$ already begin to coincide with corresponding points on the graph of

Table 9

$n = 25$				
m	x	$P_n(m)$	$\sqrt{npq}\, P_n(m)$	$\varphi(x)$
8	−3.00	0.0006	0.0023	0.0044
9	−2.75	0.0015	0.0059	0.0091
10	−2.50	0.0034	0.0134	0.0175
11	−2.25	0.0069	0.0275	0.0317
12	−2.00	0.0127	0.0510	0.0540
13	−1.75	0.0216	0.0863	0.0862
14	−1.50	0.0335	0.1341	0.1295
15	−1.25	0.0481	0.1923	0.1826
16	−1.00	0.0638	0.2553	0.2420
17	−0.75	0.0788	0.3154	0.3011
18	−0.50	0.0909	0.3636	0.3521
19	−0.25	0.0981	0.3923	0.3867
20	0.00	0.0993	0.3972	0.3989
21	0.25	0.0946	0.3783	0.3867
22	0.50	0.0849	0.3396	0.3521
23	0.75	0.0720	0.2879	0.3011
24	1.00	0.0577	0.2309	0.2420
25	1.25	0.0439	0.1755	0.1826
26	1.50	0.0316	0.1266	0.1295
27	1.75	0.0217	0.0867	0.0862
28	2.00	0.0141	0.0565	0.0540
29	2.25	0.0088	0.0351	0.0317
30	2.50	0.0052	0.0208	0.0175
31	2.75	0.0029	0.0117	0.0091
32	3.00	0.0016	0.0063	0.0044

the function $y = \varphi(x)$. This agreement becomes even more pronounced for values of n larger than 25.

In order to get a clear idea of the extent to which the asymptotic DeMoivre-Laplace formula can be effectively used for finite values of n, let us consider an example. For simplicity, let us examine the case $p = q = 1/2$ and let us take only those values of n for which the value $x_{nm} = 1$ is possible. For example, such values could be $n = 25$, 100, 400, and 1156. For these n, $x_{nm} = 1$ when $n = 15$, 55, 210, and 595.

For brevity, let us set $P_n(m) = P_n$ and

$$\frac{1}{\sqrt{2\pi npq}}\, e^{-(x_{nm}^2/2)} = Q_n$$

for $p = q = 1/2$ and $x_{nm} = 1$.

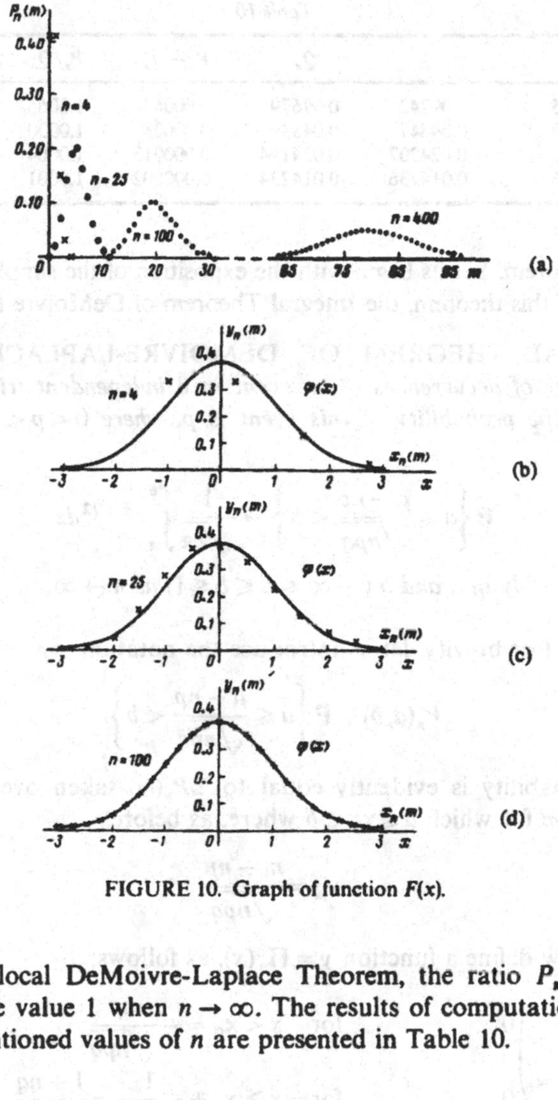

FIGURE 10. Graph of function $F(x)$.

By the local DeMoivre-Laplace Theorem, the ratio P_n/Q_n must tend to the value 1 when $n \to \infty$. The results of computation for the above-mentioned values of n are presented in Table 10.

12 THE INTEGRAL LIMIT THEOREM

The Local Limit Theorem proved in the preceding section will now be used to establish another limit relation of probability theory, the Integral

Table 10

n	P_n	Q_n	$P_n - Q_n$	P_n/Q_n
25	0.09742	0.09679	0.00063	1.0065
100	0.04847	0.04839	0.00008	1.0030
400	0.024207	0.024194	0.000013	1.0004
1156	0.014236	0.014234	0.000002	1.0001

Limit Theorem. Let us begin with the exposition of the simplest particular case of this theorem, the Integral Theorem of DeMoivre-Laplace.

INTEGRAL THEOREM OF DEMOIVRE-LAPLACE. *If* μ *is the number of occurrences of an event in n independent trials in each of which the probability of this event is p, where* $0 < p < 1$, *then the relation*

$$\mathbf{P}\left\{a \leqslant \frac{\mu - np}{\sqrt{npq}} < b\right\} \to \frac{1}{\sqrt{2\pi}} \int_a^b e^{-z^2/2} dz$$

holds uniformly in a and b $(-\infty \leqslant a \leqslant b \leqslant 1)$ *as* $n \to \infty$.

PROOF. For brevity, let us introduce the notation

$$P_n(a, b) = \mathbf{P}\left\{a \leqslant \frac{\mu - np}{\sqrt{npq}} < b\right\}.$$

This probability is evidently equal to $\Sigma P_n(m)$ taken over all those values of m for which $a \leqslant x_m \leqslant b$ where, as before,

$$x_m = \frac{m - np}{\sqrt{npq}}$$

Let us now define a function $y = \Pi_n(x)$, as follows:

$$y = \prod_n(x) = \begin{cases} 0 & \text{for} \quad x < x_0 = -\dfrac{np}{\sqrt{npq}} \\ 0 & \text{for} \quad x \geqslant x_n + \dfrac{1}{\sqrt{npq}} = \dfrac{1 + nq}{\sqrt{npq}}, \\ \sqrt{npq}\, P_n(m) & \text{for} \quad x_m \leqslant x < x_{m+1} \; (m = 0, 1, \ldots, n). \end{cases}$$

The probability $P_n(m)$ is obviously equal to the area bounded by the curve $y = \Pi_n(x)$, the x-axis, and the ordinates $x = x_m$ and $x = x_{m+1}$,

$$P_n(m) = \sqrt{npq}\, P_n(m)(x_{m+1} - x_m) = \int_{x_m}^{x_{m+1}} \prod_n(x)\,dx.$$

It follows that the desired probability $P_n(a, b)$ equals the area bounded by the curve $y = \Pi_n(x)$, the x-axis, and the ordinates

$$x_{\underline{m}} \quad \text{and} \quad x_{\bar{m}}$$

where $\bar{m}$ and $\bar{m}$ are defined by the inequalities

$$a \leqslant x_{\bar{m}} \leqslant a + \frac{1}{\sqrt{npq}}, \quad b \leqslant x_{\underline{m}} < b + \frac{1}{\sqrt{npq}}.$$

Thus,

$$\begin{aligned} P_n(a, b) &= \int_{x_{\underline{m}}}^{x_{\bar{m}}} \prod_n(x)\,dx \\ &= \int_a^b \prod_n(x)\,dx + \int_b^{x_{\bar{m}}} \prod_n(x)\,dx - \int_a^{x_{\underline{m}}} \prod_n(x)\,dx. \end{aligned}$$

Since the maximum value of the probability $P_n(m)$ occurs at $m_0 = [(n+1)p]$, the maximum value of $\Pi_n(x)$ lies in the interval

$$0 \leqslant \frac{m_0 - np}{\sqrt{npq}} \leqslant x < \frac{m_0 + 1 - np}{\sqrt{npq}} \leqslant \frac{2}{\sqrt{npq}}.$$

The Local DeMoivre-Laplace Theorem is applicable in this interval and one can therefore conclude that for all sufficiently large values of n

$$\max \prod_n(x) < 2\frac{1}{\sqrt{2\pi}} \max e^{-x^2/2} = \sqrt{2/\pi}.$$

From this, one can deduce first of all that

$$\begin{aligned} |\rho_n| = \left| \int_b^{x_m} \prod_n(x)\,dx - \int_a^{x_m} \prod_n(x)\,dx \right| &\leqslant \int_b^{x_m} \max \prod_n(x)\,dx \\ + \int_a^{x_m} \max \prod_n(x)\,dx &< \sqrt{\frac{2}{\pi}}(-b + x_m + x_m - a) \leqslant 2\sqrt{\frac{2}{\pi npq}} \end{aligned}$$

and therefore that

$$\lim_{n\to\infty} \rho_n = 0.$$

Thus, $P_n(a, b)$ only differs from

$$\int_a^b \prod_n(x)\,dx$$

by an infinitesimal value.

First assume that a and b are finite. Under this assumption, by the local limit theorem for $a \leq x_m < b$, one has

$$\prod_n(x_m) = \frac{1}{\sqrt{2\pi}} e^{-x_m^2/2}\,[1 + \alpha_n(x_m)],$$

where $\alpha_n(x_m) \to 0$ uniformly in x_m as $n \to \infty$. Obviously, for intermediate values of the argument as well,

$$\prod_n(x) = \frac{1}{\sqrt{2\pi}} e^{-x^2/2}\,[1 + \alpha_n(x)],$$

where

$$\lim_{n\to\infty} \lim_{a \leqslant x < b} \alpha_n(x) = 0.$$

Indeed, for any m in the interval $x_m \leqslant x < x_{m+1}$ one has

$$\prod_n(x) = \prod_n(x_m) = \frac{1}{\sqrt{2\pi}} e^{-x^2/2}\,[1 + \alpha_n(x)],$$

where

$$\alpha_n(x) = e^{(x^2 - x_m^2)/2}\,[\alpha_n(x_m) + 1] - 1.$$

Since

$$\frac{x^2 - x_m^2}{2} \leqslant |x| \cdot |x - x_m| < \frac{\max(|a|, |b|)}{\sqrt{npq}},$$

it is clear that

$$\lim_{n\to\infty} \max_{a \leqslant x < b} \alpha(x) = 0.$$

Collecting these estimates one obtains

$$P_n(a,b) = \frac{1}{\sqrt{2\pi}}\int_a^b e^{-x^2/2}\,dx + R_n,$$

where

$$R_n = \frac{1}{\sqrt{2\pi}}\int_a^b e^{-x^2/2}\alpha_n(x)dx + \rho_n.$$

Since

$$|R_n| \leqslant \max_{a \leqslant x < b} |\alpha_n(x)| \cdot \frac{1}{\sqrt{2\pi}}\int_a^b e^{-x^2/2}\,dx + \rho_n,$$

it follows that

$$\lim_{n\to\infty} R_n = 0.$$

The theorem is now proved, under the assumption made above. Let us omit this restriction.

First of all one observes that

$$\frac{1}{\sqrt{2\pi}}\int e^{-z^2/2}\,dz = 1.$$

Therefore, for any $\varepsilon > 0$ it is possible to choose A sufficiently large that

$$\frac{1}{\sqrt{2\pi}}\int_{-A}^{A} e^{-z^2/2}\,dz > 1 - \frac{\varepsilon}{4},$$

$$\frac{1}{\sqrt{2\pi}}\int_{-\infty}^{A} e^{-z^2/2}\,dz = \frac{1}{\sqrt{2\pi}}\int_{A} e^{-z^2/2}\,dz < \frac{\varepsilon}{8}.$$

Then one, in accordance with what has been proved above, chooses n so large that for $-A \leqslant a \leqslant b < A$

$$\left|P_n(a,b) - \frac{1}{\sqrt{2\pi}}\int_a^b e^{-z^2/2}\,dz\right| < \frac{\varepsilon}{4}.$$

It is then obvious that

$$P_n(-A, A) > 1 - \varepsilon/2,$$

$$P(-\infty, -A) + P(A, +\infty) = 1 - P(-A, A) < \varepsilon/2.$$

Let us now prove that, for any a and $b(-\infty \leqslant a \leqslant b \leqslant +\infty)$,

$$\left| P_n(a, b) - \frac{1}{\sqrt{2\pi}} \int_a^b e^{-z^2/2}\, dz \right| < \varepsilon,$$

whereupon the proof of the DeMoivre-Laplace Theorem will clearly be complete.

To do this, it is necessary to consider separately each of the different ways in which the points a and b can be situated with respect to the interval $(-A, A)$. For example, let us consider the case $a \leqslant -A$, $b \geq A$ (the remaining cases being left to the reader to prove).

In the case under consideration,

$$\frac{1}{\sqrt{2\pi}} \int_a^b e^{-z^2/2}\, dz = \frac{1}{\sqrt{2\pi}} \left(\int_a^{-A} + \int_{-A}^{A} + \int_A^b e^{-z^2/2}\, dz \right),$$

$$P_n(a, b) = P_n(a, -A) + P_n(-A, A) + P_n(A, b).$$

Therefore,

$$\left| P_n(a, b) - \frac{1}{\sqrt{2\pi}} \int_a^b e^{-z^2/2}\, dz \right| \leqslant \left| P_n(a, -A) - \frac{1}{\sqrt{2\pi}} \int_a^{-A} e^{-z^2/2}\, dz \right|$$

$$+ \left| P_n(-A, A) - \frac{1}{\sqrt{2\pi}} \int_{-A}^{A} e^{-z^2/2}\, dz \right| + \left| P_n(A, b) \right.$$

$$\left. - \frac{1}{\sqrt{2\pi}} \int_A^b e^{-z^2/2}\, dz \right| \leqslant P_n(-\infty, -A) + \frac{1}{\sqrt{2\pi}} \int_{-\infty}^{-A} e^{-z^2/2}\, dz$$

$$+ \left| P_n(-A, A) - \frac{1}{\sqrt{2\pi}} \int_{-A}^{A} e^{-z^2/2}\, dz \right| + P_n(A, +\infty)$$

$$+ \frac{1}{\sqrt{2\pi}} \int_A^{\infty} e^{-z^2/2}\, dz < \frac{\varepsilon}{2} + \frac{\varepsilon}{4} + \frac{\varepsilon}{8} + \frac{\varepsilon}{8} = \varepsilon.$$

Let us now proceed to the formulation of the Integral Limit Theorem for the general case of repeated independent trials. As before,

$\mu_i (i = 1, 2, \ldots, k)$ is the number of occurrences of the event $A_i^{(s)} (s = 1, 2, \ldots, n)$ in n sequentially repeated trials. Depending on chance, the numbers μ_i can take values $0, 1, 2, \ldots, n$ in such a manner that only one of k outcomes is possible and all of these outcomes are mutually exclusive. Then the following equality must take place

$$\mu_1 + \mu_2 + \cdots + \mu_k = n \tag{1}$$

The quantities $\mu_1, \mu_2, \ldots, \mu_k$ might be considered as the rectangular coordinates of a point in k-dimensional Euclidean space.

In this way, the outcomes of the n trials are reflected by a point whose coordinates are integers not less than 0 and not greater than n. From now on we shall call these points lattice points. Equation (1) shows that the outcomes of the trials are reflected not by any lattice points whatever within the hypercube $0 \leqslant \mu_i \leqslant n (i = 1, 2, \ldots, k)$ but only by those points that lie on the hyperplane (1). Figure 11 depicts the positions of the possible outcomes of the trials on the hyperplane (1) for the case $n = 3$, $k = 3$.

Let us now make a transformation of coordinates by means of the formulas

$$x_i = \frac{\mu_i - np_i}{\sqrt{np_i q_i}} \qquad (i = 1, 2, \ldots, k; q_i = 1 - p_i).$$

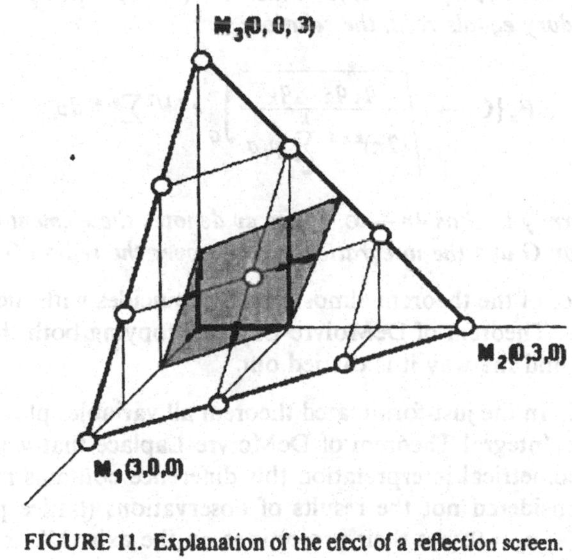

FIGURE 11. Explanation of the efect of a reflection screen.

In the new coordinates, the equation of the hyperplane (1) takes on the following form:

$$\sum_{i=1}^{k} x_i\sqrt{np_i q_i} = 0. \tag{2}$$

Let us agree that the points of the hyperplane (2) which correspond to the lattice points of the hyperplane (1) are also to be called "lattice points."

Denote $P_n(G)$ the probability that as the result of n trials the number of occurrences $\mu_i = (i = 1, 2, \ldots, k)$ of each of the possible outcomes will be such that the point with the coordinates

$$x_i = \frac{\mu_i - np_i}{\sqrt{np_i q_i}}$$

falls inside some region G.

Then the following theorem takes place.

THEOREM. *If in a scheme of repeated independent trials each of which has k possible outcomes, the probability of each outcome does not depend on the number of the trial and differs from* 0 *and* 1, *then for any region G of the hyperplane* (2) *for which the* $(k-1)$*-dimensional volume of its boundary equals zero, the relation*

$$P_n\{G\} \to \sqrt{\frac{q_1 q_2 \ldots q_k}{(2\pi)^{k-1} \sum_{i=1}^{k} p_i q_i}} \int_G e^{-1/2 \sum q_i x_i^2}\, dv,$$

holds uniformly in G as $4n \to \infty$, *where dv denotes the element of volume of the region G and the integration is taken over the region G.*

The proof of the theorem almost exactly coincides with the proof of the Integral Theorem of DeMoivre-Laplace, copying both the idea of that proof and the way it is carried out.

REMARK. In the just-formulated theorem all variables play a similar role. In the Integral Theorem of DeMoivre-Laplace that was not the case. In geometrical interpretation this difference contains in the fact that we considered not the results of observations (lattice points on the line $x_1 + x_2 = 0$) but their projections on the axis OX. In an analogous way one can consider not integration in the region G but in its projection on some coordinate hyperplane, say, on plane $x_k = 0$. The elementary volume dv' on hyperplane $x_k = 0$ corresponds to the elementary volume dv of hyperplane (2) via the expression

$$dv' = dv \cos \varphi,$$

where φ is an angle between the hyperplanes mentioned above. It is easy to calculate that

$$\cos \varphi = \frac{\sqrt{p_k q_k}}{\sqrt{\Sigma p_i q_i}}.$$

In the coordinate hyperplane the unit of elementary volume is $dv' = dx_1 \, dx_2 \ldots dx_{k-1}$ therefore the following equality

$$\sqrt{\frac{q_1 q_2 \cdots q_{k1}}{(\pi)^k}} \Sigma p_i q_i \int_G e^{-(1/2)\Sigma q_i x_i^2}$$

$$= \sqrt{\frac{q_1 q_2 \cdots q_{k-1}}{(2\pi)^{k-1} p_k}} \int_G e^{-(1/2)\Sigma_1^k q_i x_i^2} dx_1 \cdots dx_{k-1}$$

holds. In the function under the integral, the following substitution

$$x_k = -\frac{1}{\sqrt{p_k q_k}} \sum_{i=1}^{k-1} \sqrt{p_i q_i} x_i.$$

must be done where x_k is expressed via $x_1, x_2, \ldots, x_{k-1}$. After this substitution, one has

$$\sum_{i=1}^{k} q_i x_i^2 = \sum_{i=1}^{k-1} q_i \left(1 + \frac{p_i}{p_k}\right)^2 x_i^2 + 2 \sum_{1 \leq i < j \leq k-1} x_i x_j \frac{\sqrt{p_{qi} p_j q_j}}{p_k}$$

$$= Q(x_1, x_2, \ldots, x_{k-1}).$$

Thus the Integral Limit Theorem can be reformulated:
Under the conditions of the Integral Limit Theorem for $n \to \infty$

$$P(G) \to \sqrt{\frac{q_1 q_2 \cdots q_{k-1}}{(2\pi)^{k-1} p_k}} \int_{G'} e^{-(1/2) Q(x_1, x_2, \ldots, x_{k-1})} dx_1 \, dx_2 \cdots dx_{k-1}. \quad (3)$$

It is clear that the Integral Limit DeMoivre-Laplace Theorem is a particular case of the just-proved theorem because it could be easily obtained from formula (3). For this purpose, it is enough to notice that for Bernoulli trials $k=2$, $p_1=p2$, $q=p_2=1-p$.

For $k=3$, formula (3) takes the form

$$P(G)\to\sqrt{\frac{q_1 q_2}{(2\pi)^2 p_3}}\int_{G'} e^{-(1/2)Q(x_1,x_2)}\,dx_1\,dx_2,$$

where

$$P_3=1-p_1-p_2,$$

$$Q(x_1,x_2)=q_1\left(1+\frac{p_1}{p_3}\right)x_1^2+q_2\left(1+\frac{p_2}{p_3}\right)x_2^2+2\frac{\sqrt{p_1 q_1 p_2 q_2}}{p_3}x_1 x_2$$

$$=\frac{q_1 q_2}{p_3}\left(x_1^2+x_2^2+2\sqrt{\frac{p_1 p_2}{q_1 q_2}}\,x_1 x_2\right).$$

A simple calculation shows that $p_3=1-p_1-p_2=q_1 q_2-p_1 p_2$, therefore

$$Q(x_1,x_2)=\frac{1}{1-\dfrac{p_1 p_2}{q_1 q_2}}\left(x_1^2+x_2^2+2\sqrt{\frac{p_1 p_2}{q_1 q_2}}\,x_1 x_2\right).$$

13 APPLICATIONS OF THE INTEGRAL THEOREM OF DE MOIVRE-LAPLACE

As a first application of the Integral Theorem of DeMoivre-Laplace, let us estimate the probability of the inequality

$$\left|\frac{\mu}{n}-p\right|<\varepsilon,$$

where $\varepsilon>0$ is a constant.

One has

$$\mathbf{P}\left\{\left|\frac{\mu}{n}-p\right|<\varepsilon\right\}=\mathbf{P}\left\{-\varepsilon\sqrt{\frac{n}{pq}}<\frac{\mu-np}{\sqrt{npq}}<\varepsilon\sqrt{\frac{n}{pq}}\right\}$$

and this implies, by virtue of the Integral Theorem of DeMoivre-Laplace, that

$$\lim_{n\to\infty} \mathbf{P}\left\{\left|\frac{\mu}{n} - p\right| < \varepsilon\right\} = \frac{1}{\sqrt{2\pi}} \int e^{-z^2/2}\, dz = 1.$$

Thus, for any constant ε, the probability of the inequality

$$\left|\frac{\mu}{n} - p\right| < \varepsilon$$

Approaches 1.

This fact was first discovered by Jacob Bernoulli. It is called the Law of Large Numbers, or Bernoulli's Theorem. Bernoulli's Theorem and its multiple generalizations are among the most important theorems of probability theory. Through Bernoulli Theorem and its generalizations, the probability theory comes into contact with practice. They open the door for the successful applications of probability theory to diverse scientific and engineering problems. This will be discussed in detail in the chapter devoted to the Law of Large Numbers. There will be given a proof of Bernoulli's Theorem using simpler methods distinct both from the one just presented and from the one given by Bernoulli himself.

Consider now some typical problems leading to the DeMoivre-Laplace Theorem.

Let n independent trials be performed in each of which the probability of occurrence of an event A is p.

I. A question arising is: What is the probability that the relative frequency of occurrence of the event A will differ from the probability p by not more than α? This probability equals

$$\mathbf{P}\left\{\left|\frac{\mu}{n} - p\right| \leqslant \alpha\right\} = \mathbf{P}\left\{-\alpha\sqrt{\frac{n}{pq}} \leqslant \frac{\mu - np}{\sqrt{npq}} \leqslant \alpha\sqrt{\frac{n}{pq}}\right\}$$

$$\sim \frac{1}{\sqrt{2\pi}} \int_{-\alpha\sqrt{n/pq}}^{\alpha\sqrt{n/pq}} e^{-x^2/2}\, dx = \frac{2}{\sqrt{2\pi}} \int_0^{\alpha\sqrt{n/pq}} e^{-x^2/2}\, dx.$$

II. What is the smallest number of trials one must carry out in order that, with a probability not less than β, the relative frequency differs from p by no more than α? To answer the question, one has to determine n from the inequality

$$\mathsf{P}\left\{\left|\frac{\mu}{n}-p\right|\leqslant\alpha\right\}\geqslant\beta.$$

One replaces the probability appearing on the left-hand side of this inequality by the approximate value of the integral given by the DeMoivre-Laplace Integral Theorem. To obtain n, as a result, one obtains the inequality

$$\frac{2}{\sqrt{2\pi}}\int_0^{\alpha\sqrt{n/pq}} e^{-(x^2/2)}\,dx\geqslant\beta.$$

III. For a given probability β and a given number of trials n, it is required to find a bound for the possible values of $|(\mu/n)-p|$. In other words, knowing β and n, it is necessary to determine an α for which

$$\mathsf{P}\left\{\left|\frac{\mu}{n}-p\right|<\alpha\right\}=\beta.$$

The application of the Integral Theorem yields the equation

$$\frac{2}{\sqrt{2\pi}}\int_0^{\alpha\sqrt{n/pq}} e^{-x^2/2}\,dx=\beta$$

for the determination of α.

The numerical solution of each of these problems requires the computation of the values of the integral

$$\Phi(x)=\frac{1}{\sqrt{2\pi}}\int_0^x e^{-z^2/2}\,dz \tag{1}$$

for arbitrary values of x, and also the solution of the inverse problem: given the value of the integral $\Phi(x)$, find the corresponding value of the argument x. These computations necessitate special tables, since for $0<x<\infty$ the integral (1) is not expressible in closed form in terms of elementary functions. Such tables have been compiled, and can be found at the end of the book.

Figure 12 gives a graphical representation of the function $\Phi(x)$. By use of the table of values of the function $\Phi(x)$ one can also compute

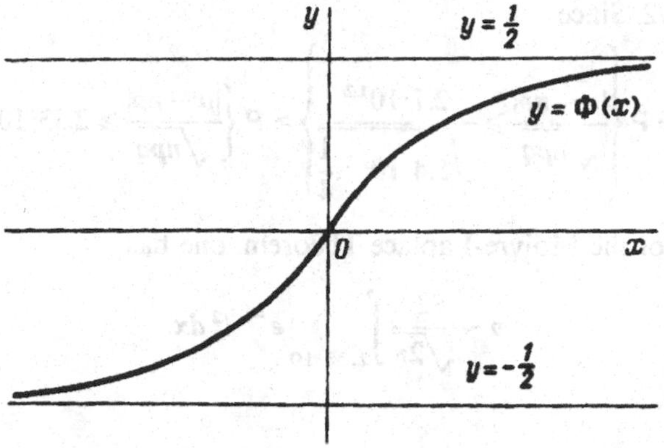

FIGURE 12. Illustration of two symmetrical paths.

the value of the integral

$$J(a,b) = \frac{1}{\sqrt{2\pi}} \int_a^b e^{-z^2/2} dz$$

with the formula $J(a,b) = \Phi(b) - \Phi(a)$.

The table of values of $\phi(x)$ are compiled for positive x only. For negative x, the function $\Phi(x)$ can be determined as

$$\Phi(-x) = -\Phi(x).$$

It is now possible to complete the solution of Example 2 of Section 9.

Example 1. In Example 2 of Section 9, it was necessary to find the probability

$$p = \Sigma \mathbf{P}\{\mu = m\},$$

where the summation extends over those values of m for which

$$|m - 2.7 \cdot 10^{22}| \geqslant 2.7 \cdot 10^{12},$$

under the condition that the total number of trials is $n = 5.44 \cdot 10^{22}$

and $p=1/2$. Since

$$p=\mathbf{P}\left\{\frac{|\mu-np|}{\sqrt{npq}}\geqslant\frac{2.7\cdot 10^{12}}{\sqrt{5.4\cdot 10^{22}\cdot\frac{1}{4}}}\right\}\approx\mathbf{P}\left\{\frac{|\mu-np|}{\sqrt{npq}}\geqslant 2.33\cdot 10\right\},$$

by virtue of the Moivre-Laplace Theorem, one has

$$p\sim\frac{2}{\sqrt{2\pi}}\int_{2.33\cdot 10}e^{-x^2/2}\,dx.$$

Because

$$\int_z e^{-x^2/2}\,dx<\frac{1}{z}\int_z^{\infty}xe^{-x^2/2}\,dx=\frac{1}{z}e^{-z^2/2},$$

it follows that

$$p<\frac{1}{\sqrt{2\pi}\cdot 10^5}e^{-2.7\cdot 100}<10^{-100}.$$

One may judge how small this probability is by the following com parison. Suppose that a sphere of radius 6,000 km is filled with white sand in which there is a single black grain. Each grain has a volume of 1 mm^3. One grain is taken at random from the entire mass. What is the probability that it will be black? It is easily to calculate that the volume of a sphere of radius 6,000 km is a little less than 10^{30} mm^3, and consequently the probability of choosing the black grain is a little more than 10^{-30}.

Example 2. In Example 1 of Section 9, one had to determine the probability that the number of defective items would be no larger than seventy if the probability of each item being defective was $p=0.005$, and if the total number of items was 10,000. By the DeMoivre-Laplace Integral Theorem, this probability is given by

$$\mathbf{P}\{\mu\leqslant 70\}=\mathbf{P}\left\{-\frac{50}{\sqrt{49.75}}\leqslant\frac{\mu-np}{\sqrt{npq}}\leqslant\frac{20}{\sqrt{49.75}}\right\}$$

$$=\mathbf{P}\left\{-7.09\leqslant\frac{\mu-np}{\sqrt{npq}}\leqslant 2.84\right\}\sim\frac{1}{\sqrt{2\pi}}\int_{-7.09}^{2.84}e^{-z^2/2}\,dz$$

$$=\Phi(2.84)-\Phi(-7.09)=\Phi(2.84)+\Phi(7.09)=0.9975.$$

The value of $\Phi(x)$ is not tabulated for $x=7.09$. We have approximated it as 1/2, committing, in so doing, an error less than 10^{-10}.

Naturally, in the examples of this and the preceding sections, as well as in any other problem involving the determination of the probabilities $P_n(m)$ for arbitrary finite values of n and m by means of the asymptotic formula of DeMoivre-Laplace, it is necessary to estimate the error introduced by making such a substitution. For a long time, the theorems of DeMoivre-Laplace were applied to the solution of problems of this kind without any satisfactory estimate of the remainder term. A purely empirical feeling of certainty arose that, as long as n is about some hundreds or larger and p is not too close to 0 or 1, the use of the DeMoivre-Laplace Theorem leads to practically satisfactory results. At present, there exist sufficiently good estimates of the error committed in using the asymptotic formula of DeMoivre-Laplace.

We consider one more extension of the Bernoulli Theorem to the general case of repeated independent trials. Let k different outcomes in each trial be possible with the respective probabilities $p_1, p_2, \ldots, p_k$ and the respective number of occurrences of each outcome in n repeated independent trials $\mu_1, \mu_2, \ldots, \mu_k$. Let us determine the probability of the simultaneous realization of the inequalities

$$\left|\frac{\mu_1}{n}-p_1\right|<\varepsilon_1, \left|\frac{\mu_2}{n}-p_2\right|<\varepsilon_2, \ldots, \left|\frac{\mu_k}{n}-p_k\right|<\varepsilon_k. \tag{2}$$

i.e., of the inequalities

$$|x_1|<\varepsilon_1\sqrt{\frac{n}{p_1q_1}}, |x_2|<\varepsilon_2\sqrt{\frac{n}{p_2q_2}}, \ldots, |x_k|<\varepsilon_k\sqrt{\frac{n}{p_kq_k}}.$$

The last of these inequalities is a consequence of the preceding ones, since the first $k-1$ inequalities in (2) yield the estimate

$$|x_k|=\left|-\sum_{i=1}^{k-1}\sqrt{\frac{p_iq_i}{p_kq_k}}\right|\leqslant\sum_{i=1}^{k-1}\sqrt{\frac{p_iq_i}{p_kq_k}}\,\varepsilon_i. \tag{3}$$

The probability of the first $k-1$ inequalities of (2) (and therefore the inequality (3) as well) approaches in the limit the integral

$$\sqrt{\frac{q_1 q_2 \cdots q_{k-1}}{(2\pi)^{k-1} p_k}} \int \cdots \int e^{-(1/2)Q(x_1, \ldots, x_k)} dx_1\, dx_2 \cdots dx_{k-1} = 1.$$

as $n \to \infty$.

14 POISSON'S THEOREM

Proving the Local DeMoivre-Laplace Theorem one saw that the asymptotic approximation of the probability $P_n(m)$ by means of the function

$$\frac{1}{\sqrt{2\pi}} e^{-x^2/2}$$

becomes increasingly worse as the probability p differs more and more from 1/2, i.e., the smaller the values of p or q under consideration, the poorer this approximation. However, a considerable number of practical problems call for the necessity of computing the probabilities $P_n(m)$ for just such small values of p (or, symmetrically, for small values of q). In order for the DeMoivre-Laplace Theorem to give results with negligible error in these cases, the number of trials must be very large. The problem thus arises of seeking an asymptotic formula which is especially suited to the case of small p. Such a formula was discovered by Poisson.

Let us consider a sequence of series of events

$$\begin{array}{l} E_{11}, \\ E_{21}, E_{22} \\ E_{31}, E_{32}, E_{33} \\ \cdots\cdots\cdots\cdots\cdots \\ E_{n1}, E_{n2}, E_{n3}, \ldots, E_{nn} \\ \cdots\cdots\cdots\cdots\cdots \end{array}$$

in which the events in any one of the set are mutually independent and each of them has a probability p_n depending only on the number of the set. The number of actual occurrences of an event in the n-th set is denoted by μ_n.

POISSON'S THEOREM. *If $p_n \to 0$ as $n \to \infty$, then*

$$\mathbf{P}(\mu_n = m) - \frac{a_n^m}{m!} e^{-a_n} \to 0, \tag{1}$$

where

$$a_n = np_n.$$

PROOF. It is obvious that

$$P_n(m) = \mathbf{P}\{\mu_n = m\} = C_n^m p_n^m (1 - p_n)^{n-m}$$

$$= \frac{n!}{m!(n-m)!}\left(\frac{a_n}{n}\right)^m \left(1 - \frac{a_n}{n}\right)^{n-m}$$

$$= \frac{a_n^m}{m!}\left(1 - \frac{a_n}{n}\right)^n \frac{\left(1 - \frac{1}{n}\right)\left(1 - \frac{2}{n}\right)\cdots\left(1 - \frac{m-1}{n}\right)}{\left(1 - \frac{a_n}{n}\right)^m}. \tag{2}$$

Let m be fixed. Let us choose an arbitrary $\varepsilon > 0$. Then, one can choose $A = A(\varepsilon)$ so large that for $a \geqslant A$ the inequality

$$\frac{a^m}{m!} e^{-a/2} \leqslant \frac{\varepsilon}{2}$$

holds.

Let us first consider those values of n for which $a_n \geqslant A$. For these n, by the inequality $1 - x < e^{-x}$ for $0 \leqslant x \leqslant 1$, one gets

$$P_n(m) \leqslant \frac{a_n^m}{m!} e^{-[(n-m)/n]a_n} \leqslant \frac{\varepsilon}{2} \quad \text{for} \quad \geqslant 2m$$

$$\frac{a_n^m}{m!} e^{-a_n} < \frac{\varepsilon}{2}.$$

Therefore, for the indicated values of n

$$\left| P_n(m) - \frac{a_n^m}{m!} e^{-a_n} \right| < \frac{\varepsilon}{2} + \frac{\varepsilon}{2} = \varepsilon.$$

Let us now consider those values of n for which $a_n \leqslant A$. Since for $a_n \leqslant A$ and fixed m

$$\lim_{n \to \infty} \left\{ \left(1 - \frac{a_n}{n}\right)^n - e^{-a_n} \right\} = 0$$

it follows that

$$\lim_{n\to\infty} \frac{\left(1-\frac{1}{n}\right)\left(1-\frac{2}{n}\right)\ldots\left(1-\frac{m-1}{n}\right)}{\left(1-\frac{a}{n}\right)^m} = 1,$$

and by formula (2) for $n \geqslant n_0(\varepsilon)$, one has

$$\left|P_n(m) - \frac{a_n^m}{m!} e^{-a_n}\right| < \varepsilon,$$

Q.E.D.

Notice that the Poisson theorem is also valid for the case where the probability of the event A equals 0 in every trial. In this case, $a_n = 0$.

Let us denote

$$P(m) = \frac{a^m}{m!} e^{-a}.$$

This probability distribution is called the Poisson law.

It is easily to calculate that the quantities $P(m)$ satisfy the relation

$$\sum_m P(m) = 1.$$

Let us study the behavior of $P(m)$ as a function of m. For this purpose, we consider the ratio

$$\frac{P(m)}{P(m-1)} = \frac{a}{m}.$$

If $m > a$ then $P(m) < P(m-1)$; however, if $m < a$, then $P(m) > P(m-1)$; and finally, if $m = a$, then $P(m) - P(m-1)$. From this one concludes that $P(m)$ increases from $m = 0$ up to $m_0 = [a]$, and with the further increase of m decreases monotonically. If a is an integer, then $P(m)$ has two maximum values, at $m_0 = a$ and at $m_0 = a - 1$.

Let us now discuss some examples.

Example 1. At each firing the probability of hitting a target equals 0.001. Find the probability of hitting a target with two or more bullets if the total number of shots fired is 5,000.[1]

Considering each firing as one trial and a target hit as the event of interest, one can use Poisson's Theorem for calculating the probability

$P(\mu_n \geqslant 2)$. In the example $a_n = np = 0.001 \cdot 5000 = 5$. Consequently the required probability is equal to

$$\mathsf{P}\{\mu_n \geqslant 2\} = \sum_{m=2} P_n(m) = 1 - P_n(0) - P_n(1).$$

By Poisson's Theorem,

$$P_n(0) \approx e^{-5},\ P_n(1) \approx 5e^{-5}.$$

Therefore,

$$\mathsf{P}\{\mu_n \geqslant 2\} \approx 1 - 6e^{-5} \approx 0.9596.$$

The probability $P(m)$ takes on its maximum value for $m = 4$ and $m = 5$.

These probabilities with accuracy of four decimals equal

$$P(4) = P(5) \approx 0.1751.$$

Computations by the exact formula give the values $P_{5000}(0) = 0.0067$ and $P_{5000}(1) = 0.0336$ with the same accuracy, and therefore

$$\mathsf{P}\{\mu_n \geqslant 2\} = 0.9575.$$

Example 2. In a spinning factory, a worker attends several hundred spindles each of which spins its own skein. Because of irregularity in the thread tension, its unevenness, and other reasons, the yarn breaks at random moments. It is important to know how frequently breaks can occur under a variety of operational conditions (quality of yarn, spindle speed, etc.).

Assuming that the worker attends 800 spindles and that the probability of a break in the yarn during some fixed interval τ equals 0.005 for each spindle, find the most probable number of breaks and the probability that no more than 10 breaks will occur during this interval of time.

Since

$$a_n = np = 0.005 \cdot 800 = 4,$$

[1] In World War II, analogous conditions were actually realized in the use of small-arms fire against airplanes. An airplane could be shot down by bullets only if they reached one of a few vulnerable spots: the motor, the fuel tank, the pilot himself, etc. The probability of a hit in these vulnerable spots was extremely small but, as a rule, an entire division fired at a plane at once and the entire number of shots fired at the airplane was considerably large. As a result, the probability of hitting the plane with one or two bullets was relatively high. This fact was observed in practice.

the most probable number of breaks in the interval τ is two, namely, 3 and 4. Their probabilities are

$$P_{800}(3) = P_{800}(4) = \binom{800}{4} \cdot 0.005^4 \cdot 0.995^{796}.$$

By Poisson's formula, one has

$$P_{800}(3) = P_{800}(4) \sim \frac{4^3}{3!} e^{-4} = \frac{32}{3} \cdot e^{-4} = 0.1954.$$

The exact value $P_{800}(3) = P_{800}(4) = 0.1945$. The probability that no more than 10 breaks will have occurred in an interval of time τ is equal to

$$\mathsf{P}\{\mu_n \leqslant 10\} = \sum_{m=0}^{10} P_{800}(m) = 1 - \sum_{m=11}^{\infty} P_{800}(m).$$

By virtue of Poisson's Theorem,

$$P_{800}(m) \sim \frac{4^m}{m!} e^{-4} (m = 0, 1, 2, \ldots),$$

therefore

$$\mathsf{P}\{\mu_n \leqslant 10\} = 1 - \sum_{m=11}^{\infty} \frac{4^m}{m!} e^{-4}.$$

But

$$\sum_{m=11}^{\infty} \frac{4^m}{m!} e^{-4} > \left(\frac{4^{11}}{11!} + \frac{4^{12}}{12!} + \frac{4^{13}}{13!}\right) e^{-4} = \frac{4^{12} \cdot 14}{11!39} e^{-4} = 0.00276.$$

On the other hand,

$$\sum_{m=11}^{\infty} \frac{4^m}{m!} e^{-4} < \frac{4^{11}}{11!} e^{-4} + \frac{4^{12}}{12!} e^{-4}$$

$$+ e^{-4} \frac{4^{13}}{13!} \left[1 + \frac{4}{14} + \left(\frac{4}{14}\right)^2 + \cdots\right] = \frac{4^{12} \cdot 24}{11!35} e^{-4} = 0.00284.$$

Thus

$$0.99716 \leqslant \mathsf{P}\{\mu_n \leqslant 10\} \leqslant 0.99724.$$

Analogously to the application of the DeMoivre-Laplace Theorem, the question arises how to estimate the error made in the replacement of the exact formula for calculating $P_n(m)$ by the asymptotic formula of Poisson.

From the equality

$$P_n(0) = \left(1 - \frac{a_n}{n}\right)^n = e^{n\ln(1-(a_n/n))} = \exp\left\{-n \sum_{k=1} \frac{1}{k}\left(\frac{a_n}{n}\right)^k\right\} = e^{-a_n}(1 - R_n),$$

where

$$R_n = 1 - \exp\left\{-n \sum_{k=2} \frac{1}{k}\left(\frac{a_n}{n}\right)^k\right\},$$

one can easily find this estimate for the case $m = 0$. Indeed, since for any positive x one has $0 < 1 - e^{-x} < x$, then for any a_n and n the following inequality can be written

$$0 < R_n < n \sum_{k=2} \frac{1}{k}\left(\frac{a_n}{n}\right)^k.$$

Since

$$\sum_{k=2}^{\infty} \frac{1}{k}\left(\frac{a_n}{n}\right)^k \leqslant \frac{a_n^2}{2n^2} + \frac{1}{3}\sum_{k=3}^{\infty}\left(\frac{a_n}{n}\right)^k = \frac{a_n^2}{2n^2} + \frac{a_n^3}{3n^3\left(1 - \frac{a}{n}\right)}$$

$$= \frac{a_n^2}{6n^2} \cdot \frac{3n - a_n}{n - a_n} < \frac{a_n^2}{2n(n - a_n)},$$

then

$$0 < R_n < \frac{a^2 n}{2(n - a_n)}.$$

From the fact that R_n is non-negative, one concludes that for the substitution of $P_n(0)$ by $\exp(-a_n)$ the value of the probability $P_n(0)$ is slightly increased.

15 ILLUSTRATION OF THE SCHEME OF INDEPENDENT TRIALS

As an illustration using the preceding results for applied purposes, let us consider, in schematic form, the problem of the random walk of a particle on a straight line. This problem may be regarded as the modeling prototype of actual physical problems in the diffusion theory, Brownian motion, etc.

Imagine that a particle, initially located at the point $x=0$, is subjected to random collisions at some moments of time. In consequence of each of such collision the particle moves one unit of length to the right or to the left depending on the direction of the impact. Hence, at each moment of collision the particle is displaced one unit to the right with probability 1/2 or one unit to the left with probability 1/2. As the result of n impacts, the particle will be displaced a distance μ. It is clear that this problem is equivalent to Bernoulli trials in their purest form. It follows that for any n and m one can calculate the probability that $m=\mu$:

$$\mathsf{P}\{\mu=m\}=\begin{cases} C_2^{(m+n)/2}\left(\frac{1}{2}\right)^n, & \text{if } -n\leqslant m\leqslant n,\\ 0, & \text{if } |m|>n. \end{cases}$$

For large values of n, the Local DeMoivre-Laplace Theorem can be applied

$$\mathsf{P}\{\mu=m\}\approx\frac{1}{\sqrt{2\pi n}}e^{-m^2/2n}. \quad (1)$$

The formula obtained can be interpreted as follows. Let a large number of particles initially be located at $x=0$. All particles begin to move along the line independently under influences of the random impacts. Then the fraction of particles that has been displaced a distance m after n impacts is given by formula (1).

Of course, idealized conditions for the motion of the particles were considered above. The conditions under which molecules actually move is far more complex. The result obtained, however, gives an accurate qualitative picture of the phenomenon.

In physics, one has to consider more complicated examples of random walks. This consideration is confined by a schematic treatment of the effect on a particle of 1) a reflecting barrier and 2) an absorbing barrier.

Let us imagine that a reflecting barrier is located s units to the right of the point $x=0$ so that if at any time a particle reaches the barrier, the next impact reflects it back with probability 1.

For clarity, let us plot the position of the particle in the plane (x, t). The path of the particle will be represented in the form of a polygonal line. The particle moves under each impact one unit "upward" and one unit to the right or left (with probability 1/2 in each case if $x < s$). If $x = s$, however, the particle moves one unit 'upward' and one unit to the left.

For computing the probability $\mathbf{P}\{\mu = m\}$, one proceeds as follows. Imagine that the barrier is removed and the particle is allowed to move freely. Figure 13 shows such idealized paths leading to the points A and A' which are symmetrically located with respect to the barrier. It is clear one of that a physical particle undergone reflections reaches point A by one of two ways (1) directly without interaction with a reflecting barrier or (2) after reflection if it should reach the imaginary point, A', beyond the reflecting barrier. The probability of the particle reaching point A directly is obviously equal to

$$\mathbf{P}\{\mu = m\} = \frac{n!}{\left(\frac{m+n}{2}\right)!\left(\frac{n-m}{2}\right)!}\left(\frac{1}{2}\right)^n.$$

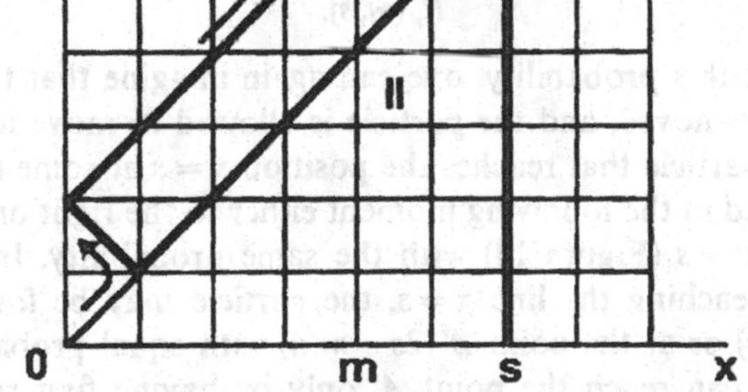

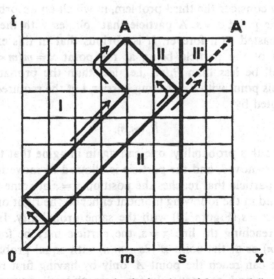

FIGURE 13. Graph of different density functions of the normal distribution.

Similarly, the probability of the particle reaching the point A' (with the abscissa $2s-m$) equals

$$\mathsf{P}\{\mu=2s-m\}=\frac{n!}{\left(s+\frac{n-m}{2}\right)!\left(\frac{n+m}{2}-s\right)!}\left(\frac{1}{2}\right)^n.$$

The required probability is therefore equal to

$$p_n\{m;s\}=\mathsf{P}\{\mu=m\}+\mathsf{P}\{\mu=2s-m\}.$$

Using the Local DeMoivre-Laplace Limit Theorem, one finds:

$$p_n\{m,s\}\sim\frac{1}{\sqrt{2\pi n}}\{e^{-m^2/2n}+e^{-(2s-m)^2/2n}\}$$

This is the well-known formula in the theory of Brownian motion. It can be put in more symmetric form by shifting the origin to the point $x=s$. It can be done by making the change of variable z for $z=x-s$. Finally, one obtains:

$$P_n(z=k)=P_n\{k+s,s\}=\frac{1}{\sqrt{2\pi n}}\{e^{-(k+s)^2/2n}+e^{-(k-s)^2/2n}\}.$$

Let us now consider the third problem, in which an absorbing barrier exists at the point $z=s$. A particle that collides with the absorbing barrier is pasted to it forever. It is obvious that in this example the probability of the particle being at the point $x=m\,(m<s)$ after n impacts will be less than $P_n(m)$,i.e., less than the probability of its reaching this point with a reflecting barrier. Let the required probability be denoted by

$$\bar{P}_n(m,s).$$

To compute this probability, one can again imagine that the absorbing barrier removed, and the particle is allowed to move freely along the line. A particle that reaches the position $x=s$ at some moment of time, is found in the following moment either to the right or to the left of the line $x=s$ (Figure 14) with the same probability. In a similar way, after reaching the line $x=s$, the particle may be found at the point $A(m,n)$ or at the point $A'(2s-m,n)$ with equal probability. But the particle can reach the point A' only by having first reached the position $x=s$, and, consequently, for any path leading to the point A', there exists another path which is symmetric to it with respect to the line $x = s$ and leads to the point A (Figure 14). Note that the symmetry of the paths is considered only from the moment the particle reaches the line $x = s$. This reasoning shows that the required probability equals

$$\bar{P}_n(m, s) = \mathbf{P}\{\mu = m\} - \mathbf{P}\{\mu = 2s - m\}.$$

By the Local DeMoivre-Laplace Theorem, one has

$$\bar{P}_n(m, s) \sim \frac{1}{\sqrt{2\pi n}}\{e^{-m^2/2n} - e^{-(2s-m)^2/2n}\}.$$

16 EXERCISES

1. A workman attends 12 identical machines. The probability that a machine will require his attention during an time interval of length τ equals 1/3. Find the probability that

a) 4 machines will require the workman's attention during interval τ;

b) the number of times his attention is required during this interval will be between 3 and 6 (including limits).

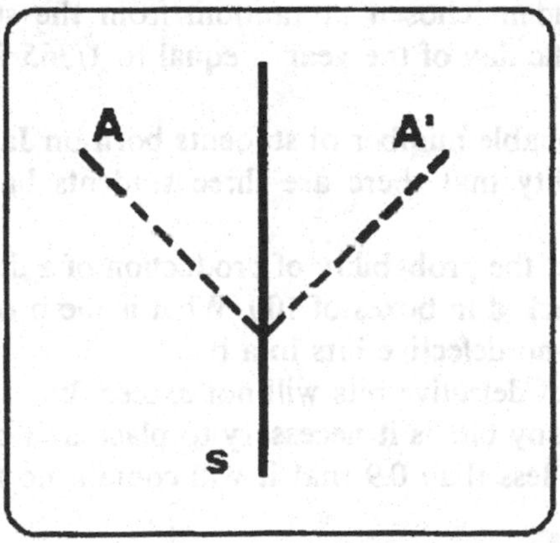

FIGURE 14. Geometrical interpretation for Example 1.

2. There are 10 children in a certain family. Assuming that the probability of the birth of a boy, or of a girl, is 1/2, determine the probability that

a) this family has 5 boys and 5 girls;

b) the number of boys lies between 3 and 8, inclusive.

3. In a group of 4 people, the birthdays of three occur in the same month and that of the fourth, in one of the remaining eleven months. Assuming that the probability that anyone's birthday is in any particular month is equal to 1/12, find the probability that

a) the three designated persons were born in January, and the fourth in October;

b) the three persons were born in one of any month and the fourth in some other month.

4. In 14,000 tosses of a coin, heads has come up 7,428 times. What is the probability of the number of heads differing from the quantity np by an amount equal to or greater than the difference in this experiment? The coin is fair, i.e., the probability of tossing a head in each trial is 1/2.

5. There are n electrical appliances plugged into an electric circuit. Each consumes a kW. An appliance switches on randomly, i.e., consumes the power at any specified moment with the probability p. Find the probability that at a given moment the power used

a) will be less than $rnap$ $(r > O)$;

b) will exceed $rnap$ under the condition that np is large.

6. There are 730 students at a university. The probability that the birthday of a student chosen at random from the student register occurs on a specific day of the year is equal to 1/365 for each day of the year. Find

a) the most probable number of students born on January 1;

b) the probability that there are three students having the same birthday.

7. It is known that the probability of production of a defective drill bit is 0.02. Bits are packed in boxes of 100. What is the probability that

a) there will be no defective bits in a box;

b) the number of defective bits will not exceed 3;

c) Also, how many bits is it necessary to place in a box so that the probability is not less than 0.9 that it will contain no fewer than 100 good ones?

Hint: Use the Poisson distribution.

8. Ten thousand persons of the same age and social group have policies with an insurance company. The probability of death during the year is 0.006 for each one of them. Each insured pays a premium of \$200 on January 1 of each year. In the event he dies, his beneficiaries receive \$10,000 from the company. What is the probability that

a) the company will lose money;

b) the company will make a profit of not less than \$400,000; \$600,000; \$800,000.

9. Prove the following theorem: If P and P' are the probabilities of the most probable number of occurrences of an event A in n and $n+1$ independent trials, respectively, then $P' \leqslant P$. Equality is excluded if $(n+1)p$ is not an integer. (In each trial $P(A) = p$.)

10. In Bernoulli trials $p = 1/2$. Prove that

(a)

$$\frac{1}{2\sqrt{n}} \leqslant P_{2n}(n) \leqslant \frac{1}{\sqrt{n+1}};$$

(b)

$$\lim_{n\to\infty} \frac{P_{2n}(n \pm h)}{P_{2n}(n)} = e^{-z^2},$$

if $z = h/\sqrt{n}$ $(0 \leqslant z < \infty)$.

11. Prove that for $npq \geqslant 25$

$$P_n(m) = \frac{1}{\sqrt{2npq}} e^{-z^2/2} \left[1 + \frac{(q-p)(z^3 - 3z)}{6\sqrt{npq}} \right] + \Delta,$$

$$z = \frac{m - np}{\sqrt{npq}}, \quad |\Delta| < \frac{0,15 + 0,25|p-q|}{\sqrt{(npq)^3}} |z| e^{-(3/2)\sqrt{npq}}.$$

12. There are made n independent trials. The probability that an event A occurs in trial i is p_i; $P_n(m)$ is the probability of the m-fold occurrence of the event A in the n trials. Prove that

(a)

$$\frac{P_n(1)}{P_n(0)} \geqslant \frac{P_n(2)}{P_n(1)} \geqslant \cdots \geqslant \frac{P_n(n)}{P_n(n-1)};$$

(b) $P_n(m)$ at first increases and then decreases (except the situation when $P_n(0)$ or $P_n(n)$ are themselves maximum values).

13. Prove that for $x>0$ the function

$$\int_0^\infty \exp\left(-\frac{z^2}{2}\right)dz$$

satisfies the inequalities

$$\frac{x}{1+x^2}e^{-(1/2)x^2} \leqslant \int_x^\infty e^{-(1/2)z^2}\,dz \leqslant \frac{1}{x}e^{-(1/2)x^2}.$$

14. *Banach's match box problem.* A mathematician carries two match boxes in his pocket. Each time he wants to use a match, he selects one of the boxes at random. Find the probability that when the mathematician discovers that one of the boxes is empty, the other box contains r matches ($r=0,1,2,\ldots,n$; where n is the number of matches initially contained in each box).

15. There are n machines hooked up to an electrical circuit. The probability that a machine consuming the power from the circuit at time t will stop doing so before time $t+\Delta t$ equals $\alpha\Delta t+o(\Delta t)$. If a machine is not consuming power at time t, then the probability that it will begin to do so before time $t+\Delta t$ is $\beta\Delta t+o(\Delta t)$ independent of the operation of the other machines. Derive the differential equations satisfied by $P_r(t)$, the probability that r machines are consuming power at time t.

REMARK. One can easily find concrete situations in which such a situation occurs: trolley buses, electric welding, the consumption of power by machines with automatic cutoffs, etc.

16. A single workman attends n identical automatic machines. If a machine is operating at time t, then the probability that it will require his attention before time $t+\Delta t$ is $\alpha\Delta t+o(\Delta t)$. If the workman is attending one of the machines at time t, then the probability that he will finish attending to it before time $t+\Delta t$ is $\alpha\Delta t+o(\Delta t)$. Derive the differential equations that are satisfied by $P_r(t)$, the probability that there are $n-r$ machines operating at time t, one is being attended to and $r-1$ are awaiting his attention. (Denote by $P_0(t)$ the probability that all of the machines are in operation.)

Note: Analogously, it is not difficult to derive the differential equations for the more complicated problem in which N machines are attended by a team of k workmen. For practical reasons, it is important to compare the efficiency of the various systems of work organization. For this purpose, it is useful to invesstigate the steady-state condition, i.e., to consider the probabilities $P_r(t)$ as $t \to \infty$.

It appears that the work of a team attending to kn machines is more effective than the maintenance of n machines by a single workman, both in the sense of better utilization of the running time of the machine and the work time of the workmen.

Chapter 4

MARKOV CHAINS

17 DEFINITION OF A MARKOV CHAIN

A straightforward generalization of the scheme of independent trials is a so-called Markov chain, which was introduced and systematically investigated by the famous Russian mathematician A. A. Markov. We shall restrict the presentation of the theory to an elementary level.

Imagine that a sequence of trials is performed in each of which one of k mutually exclusive or disjoint events $A_1^{(s)}, A_2^{(s)}, \ldots, A_k^{(s)}$ may be realized (as before, the superscript denotes the number of the trial). One says that the sequence of trials forms a Markov chain, or more precisely, a simple Markov chain, if the conditional probability of occurrence of the event $A_i^{(s+1)} (i = 1, 2, \ldots, k)$ in trial $s + 1 (s = 1, 2, 3, \ldots)$ depends only on which event has occurred in trial s and is not affected by further information as to which events have occurred in the previous trials.

One often uses a different terminology in the theory of Markov chains. One speaks of a physical system S which at any time can be in one of the states $A_1, A_2, \ldots, A_k$ and which changes its state only at the times $t_1, t_2, \ldots, t_n, \ldots$. For Markov chains, the probability of the system transition into some state $A_i (i = 1, 2, \ldots, k)$ at time $\tau (t_s < \tau < t_{s+1})$ depends only on the state in which the system was at time $t (t_{s-1} < t < t_s)$ and is uninfluenced by anything that may be known about its states at previous moments of time.

Let us consider two illustrative examples.

Example 1. Imagine a particle moving along a line under the influence of random impacts occurring at moments $t_1, t_2, t_3, \ldots$. The particle may be at the points with integer coordinates $a, a + 1, a + 2, \ldots, b$). The points a and b are reflecting barriers. Each impact moves the particle to the right with probability p or to the left with probability

$q = 1 - p$, until the particle does not reach one of the barriers. However, if the particle is at a barrier then the impact moves it towards the interior of the interval between the barriers, i.e. the particle reflects from the barrier. We see that this example of the particle's random walking is a typical Markov chain. In an analogous way, one may consider the cases in which the particle is adhered to one or both of the barriers.

Example 2. In the Bohr's model of the hydrogen atom, the electron may be found in one of certain permissible orbits. Let A_i denote the event consisting in the electron being found in orbit i. Suppose further that a change in the state of the atom can only happen at moments $t_1, t_2, t_3, \ldots$ (in reality, they are random). The probability of a transition from orbit i to orbit j at time t_s, depends only on i and j (the difference $j - i$ depends on the energy increment the atom has attained at time t_s) and not on the previous orbits of the electron.

The latter example represents a Markov chain with an infinite number of states. Notice that this example would be far closer to real conditions if the time of transition of the system from one state to a new state could vary continuously.

18 TRANSITION MATRIX

Further we restrict our presentation by the simplest facts relating to *homogeneous Markov chains*, in which the conditional probability of occurrence of the event $A_j^{(s+1)}$ in trial $s + 1$, given that the event $A_i^{(s)}$ has been realized in trial s, is independent of the number of the trial. One calls this probability the *probability of transition* and denotes it by p_{ij}. In this notation, the first subscript always indicates the outcome of the preceding trial and the second subscript indicates the state into which the system passes in the following moment of time.

The complete picture of the possible changes in state which might occur is given by the matrix

$$\prod_1 = \begin{pmatrix} p_{11} & p_{12} & \cdots & p_{1k} \\ p_{21} & p_{22} & \cdots & p_{2k} \\ \cdots & \cdots & \cdots & \cdots \\ p_{k1} & p_{k2} & \cdots & p_{kk} \end{pmatrix},$$

constructed from the transition probabilities. One calls this matrix the *matrix of transition probabilities.*

Notice what conditions must be satisfied for the elements of this matrix. First of all, they are probabilities, so $0 \leqslant p_{ij} \leqslant 1$. Furthermore, in trial s the system from state $A_i(s)$ necessarily moves to one of other states $A_j^{(s+1)}$ which implies the condition

$$\sum_{j=1}^{k} p_{ij} = 1 \qquad (i = 1, 2, \ldots, k).$$

This means that the sum of all the elements of any row of the transition matrix equals 1.

Our first task in the theory of Markov chains is the determination of the transition probability from state $A_i^{(s)}$ in trial s into the state $A_j^{(s+n)}$ in n subsequent steps. Denote this probability by $P_{ij}(n)$. Consider any intermediate trial, say $s+m, m<n$. In this trial, one of the possible outcomes $A_r^{(s+m)} (1 \leqslant r \leqslant k)$ will occur. The probability of transition into this state is $P_{ir}(m)$. The probability of the transition from state $A_r^{(s+m)}$ into state $A_j^{(s+n)}$ is $P_{rj}(n-m)$. By the formula of total probability, one obtains the *Markov's equation*:

$$P_{ij}(n) = \sum_{r=1}^{k} P_{ir}(m) \cdot P_{rj}(n-m). \tag{1}$$

Let π_n denote the matrix of transition probabilities after n steps

$$\pi_n = \begin{pmatrix} P_{11}(n) & P_{12}(n) & \cdots & P_{1k}(n) \\ \cdots & \cdots & \cdots & \cdots \\ P_{k1}(n) & P_{k2}(n) & \cdots & P_{kk}(n) \end{pmatrix}.$$

In accordance with (1), between matrices π_n with different subscripts the following relation is satisfied

$$\pi_n = \pi_m \cdot \pi_{n-m} \qquad (0 < m < n).$$

In particular, for $n = 2$, one finds:

$$\pi_2 = \pi_1 \cdot \pi_1 = \pi_1^2;$$

for $n = 3$,

$$\pi_3 = \pi_1 \cdot \pi_2 = \pi_2 \cdot \pi_1 = \pi_1^3;$$

and for any n whatever, one has

$$\pi_n = \pi_1^n.$$

We note a special case of formula (1) for $m = 1$

$$P_{ij}(n) = \sum_{r=1}^{n} p_{ir} P_{rj}(n-1).$$

The reader may consider the following as an exercise: Write the transition matrix for the first example of the previous section.

19 THEOREM ON LIMITING PROBABILITIES

THEOREM. *If for some value of $s > 0$ all elements of the transition matrix π_s are positive, then there exist constants p_j $(j = 1, 2, \ldots, k)$ such that the limit relations*

$$\lim_{n\to\infty} P_{ij}(n) = p_j.$$

hold independently of subscript i.

PROOF: The main idea of the proof of this theorem is very simple. One first establishes that the largest of the probabilities $P_{ij}(n)$ cannot increase and the smallest of the probabilities cannot decrease as n increases. Then one shows that the maximum of the difference $|P_{ij}(n) - P_{lj}(n)|$ $(i, l = 1, 2, \ldots, k)$ approaches zero as $n \to \infty$. This completes the proof of the theorem. Indeed, from the known theorem of the limit of monotonic sequences, one concludes from the first two properties mentioned above of the probabilities $P_{ij}(n)$ that the following limits exist:

$$\lim_{n\to\infty} \min_{1\leqslant i\leqslant k} P_{ij}(n) = \bar{p}_j$$

and

$$\lim_{n\to\infty} \max_{1\leqslant i\leqslant k} P_{ij}(n) = \bar{\bar{p}}_j.$$

And since by the third property mentioned

$$\lim_{n\to\infty} \max_{1\leqslant i,l\leqslant k} |P_{ij}(n) - P_{lj}(n)| = 0,$$

it follows that

$$\underline{p}_j = \overline{p}_j = p_j.$$

We now proceed to carry out the above plan. We observe, first of all, that for $n > 1$ the inequality

$$P_{ij}(n) = \sum_{l=1}^{k} p_{il} P_{lj}(n-1) \geqslant \min_{1\leqslant l\leqslant k} P_{lj}(n-1) \sum_{l=1}^{k} p_{il} = \min_{1\leqslant l\leqslant k} P_{lj}(n-1)$$

holds. This inequality is valid for every i and, in particular, for the value for which

$$P_{ij}(n) = \min_{1\leqslant i\leqslant k} P_{ij}(n).$$

Thus, one has

$$\min_{1\leqslant i\leqslant k} P_{ij}(n) \geqslant \min_{1\leqslant i\leqslant k} P_{ij}(n-1).$$

In a similar way, it is easy to show that

$$\max_{1\leqslant i\leqslant k} P_{ij}(n) \leqslant \max_{1\leqslant i\leqslant k} P_{ij}(n-1).$$

One may assume that $n > s$ and therefore, by formula (1) it is permissible to write:

$$P_{ij}(n) = \sum_{r=1}^{k} P_{ir}(s)\cdot P_{rj}(n-s).$$

Consider the difference

$$P_{ij}(n) - P_{lj}(n) = \sum_{r=1}^{k} P_{ir}(s) P_{rj}(n-s) - \sum_{r=1}^{k} P_{lr}(s)\cdot P_{rj}(n-s)$$

$$= \sum_{r=1}^{k} [P_{ir}(s) - P_{lr}(s)] P_{rj}(n-s).$$

Denote those differences $P_{ir}(s) - P_{lr}(s)$ that are positive by the symbol $\beta_{il}^{(r)}$ and those that are non-positive by $\beta_{il}^{\prime(r)}$. Since

$$\sum_{r=1}^{k} P_{ir}(s) = \sum_{r=1}^{k} P_{lr}(s) = 1,$$

it follows that

$$\sum_{r=1}^{k} [P_{ir}(s) - P_{lr}(s)] = \sum_{(r)} \beta_{il}^{(r)} - \sum_{(r)} \beta_{il}^{\prime(r)} = 0. \tag{2}$$

From this equality, one concludes that

$$h_{il} = \sum_{(r)} \beta_{il}^{(r)} = \sum_{(r)} \beta_{il}^{\prime(r)}.$$

Since, by assumption, $P_{ir}^{(s)} > 0$ for all i and r $(i, r = 1, 2, \ldots, k)$, one has

$$\sum_{(r)} \beta_{il}^{(r)} < \sum_{r=1}^{k} P_{il}(s) = 1.$$

Thus

$$0 \leqslant h_{il} < 1.$$

Let

$$h = \max_{1 \leqslant i,l \leqslant k} h_{il}.$$

Since the number of possible outcomes is finite, the quantity h along with h_{ij} satisfied the inequality

$$0 \leqslant h < 1. \tag{3}$$

From (1), one finds that for any i and j $(i, j = 1, 2, \ldots, k)$

$$\left| P_{ij}(n) - P_{lj}(n) \right| = \left| \sum_{(r)} \beta_{il}^{(r)} P_{rj}(n-s) - \sum_{(r)} \beta_{il}^{\prime(r)} P_{rj}(n-s) \right|$$

$$\leqslant \left| \max_{1 \leqslant r \leqslant k} P_{rj}(n-s) \sum_{(r)} \beta_{il}^{(r)} - \min_{1 \leqslant r \leqslant k} P_{rj}(n-s) \sum_{(r)} \beta_{il}^{\prime(r)} \right|$$

$$\leqslant h \left| \max_{1 \leqslant r \leqslant k} P_{rj}(n-s) - \min_{1 \leqslant r \leqslant k} P_{rj}(n-s) \right|$$

$$\leqslant h \max_{1 \leqslant i,l \leqslant k} \left| P_{ij}(n-s) - P_{lj}(n-s) \right|$$

and, therefore, also that

$$\max_{1\leqslant i,l\leqslant k} |P_{ij}(n)-P_{lj}(n)| \leqslant h \max_{1\leqslant i,l\leqslant k} |P_{ij}(n-s)-P_{lj}(n-s)|.$$

By applying this inequality [n/s] times, one finds:

$$\max_{1\leqslant i,l\leqslant k}\left|P_{ij}(n)-P_{lj}(n)\right| \leqslant h^{[n/s]} \max_{1\leqslant i,l\leqslant k}\left|P_{ij}\left(n-\left[\frac{n}{s}\right]s\right)-P_{lj}\left(n-\left[\frac{n}{s}\right]s\right)\right|.$$

Since $|P_{ij}(n)-P_{lj}(n)|$ is always less than 1, it follows that

$$\max_{1\leqslant i,l\leqslant k} |P_{ij}(n)-P_{lj}(n)| \leqslant h^{[n/s]}.$$

As $n\to\infty, [n/s]\to\infty$ also, and therefore, by (3), it follows from this that

$$\lim_{n\to\infty}\max_{1\leqslant i,l\leqslant k} |P_{ij}(n)-P_{lj}(n)| = 0.$$

The proof also implies that

$$\sum_{j=1}^{k} p_j = 1.$$

In fact,

$$\sum_{j=1}^{k} p_j = \lim_{n\to\infty}\sum_{j=1}^{k} P_{ij}(n) = 1.$$

Thus, the probability p_j may be interpreted as the probability of occurrence of the outcome $A_j^{(n)}$ in trial n when n is large.

The physical meaning of this theorem is clear: The probability that the system is in state ij is practically independent of what happened with the system in the distant past.

The theorem laid out above was first proved by the creator of the concept of chain dependence, A. A. Markov. It was the first rigorously proved statement among the so-called ergodic theorems that play such an important role in modern physics and engineering.

20 EXERCISES

1. Given the matrix of transition probabilities

$$\pi_1 = \begin{pmatrix} 1/2 & 1/3 & 1/6 \\ 1/2 & 1/3 & 1/6 \\ 1/2 & 1/3 & 1/6 \end{pmatrix}.$$

How many states does the system have? Find the two-step transition probabilities.

2. An electron may be found in one of a counting number of orbits depending on the available energy. The transition from orbit i to orbit j occurs in one second with probability $c_i e^{-a|i-j|}$.

Find (a) the transition probabilities for two seconds, and (b) the constant, c_i.

3. Given the matrix of transition probabilities

$$\pi_1 = \begin{pmatrix} 0 & 1/2 & 1/2 \\ 1/2 & 0 & 1/2 \\ 1/2 & 1/2 & 0 \end{pmatrix}.$$

Is the ergodic Markov's Theorem applicable in this case? If so, find the limiting probabilities.

Chapter 5

RANDOM VARIABLES AND DISTRIBUTION FUNCTIONS

21 FUNDAMENTAL PROPERTIES OF DISTRIBUTION FUNCTIONS

The concept of the random variable is one of the fundamental concepts of probability theory. Before we present the formal definition of this concept, let us consider some examples.

The number of cosmic particles that strike a specific square of the earth's surface in a specified time interval is subject to significant fluctuations depending on many random factors.

The number of calls received from subscribers at a telephone central office in a given time interval is also a random variable, not a constant in time.

The distance between the point of impact of a shell and the center of the target is determined by a number of varied causes of a random nature. Thus, in the theory of firing, one must regard the phenomenon of shell dispersion about the target as random. The deviation from the center of the target is assumed as a random variable.

The velocity of a gas molecule does not remain constant but changes depending on its collisions with other molecules. There are a great many such collisions, even in a very short time interval. Even if one knows the exact velocity of a molecule at a fixed moment of time, this does not make it possible to specify its value with complete definiteness after, say, 0.01 or 0.001 seconds. The velocity of a molecule is uncertain, it has a random character.

These examples show quite clearly that one deals with random variables in the varied fields of science and engineering. The natural and at the same time important problem of creating methods for studying random variables arises.

Despite the differences in the specific content of the examples considered above, from the mathematical standpoint all of them present a similar phenomenon. In each example, one deals with a variable which characterizes the phenomenon concerned, and this variable is capable of taking on various values, under the influence of chance. To say in advance which value the variable will take on with absolute accuracy is impossible, since it randomly changes from experiment to experiment.

Thus, in order to know a random variable, it is first of all necessary to know which values it can assume. However, a mere listing of the values of a random variable is insufficient for making any essential judgment. Indeed, if in the third example, one considers the gas for different temperatures, the range of possible values of the molecular velocities will be the same, although the states of the gas will be different. Thus, to assign a random variable, it is necessary to know not only what values it can take on but also, how often, i.e., with what probability these values are assumed.

The variety of random variables is very large. The number of values they take on may be finite, enumerable, or non-enumerable. These values may be distributed discretely or continuously in some interval, or, without filling out an interval, they may be everywhere dense. In order to assign probabilities to the values of random variables of such a different nature and yet assign them in a consistent way, one introduces the concept of a *distribution function of a random variable*.

Let ξ be a random variable and x be an arbitrary real number. The probability that ξ will assume a value less than x is called the distribution function of the random variable ξ

$$F(x) = \mathbf{P}\{\xi < x\}.$$

Henceforth, as a rule, random variables are denoted by Greek letters and the values which they take on by lowercase italic letters.

Thus, a random variable is a variable whose values depend on chance and for which there exists a distribution function. (A formal definition of a random variable is found on page 116.)

Consider some examples of distribution functions. Example 1. Let μ denote the number of occurrences of an event A in a sequence of n independent trials in each of which the probability of its occurrence equals p. Depending on chances μ may assume any integer value between 0 and n (inclusive). According to the results of Chapter 2, one has

$$P_n(m) = \mathbf{P}\{\mu = m\} = \binom{n}{m} p^m q^{n-m}.$$

The distribution function of the variable μ is defined as follows:

$$F(x) \begin{cases} 0 & \text{for} \quad x \leqslant 0, \\ \sum\limits_{k<x} P_n(k) & \text{for} \quad 0 < x \leqslant n, \\ 1 & \text{for} \quad x > n. \end{cases}$$

The distribution function is a step function with jumps at the points $x = 0, 1, 2, \ldots, n$. The jump at the point $x = k$ is $P_n(k)$.

This example shows that the so-called Bernoulli scheme can be included in the general theory of random variables.

Example 2. Let the random variable ξ take on the values $0, 1, 2, \ldots$ with the probabilities

$$p_n = \mathbf{P}\{\xi = n\} = \frac{\lambda^n e^{-\lambda}}{n!} \qquad (n = 0, 1, 2, \ldots),$$

where $\lambda > 0$ is a constant. The distribution function of the variable ξ resembles a staircase with an infinite number of steps with jumps at all of the non-negative integer points. The height of the jump at the point $x = n$ equals p_n and for $x \leqslant 0$, $F(x) = 0$. The distribution of such a random variable is called the Poisson distribution.

Example 3. A random variable is said to be normally distributed if its distribution function is given by

$$\Phi(x) = C \int_{-\infty}^{x} e^{-(z-a^2)/(2\sigma^2)} dz,$$

where $C > 0$, $\sigma > 0$ and a are constants. Later on, a connection between the constants σ and C will be established and the probabilistic meaning of parameters a and σ will be explained. Normally distributed random variables play an exclusively important role in probability theory and its applications. In the sequel to this book, readers will have many opportunities to be convinced of this.

Note that while the random variables in our first two examples were discrete and could only assume an finite or an enumerable number of values, a normally distributed random variable may take on the values in any interval. Really, as it will be shown below, the probability that a normally distributed random variable takes on values in the interval $x_1 \leqslant \xi x_2$ is

$$\Phi(x_2) - \Phi(x_1) = C\int_{x_1}^{x_2} e^{-(z-a)^2/2\sigma^2} dz$$

and therefore, for any x_1 and x_2 $(x_1 \neq x_2)$ this probability is positive.

After these preliminary intuitive remarks, one now can proceed to a strong formal definition of the random variable.

In accordance with the general concept of the random event, let us consider a set of elementary events Ω. A certain value

$$\xi = f(\omega)$$

corresponds to each elementary event ω.

It is said that 3ξ is a random variable if the function $f(\omega)$ is measurable in probability introduced in the set ω. In other words, it is necessary that for each set $A_\xi(\xi \in A_\omega)$ measurable in the Borel sense, the set A_ω of such ω for which $f(\omega) \subset A_\xi$ belongs to the set of random events F and, consequently, for this set the probability should be defined as

$$\mathbf{P}\{\xi \subset A_\xi\} = \mathbf{P}\{A_\omega\}.$$

In particular, if the set A_ω is defined as $\xi < x$, then the probability $\mathbf{P}\{A_\omega\}$ is a function of x

$$\mathbf{P}\{\xi < x\} = \mathbf{P}\{A_\omega\} = F(x),$$

which is called the distribution function of a random variable ξ.

Example 4. Consider a sequence of n independent trials in each of which the probability of occurrence of an event A equals p. The elementary events consist of sequences of occurrences and non-occurrences of the event A in the n trials. For instance, one of the elementary events is the occurrence of the event A in every trial. It clear that there are 2^n elementary events in total.

Now define a function $\mu = f(\omega)$ of the elementary event ω as follows: It is equal to the number of occurrences of the event A in the elementary event ω. In accordance with the results of Chapter 2, one has:

$$\mathbf{P}\{\mu = k\} = P_n(k) = \binom{n}{k} p^k q^{n-k}.$$

The measurability of the function $\mu = f(\omega)$ in the field of probabilities is immediately evident. From this, by the definition, one concludes that μ is a random variable.

Example 5. Three observations are made of the position of a molecule moving on a straight line. The set of elementary events consists of the points in three-dimensional Euclidian space R^3. The set of random events F consists of all possible Borel sets in the space R^3.

For every random event A, the probability $\mathbf{P}\{A\}$ can be found as

$$\mathbf{P}\{A\} = \frac{1}{(\sigma\sqrt{2\pi})^3} \iiint_A e^{-1/2\sigma^2 [(x_1 - a)^2 + (x_2 - a)^2 + (x_3 - a)^2]} dx_1 dx_2 dx_3.$$

Let us now consider the function $\xi = f(\omega)$ of the elementary event defined by the equation

$$\xi = \frac{1}{3}(x_1 + x_2 + x_3).$$

This function is measurable with respect to the probability that has been introduced and hence ξ is a random variable. Its distribution function is

$$F(x) = \mathbf{P}\{\xi < x\} = \frac{1}{(\sigma\sqrt{2\pi})^3} \iiint_{x_1 + x_2 + x_3 < 3x} e^{-(1/2\sigma^2)\Sigma_{k=1}^{3}(x_k - a)^2}$$

$$\times\, dx_1 dx_2 dx_3 \quad = \frac{1}{\sigma\sqrt{\frac{2}{3}\pi}} \int_{-\infty}^{x} e^{-3(z-a)^2/2\sigma^2} dz.$$

From the point of view just elaborated, any operation on random variables can be reduced to familiar operations on functions. Thus, if ξ_1 and ξ_2 are random variables, i.e., if they are measurable functions with respect to the probability defined

$$\xi_1 = f_1(\omega), \qquad \xi_2 = f_2(\omega),$$

then any Borel function of these variables is also a random variable. For a example,

$$\zeta = \xi_1 + \xi_2$$

is measurable with respect to the probability defined and is therefore a random variable.

Later on, the remark just made will be elaborated and a number of important results will be obtained. In particular, a formula for the distribution function of a sum and of a quotient based on the distribution functions of their terms will be derived.

The distribution function of the random variable ξ allows us to determine the probability of the inequality $x_1 \leqslant \xi < x_2$ for any x_1 and x_2. Indeed, if A denotes the event that $\xi < x_2$, B denotes the event that $\xi < x_1$, and finally, C denotes the event that $x_1 \leqslant \xi < x_2$, then the following relation obviously holds:

$$A = B + C.$$

Since the events B and C are mutually exclusive, then

$$\mathsf{P}(A) = \mathsf{P}(B) + \mathsf{P}(C).$$

But

$$\mathsf{P}(A) = F(x_2), \mathsf{P}(B) = F(x_1), \mathsf{P}(C) = \mathsf{P}\{x_1 \leqslant \xi < x_2\},$$

therefore

$$\mathsf{P}\{x_1 \leqslant \xi < x_2\} = F(x_2) - F(x_1). \qquad (1)$$

Since by definition a probability is a non-negative value, it follows from equation (1) that the inequality

$$F(x_2) \geqslant F(x_1),$$

holds for any values of x_1 and x_2 $(x_2 > x_1)$, i.e., the distribution function of every random variable is a non-decreasing function.

It is evident that, for any x, the distribution function $F(x)$ satisfies the inequality

$$0 \leqslant F(x) \leqslant 1.$$

It is said that a distribution function $F(x)$ has a jump at the point $x = x_0$ if

$$F(x_0+0) - F(x_0-0) = C_0 > 0.$$

A distribution function can have at most an enumerable number of jumps. For instance, a distribution function can have at most one jump of magnitude greater than 1/2, at most three jumps of magnitude between 1/4 and 1/2. In general, there can be at most $2^n - 1$ jumps of magnitude between 2^{-n} and 2^{n-1}.

Let us establish a few additional general properties of distribution functions. Let us define $F(-\infty)$ and $F(+\infty)$ by the following equalities

$$F(-\infty) = \lim_{n\to+} F(-n), F(+\infty) = \lim_{n\to} F(+n)$$

and prove that

$$F(-\infty) = 0, \quad F(+\infty) = 1.$$

In fact, since the inequality $\xi < \infty$ is certain, it follows that

$$\mathsf{P}\{\xi < +\infty\} = 1.$$

Denote the event $k-1 \leqslant \xi < k$ by Q_k. Because the event $\xi < \infty$ is equivalent to the sum of the events Q_k, from the extended axiom of addition one has that

$$\mathsf{P}\{\xi < +\infty\} = \sum_{k=-}^{\infty} \mathsf{P}\{Q_k\}.$$

Therefore, as $n \to \infty$

$$\sum_{k=1-n}^{n} \mathsf{P}\{Q_k\} = \sum_{k=1-n}^{n} [F(k) - F(k-1)] = F(n) - F(-n) \to 1.$$

Taking into account inequality (2), one concludes that, as $n \to \infty$,

$$F(-n) \to 0, \quad F(+n) \to 1.$$

A distribution function is continuous on the left.

Let us choose some increasing sequence $x_0 < x_1 < \cdots < x_n < \cdots$ which converges to x. Denote by A_n the event $\{x \leqslant \xi x\}$. Then it is clear that $A_i \subset A_j$ for $i > j$, and that the product of all of the events A_n is the impossible event.

By the axiom of continuity, one has

$$\lim_{n\to\infty} \mathbf{P}(A_n) = \lim_{n\to\infty} \{F(x) - F(x_n)\} = F(x) - \lim_{n\to\infty} F(x_n) = F(x) - F(x-0) = 0,$$

Q.E.D.

In an analogous way, one can show that

$$\mathbf{P}\{\xi \leqslant x\} = F(x+0).$$

Thus, every distribution function is a non-decreasing function which is continuous on the left and satisfies the conditions $F(-\infty)=0$ and $F(+\infty)=1$. The converse is also true: Every function satisfying the conditions just stated can be regarded as the distribution function of some random variable.

Note that whereas each random variable determines its distribution function uniquely, there exist random variables as distinct as we please that have the same distribution function. For instance, if ξ assumes the two values -1 and 1 each with a probability of 1/2 and if $\eta = -\xi$, then it is clear that ξ always differs from η. Nevertheless, both of these random variables have the same distribution function

$$F(x) = \begin{cases} 0 & \text{for } x \leqslant -1, \\ 1/2 & \text{for } -1 < x \leqslant 1, \\ 1 & \text{for } x > 1. \end{cases}$$

22 CONTINUOUS AND DISCRETE DISTRIBUTIONS

Sometimes the behavior of a random variable is characterized not by the assignment of its distribution function but in some other way. Any such characterization is referred to as a distribution law of the random variable, if the distribution function can be obtained from it by some prescribed rule. For instance, the interval function $P\{x_1, x_2\}$ representing the probability of the inequality $x_1 \leqslant \xi < x_2$ is such a distribution law. Indeed, knowing $P\{x_1, x_2\}$, one can find the distribution function by using the formula

$$F(x) = P\{-\infty, x\}.$$

It is already known that for any x_1 and x_2 the function $P\{x_1, x_2\}$ can be found on the basis of known $F(x)$ by using the expression

$$P\{x_1, x_2\} = F(x_2) - F(x_1).$$

It is often useful to take as a distribution law the set function $P\{E\}$ defined on all Borel sets and representing the probability that the random variable assumes a value belonging to the set E. By the extended axiom of addition, the probability $P\{E\}$ is a completely additive set function, i.e., for any set E, which is the union of a finite or countable number of disjoint sets E_k, one has:

$$P\{E\} = \sum P\{E_k\}.$$

Let us first single out those of the various possible random variables that may only take on either a finite or an enumerable number of values. Such variables will be called discrete. For a complete probabilistic characterization of a discrete random variable ξ taking on the values $x_1, x_2, x_3, \ldots$ with positive probability, it suffices to know the probabilities $p_k = \mathrm{P}\{\xi = x_k\}$. (These, and only these, values x_n will be called possible values of the discrete random variable ξ). It is evident that the collection of probabilities p_k determines the distribution function $F(x)$ by means of the relation

$$F(x) = \sum p_k,$$

in which the summation extends over all of the subscripts for which $x_k < x$.

The distribution function $F(x)$ of an arbitrary discrete variable is discontinuous and increases by jumps at those values of x which are the possible values of ξ. The size of the jump of the function $F(x)$ at the point x equals the difference $F(x+0) - F(x)$.

If two of the possible values of the variable ξ are separated by an interval in which no other possible values of ξ appear, then the distribution function $F(x)$ is constant within this interval. If the number of possible values of ξ is finite, say n, then the distribution function $F(x)$ is a step function with a constant value in each of $n+1$ intervals. If, however, the number of possible values of ξ is enumerable, these possible values may be everywhere dense, so that there may or may not be any interval in which the discrete distribution function is constant. For example, let the possible values of ξ be all of the rational numbers. Suppose that these numbers are somehow ordered $r_1, r_2, \ldots$

and that the probabilities $\mathbf{P}\{\xi = r_k\} = p_k$ are defined by means of the relation $p_k = 2^{-k}$. In this example, every rational point is a point of discontinuity of the distribution function.

Another important class of random variables is represented by distributions for which there exists a non-negative function $p(x)$ satisfying the equation

$$F(x) \int_{-\infty}^{x} p(z)\,dz$$

for any x. Any random variables possessing this property are called continuous. The function $p(x)$ is called the probability density function.

Note that a probability density function possesses the following properties:

(1) $$p(x) \geqslant 0.$$

(2) For any x_1 and x_2 it satisfies the relation

$$\mathbf{P}\{x_1 \leqslant \xi < x_2\} = \int_{x_1}^{x_2} p(x)\,dx.$$

In particular, if $p(x)$ is continuous at x then $\mathbf{P}\{x \leqslant \xi < x + dx\} = \mathrm{b}p(x)\,dx$ with accuracy up to the infinitesimally small values of the highest orders.

(3) $$\int p(x)\,dx = 1,$$

Random variables distributed according to the normal or the uniform law represent an example of continuous random variables.

EXAMPLE. Let us examine the normal distribution in more detail. Its probability density function is given by

$$p(x) = C \cdot e^{-(x-a)^2/2\sigma^2}.$$

The constant C can be determined by property 3. Indeed,

$$C \int e^{-(x-a)^2/2\sigma^2} dx = 1.$$

By the substitution of variables $(x-a)/\sigma = z$, this equality can be reduced to

$$C\sigma \int e^{-z^2/2}\, dz = 1.$$

This integral is known as the Poisson integral and its value is

$$\int e^{-z^2/2}\, dz = \sqrt{2\pi}.$$

Thus, one finds that

$$C = \frac{1}{\sigma\sqrt{2\pi}}$$

and this implies that for the normal distribution

$$p(x) = \frac{1}{\sigma\sqrt{2\pi}} e^{-(x-a)^2/2\sigma^2}.$$

The function $p(x)$ attains its maximum value at $x = a$ and has points of inflection at $x = a \pm \sigma$. The graph of this function is asymptotic to the axis x as $x \to \pm\infty$. To illustrate the effect of the parameter σ on the shape of the graph of the normal density function, the graphs of $p(x)$ for $a = 0$ and (1) $\sigma^2 = 1/4$, (2) $\sigma^2 = 1$, and (3) $\sigma^2 = 4$ are depicted in Figure 15. One sees that the smaller the value of σ, the larger the

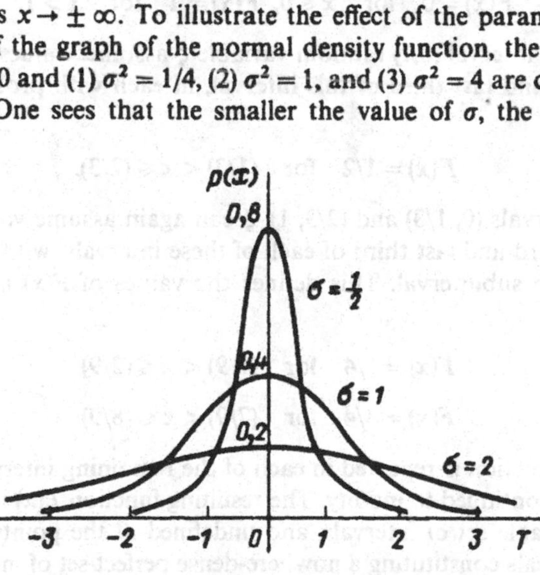

FIGURE 15. Geometrical interpretation for Example 4.

maximum value of $p(x)$ and the steeper the curve. This means, in particular, that for a normally distributed random variable ξ with parameter $a=0$ the probability of the condition $-\alpha \leqslant \xi < \alpha$ is greater if the value of σ is smaller. Therefore one may consider σ as a characteristic of the dispersion of the values of the variable ξ. For $a \neq 0$ the density curves have the same shape but they are merely relocated to the right ($a>0$) or to the left ($a<0$) depending on the sign of the parameter a.

Of course, there exist still other random variables in addition to those that are discrete or continuous. Besides those that behave like continuous variables in some intervals and like discrete variables in others, there are variables that are neither discrete nor continuous in any interval. For example, this category includes the random variables whose distribution functions are continuous but which at the same time increase only at points of a set of Lebesgue measure zero. An example of such a random variable is one which has the well known Cantor function as its distribution function. Let us recall how this function is constructed. The variable ξ takes on only values between 0 and 1. Therefore its distribution function satisfies the conditions

$$F(x)=0 \quad \text{for} \quad x \leqslant 0, \; F(x)=1 \quad \text{for} \quad x>1.$$

Within the interval (0,1) random variable ξ assumes values only in the first third and last third of this interval, in each with probability 1/2. Thus,

$$F(x)=1/2 \quad \text{for} \quad (1/3)<x \leqslant (2/3).$$

In the intervals (0, 1/3) and (2/3, 1), ξ can again assume values only in the first third and last third of each of these intervals, with probability 1/4 in each subinterval. This defines the values of $F(x)$ in two more intervals:

$$F(x)=1/4 \quad \text{for} \quad (1/9)<x \leqslant (2/9),$$

$$F(x)=3/4 \quad \text{for} \quad (7/9)<x \leqslant (8/9).$$

This construction is repeated in each of the remaining intervals and the process is continued to infinity. The resulting function $F(x)$ is defined on an enumerable set of intervals, and undefined at the points separating these intervals constituting a nowhere-dense perfect set of measure zero. On this set the function $F(x)$ is defined by continuity. The random variable ξ with the distribution function just defined is not discrete, since its distribution function is continuous; but at the same time ξ is not continuous, because its distribution function is not the integral of its derivative.

All of the definitions introduced above are readily carried over to the case of conditional probability. For example, the function $F(x/B) = \mathrm{P}\{\xi < x/B\}$ will be called the conditional distribution function of the random variable ξ under the condition B. It is obvious that $F(x/B)$ has all the properties of an ordinary distribution function.

23 MULTIDIMENSIONAL DISTRIBUTION FUNCTIONS

For further study one needs not only the concept of a random variable but also the concept of a random vector or, as it is often called, a multidimensional random variable.

Consider a probability space $\{\omega, \mathcal{J}, \mathrm{P}\}$ on which n random variables

$$\xi_1 = f_1(\omega), \quad \xi_2 = f_2(\omega), \ldots, \quad \xi_n = f_n(\omega)$$

are defined. Functions $f_1(\omega)$ are assumed to be measurable. The vector $(\xi_1, \xi_2, \ldots, \xi_n\}$ is called a random vector, or an n-dimensional random variable. Let $(\xi_1, \xi_2, \ldots, \xi_n\}$ be a random vector. Denote the set of all of elementary events ω for which all of inequalities

$$f_1(\omega) < x_1, \quad f_2(\omega) < x_2, \ldots, \quad f_n(\omega) < x_n$$

hold simultaneously by $\{\xi_1 < x_1, \xi_2 < x_2, \ldots, \xi_n < x_n\}$. Since this event is the product of the events $\{f_k(\omega) < x_k\}$ $(1 \leqslant k \leqslant n)$ it belongs to the set $\mathfrak{J}$, i.e.,

$$\{\xi_1 < x_1, \xi_2 < x_2, \ldots, \xi_n < x_n\} \in \mathfrak{F}.$$

Thus, for any collection of numbers $x_1, x_2, \ldots, x_n$ there is defined a probability $F(x_1, x_2, \ldots, x_n) = \mathrm{P}\{\xi_1 < x_1, \xi_2 < x_2, \ldots, \xi_n < x_n\}$. This function of n arguments is called the *no*-dimensional distribution function of the random vector $(\xi_1, \xi_2, \ldots, \xi_n)$.

Later on a geometrical interpretation will be used and the quantities $\xi_1, \xi_2, \ldots, \xi_n$ will be regarded as coordinates of a point in n-dimensional

Euclidean space. Obviously the position of a point $(\xi_1, \xi_2, \ldots, \xi_n)$ depends on chance and the function $F(x_1, x_2, \ldots, x_n)$, under this interpretation, yields the probability that the point $(\xi_1, \xi_2, \ldots, \xi_n)$. falls within the n-dimensional parallelepiped $\xi_1 < x_1$, $\xi_2 < x_2, \ldots, \xi_n < x_n$ whose edges are parallel to the coordinate axes.

By means of the distribution function, it is easy to compute the probability that the point $(\xi_1, \xi_2, \ldots, \xi_n)$ falls inside the parallelepiped

$$a_i \leqslant \xi_i < b_i \qquad (i = 1, 2, \ldots, n),$$

where a_i and b_i are arbitrary constants. It is not difficult to show that

$$\mathbf{P}\{a_1 \leqslant \xi_1 < b_1, a_2 \leqslant \xi_2 < b_2, \ldots, a_n \leqslant \xi_n < b_n\}$$

$$= F(b_1, b_2, \ldots, b_n) - \sum_{i=1}^{n} p_i + \sum_{i<j} p_{ij} \mp \cdots +$$

$$+(-1)^n F(a_1, a_2, \ldots, a_n), \tag{1}$$

where $p_{ij\ldots k}$ denotes the value of the function $F(c_1, c_2, \ldots, c_n)$ for $c_i = a_i$, $c_j = a_j, \ldots, c_k = a_k$ and the remaining c_s equal b_s. Proving this theorem is left to the reader. Note in particular, that $F(x_1, \ldots, x_{k-1}, +\infty, x_{k+1}, \ldots, x_n)$ gives the probability that the following system of inequalities is satisfied:

$$\xi_1 < x_1, \xi_2 < x_2, \ldots, \xi_{k-1} < x_{k-1}, \xi_{k+1} < x_{k+1}, \ldots, \xi_n < x_n.$$

Since by the extended axiom of addition for probabilities

$$\mathbf{P}\{\xi_1 < x_1, \ldots, \xi_{k-1} < x_{k-1}, \xi_{k+1} < x_{k+1}, \ldots, \xi_n < x_n\}$$

$$= \sum_{s=-\infty}^{\infty} \mathbf{P}\{\xi_1 < x_1, \ldots, \xi_{k-1} < x_{k-1}, s \leqslant \xi_k < s+1, \xi_{k+1}$$

$$< x_{k+1}, \ldots, \xi_n < x_n\} = F(x_1, \ldots, x_{k-1}, \infty, x_{k+1}, \ldots, x_n),$$

then $F(x_1, \ldots, x_{k-1}, +\infty, x_{k+1}, \ldots, x_n)$ represents the distribution function for the $(n-1)$-dimensional random variable $(\xi_1, \ldots, \xi_{k-1}, \xi_{k+1}, \ldots, \xi_n)$ By then continuing this process, one can determine the k-dimensional distribution function of any group of k quantities $\xi_{i_1}, \xi_{i_2}, \ldots, \xi_{i_k}$ from the formula

$$F_k(x_{i_1}, x_{i_2}, \ldots, x_{i_k}) = \mathbf{P}\{\xi_{i_1} < x_{i_1}, \ldots, \xi_{i_k} < x_{i_k}\} = F(c_1, c_2, \ldots, c_n),$$

where $c_s = x_s$ if $s = i_r (1 \leqslant r \leqslant k)$ and $c_s = +\infty$ in all other cases. In particular, the distribution function for the random variable ξ_k is

$$F_k(x) = F(c_1, c_2, \ldots, c_n),$$

where all $c_i (i \neq k)$ equal $+\infty$ and $c_k = x$.

The behavior of a multi-dimensional random variable, like that of a one-dimensional random variable, may be characterized by means other than its distribution function, say, by a non-negative completely additive set function $\Phi\{E\}$ defined on arbitrary Borel sets in n-dimensional space. This function is defined as the probability that the point $(\xi_1, \xi_2, \ldots, \xi_n)$ lies in the set E. This method of probabilistic characterizing an n-dimensional random variable must be regarded as most natural and most successful from a theoretical viewpoint.

Let us consider some examples.

Example 1. A random vector $(\xi_1, \xi_2, \ldots, \xi_n)$ is called uniformly distributed in the parallelepiped $a_i \leqslant \xi_i < b_i$ $(1 \leqslant i \leqslant n)$ if the probability of the point $(\xi_1, \xi_2, \ldots, \xi_n)$ lying in an arbitrary interior region of the parallelepiped is proportional to its volume and if the event consisting of the point falling inside the parallelepiped is certain.

The distribution function of the required variable has the form

$$F(x_1, \ldots, x_n) = \begin{cases} 0 & \text{if } x_i \leqslant a_i \text{ at least for one } i, \\ \prod_{i=1}^{n} \dfrac{c_i - a_i}{b_i - a_i} & \text{where } c_i = b_i \text{ if } a_i \leqslant x_i \leqslant b_i \\ & \text{and } c_i = b_i \text{ if } x_i > b_i. \end{cases}$$

Example 2. A two-dimensional random variable $(\xi_1 \, \xi_2)$ is distributed normally if its distribution function is given by

$$F(x_1, x_2) = C \int_{-\infty}^{x_1} \int_{-\infty}^{x_2} e^{-Q(x,y)} \, dx \, dy,$$

where $Q(x, y)$ is a positive-definite quadratic form.

It is known that a positive-definite quadratic form of x and y can be written in the form

$$Q(x, y) = \frac{(x-a)^2}{2A^2} - r\frac{(x-a)(y-b)}{AB} + \frac{(y-b)^2}{2B^2},$$

where A and B are positive and r, a, and b are real numbers. Inaddition r satisfies the condition $-1 < r < +1$.

It is readily apparent that for $r_2 \neq 1$, each of the random variables ξ_1 and ξ_2 is subject to a one-dimensional normal law. Indeed,

$$F_1(x_1) = \mathbf{P}\{\xi_1 < x_1\} = F(x_1, +\infty) = C \int_{-\infty}^{x_1} \int e^{-Q(x,y)}\, dx\, dy$$

$$= C \int_{-\infty}^{x_1} e^{-(x-a)^2/2A^2(1-r^2)} \int e^{-(1/2)[(y-b/B)]-r[(x-a/A)]^2}\, dy.$$

Since

$$\int e^{-(1/2)[(y-b)/B)-r(x-a)/A)^2]}\, dy = B\sqrt{\pi},$$

it follows that

$$F_1(x_1) = BC\sqrt{2\pi} \int_{-\infty}^{x_1} e^{-[(x-a)^2/2A^2](1-r^2)}\, dx. \tag{2}$$

The constant C is expressible in terms of A, B and r. This dependence can be found from the condition $F_1(+\infty) = 1$. It gives:

$$1 = BC\sqrt{2\pi} \int e^{-[(x-a)^2/2A^2](1-r^2)}\, dx = \frac{ABC\sqrt{2\pi}}{\sqrt{1-r^2}} \int e^{-z^2/2}\, dz = \frac{2ABC\pi}{\sqrt{1-r^2}}.$$

From here follows

$$C = \frac{\sqrt{1-r^2}}{2\pi AB}.$$

If $r_2 \neq 1$, then it is possible to set

$$A = \sigma_1\sqrt{1-r^2}, \qquad B = \sigma_2\sqrt{1-r^2}.$$

In this new notation the two-dimensional normal law takes the following form:

$$F(x_1, x_2) = \frac{1}{2\pi\sigma_1\sigma_2\sqrt{1-r^2}}$$

$$\times \int_{-\infty}^{x_1} \int_{-\infty}^{x_2} e^{-(1/2(1-r^2))[((x-a)^2/\sigma_1^2) - 2r((x-a)(y-b))/\sigma_1\sigma_2 + ((y-b)^2/\sigma_2^2)]}\, dx\, dy.$$

The probabilistic meaning of the parameters appearing in this formula will be explained in the following chapter.

Some properties of multidimensional distribution functions can be established in a way similar to those for the one-dimensional case. The justification of these properties is left to the reader. Here only the formulations are given. A distribution function

(1) is a non-decreasing function of each of its arguments,
(2) is continuous on the left in each of its arguments, and
(3) satisfies the relations

$$F(+\infty, +\infty, \ldots, +\infty) = 1,$$

$$\lim_{x_k \to -\infty} F(x_1, x_2, \ldots, x_n) = 0 \quad (1 \leqslant k \leqslant n)$$

for arbitrary values of the other arguments.

In the one-dimensional case, as was shown above, these properties are necessary and sufficient conditions for the function $F(x)$ to be the distribution function of some random variable. In the n-dimensional case, these properties are not sufficient. In order for the function $F(x_1, \ldots, x_n)$ to be a distribution function, it is necessary to have the three following additional properties:

(4) For any a_i and b_i $(i = 1, 2, \ldots, n)$ the expression (1) is non-negative.

This condition may not be satisfied even if the function $F(x_1, \ldots, x_n)$ obeys conditions (1)–(3). It can be shown by the following example. Let

$$F(x, y) = \begin{cases} 0 & \text{if } x \leqslant 0, \text{ or } x + y \leqslant 1, \text{ or } y \leqslant 0, \\ 1 & \text{in the remaining part of the plane.} \end{cases}$$

This function satisfies conditions (1)–(3) but

$$F(1, 1) - F(1, 1/2) - F(1/2, 1) + F(1/2, 1/2) = -1, \qquad (3)$$

and therefore the fourth condition is not satisfied.

The function $F(x, y)$ cannot be a distribution function since according to the relation (1), the expression (3) should be the probability that the point (ξ_1, ξ_2) falls in the square $(1/2) \leqslant \xi_1 < 1, (1/2) \leqslant \xi_2 < 1$.

If there exists a function $p(x_1, x_2, \ldots, x_n)$ such that the equation

$$F(x_1, x_2, \ldots, x_n) = \int_{-\infty}^{x_1} \int_{-\infty}^{x_2} \cdots \int_{-\infty}^{x_n} p(z_1, z_2, \ldots, z_n)\, dz_n \cdots dz_2\, dz_1,$$

holds for any values of $x_1, x_2, \ldots, x_n$ then this function is called the probability density function of the random vector $(\xi_1, \xi_2, \ldots, \xi_n)$. It is easy to see that a density function possesses the following properties:

$$p(x_1, x_2, \ldots, x_n) \geqslant 0. \tag{1}$$

(2) The probability that the point $(\xi_1, \xi_2, \ldots, \xi_n)$ will fall in some region G equals

$$\int_G \cdots \int p(x_1, \ldots, x_n)\, dx_n \cdots dx_1.$$

In particular, if the function $p(x_1, x_2, \ldots, x_n)$ is continuous at the point $(x_1, x_2, \ldots, x_n)$ then the probability that the point $(\xi_1, \xi_2, \ldots, \xi_n)$ will fall in the parallelepiped $x_k \leq v\xi_k < x_k + dx_k\ (k = 1, 2, \ldots, n)$ is

$$p(x_1, x_2, \ldots, x_n)\, dx_1\, dx_2 \ldots dx_n$$

with accuracy up to infinitesimally small values of higher order.

Example 3. Consider an n-dimensional random variable which is uniformly distributed in some n-dimensional region G. If V denotes the n-dimensional volume of the region G, then the density function is given by

$$p(x_1, x_2, \ldots, x_n) = \begin{cases} 0, & \text{for } p(x_1, x_2, \ldots, x_n) \bar{\in} G, \\ 1/V, & \text{for } (x_1, x_2, \ldots, x_n) \in G. \end{cases}$$

Example 4. The density function of the two-dimensional normal distribution is given by the expression

$$p(x, y) = \frac{1}{2\pi\sigma_1 \sigma_2 \sqrt{1 - r^2}} e^{-(1/2(1-r^2))\,[(x-a)^2/\sigma_1^2/ - 2r[(x-a)(y-b)/(\sigma_1\sigma_2) + (y-b)^2\sigma_2^2]}.$$

Note that the normal density function has a constant value on the ellipses

$$\frac{(x-a)^2}{\sigma_1^2} - 2r\frac{(x-a)(y-b)}{\sigma_1 \sigma_2} + \frac{(y-b)^2}{\sigma_2^2} = \lambda^2, \tag{4}$$

where λ is a constant. For this reason, the ellipses (4) are referred to as the *ellipses of equal probability*.

Let us find the probability that the point (ξ_1, ξ_2) falls inside the ellipse (4). By the definition of a density function, one can write

$$\mathsf{P}(\lambda) = \iint_{G(\lambda)} p(x, y)\,dx\,dy, \tag{5}$$

where $G(\lambda)$ denotes the region bounded by the ellipse (4). For evaluating this integral, one introduces polar coordinates

$$x - a = \rho\cos\theta, \quad y - b = \rho\sin\theta.$$

The integral in (5) takes the form

$$\mathsf{P}(\lambda) = \frac{1}{2\pi\sigma_1\sigma_2\sqrt{1-r^2}} \int_0^{2\pi\lambda/2} \int_0^{\sqrt{1-r^2}} e^{(-\rho^2/2)s^2} \rho\,d\rho\,d\theta,$$

where for brevity the following notation is introduced

$$s^2 = \frac{1}{1-r^2}\left[\frac{\cos^2\theta}{\sigma_1^2} - 2r\frac{\cos\theta\sin\theta}{\sigma_1\sigma_2} + \frac{\sin^2\theta}{\sigma_2^2}\right].$$

The integration with respect to ρ yields:

$$\mathsf{P}(\lambda) = \frac{1 - e^{-\lambda^2/2(1-r^2)}}{2\pi\sigma_1\sigma_2\sqrt{1-r^2}} \int_0^{2\pi} \frac{d\theta}{s^2}.$$

The integration with respect to θ can be carried out by the rules for integrating trigonometric functions. But one need not do so since it can be done automatically with the help of probabilistic reasoning. In fact,

$$\mathsf{P}(+\infty) = 1 = \frac{1}{2\pi\sigma_1\sigma_2\sqrt{1-r^2}} \int_0^{2\pi} \frac{d\theta}{s^2}.$$

Hence,

$$\int_0^{2\pi} \frac{d\theta}{s^2} = 2\pi\sigma_1\sigma_2\sqrt{1-r^2}$$

and therefore

$$\mathsf{P}(\lambda) = 1 - e^{-\lambda^2/2(1-r^2)}.$$

The normal distribution plays an exceptional role in varied applied problems. The distributions of many random variables playing an important practical role are proved to obey the normal distribution. For example, an enormous amount of artillery practice, carried out under dissimilar conditions, has shown that when a single gun is fired at a given target, the dispersion of the shells in the plane obeys the two-dimensional normal law. In Chapter 9 this "universality" of the normal law will be explained by the following fact. If a random variable is the sum of a very large number of independent random variables each having only a negligible effect on the sum then this resulting random variable is approximately distributed according to the normal law.

The most important concept of probability theory concerning the independence of events preserves its significance for random variables. In accordance with the definition of independence of events, random variables $\xi_1, \xi_2, \ldots, \xi_n$ are independent if for an arbitrary collection of them

$$\xi_{i_1}, \xi_{i_2}, \ldots, \xi_{i_k}$$

the following equality

$$\mathbf{P}\{\xi_{i_1} < x_{i_1}, \xi_{i_2} < x_{i_2}, \ldots, \xi_{i_k} < x_{i_k}\} = \mathbf{P}\{\xi_{i_1} < x_{i_1}\}\,\mathbf{P}\{\xi_{i_2} < x_{i_2}\} \cdots \mathbf{P}\{\xi_{i_k} < x_{i_k}\}$$

holds for arbitrary values of

$$x_{i_1}, x_{i_2}, \ldots, x_{i_k}$$

any for k $(1 \leq k \leq n)$. In particular, the relation

$$\mathbf{P}\{\xi_1 < x_1, \xi_2 < x_2, \ldots, \xi_n < x_n\} = \mathbf{P}\{\xi_1 < x_1\}\,\mathbf{P}\{\xi_2 < x_2\} \cdots \mathbf{P}\{\xi_n < x_n\}$$

is satisfied for arbitrary values of $x_1, x_2, \ldots, x_n$ or, in terms of distribution functions,

$$F(x_1, x_2, \ldots, x_n) = F_1(x_1)\,F_2(x_2) \cdots F_n(x_n),$$

where $F_k(x_k)$ denotes the distribution function of the variable ξ_k.

It is easy to see that the converse proposition is also true: If the distribution function $F(x_1, x_2, \ldots, x_n)$ of the system of random variables $\xi_1, \xi_2, \ldots, \xi_n$ is given by

$$F(x_1, x_2, \ldots, x_n) = F_1(x_1) F_2(x_2) \cdots F_n(x_n),$$

where functions $F_k(x_k)$ satisfy the relations

$$F_k(+\infty) = 1 \quad (k = 1, 2, \ldots, n),$$

then random variables $\xi_1, \xi_2, \ldots, \xi_n$ are independent, and the functions $F_1(x_1), F_2(x_2), \ldots, F_n(x_n)$ are their distribution functions.

The proof of this theorem is left to the reader.

Example 5. Consider an n-dimensional random variable whose components $\xi_1, \xi_2, \ldots, \xi_n$ are mutually independent random variables each of which is normally distributed as follows

$$F_k(x_k) = \frac{1}{\sigma_k \sqrt{2\pi}} \int_{-\infty}^{x_k} e^{-(z-a_k)^2/2\sigma_k^2} \, dz.$$

In this example, the n-dimensional distribution function is

$$F(x_1, x_2, \ldots, x_n) = (2\pi)^{-\frac{n}{2}} \prod_{k=1}^{n} \sigma_k^{-1} \int_{-\infty}^{x_k} e^{-(z-a_k)^2/2\sigma_k^2} \, dz.$$

If independent random variables $\xi_1, \xi_2, \ldots, \xi_n$ have densities $p_1(x_1), p_2(x_2), \ldots, p_n(x_n)$ then n-dimensional random variable $(\xi_1, \xi_2, \ldots, \xi_n)$ has the density function

$$p(x_1, x_2, \ldots, x_n) = p_1(x_1) p_2(x_2) \ldots p_n(x_n).$$

Example 6. If random variables $\xi_1, \xi_2, \ldots, \xi_n$ are independent and have density functions

$$p_k(x) = \frac{1}{\sigma_k \sqrt{2\pi}} e^{-(x-a_k)^2/2\sigma_k^2} (1 \leqslant k \leqslant n),$$

then n-dimensional density function of the random variable $(\xi_1, \xi_2, \ldots, \xi_n)$ is

$$p(x_1, x_2, \ldots, x_n) = \frac{(2\pi)^{-\frac{n}{2}}}{\sigma_1 \sigma_2 \ldots \sigma_n} e^{-1/2 \sum_{k=1}^{n} (x_k - a_k)^2/\sigma_k^2}. \tag{6}$$

For $n = 2$, this expression has the form

$$p(x_1, x_2) = \frac{1}{2\pi\sigma_1\sigma_2} e^{-((x_1 - a_1)^2/2\sigma_1^2) - ((x_2 - a_2)^2/2\sigma_2^2)}.$$

The comparison of this function with the two-dimensional density function (example 4) shows that for independent random variables ξ_1 and ξ_2 the parameter r equals 0.

For $n = 3$, the equation (6) can be interpreted as the density function of probabilities of the velocity components ξ_1, ξ_2, ξ_3 taken by coordinate axis only if one assume that

$$\sigma_1^2 = \sigma_2^2 = \sigma_3^2 = \frac{1}{hm},$$

where m is the molecule's mass and h is a constant.

The corresponding distribution is called the Maxwell distribution.

24 FUNCTIONS OF RANDOM VARIABLES

Now one can begin to solve the following problem: On the basis of the given distribution function $F(x_1, x_2, \ldots, x_n)$ and a collection of random variables $\xi_1, \xi_2, \ldots, \xi_n$ determine the distribution function $\Phi(y_1, y_2, \ldots, y_k)$ of the variables $\eta_1 = f(\xi_1, \xi_2, \ldots, \xi_n)$, $\eta_2 = f(\xi_1, \xi_2, \ldots, \xi_n), \ldots, \eta_k = f(\xi_1, \xi_2, \ldots, \xi_n)$.

This problem can be solved very simply in the general case but needs in extension of the concept of an integral. Let us restrict ourselves by the most important special cases: discrete and continuous random variables. The definition and the main properties of a Stieltjes integral is be presented in the next section where the general form of the most important results of the present section is also presented.

At the beginning consider the case where the n-dimensional vector $(\xi_1, \xi_2, \ldots, \xi_n)$ has a probability density function $p(x_1, x_2, \ldots, x_n)$. From the foregoing, it follows that the desired distribution function is defined by the equality

$$\Phi(y_1, y_2, \ldots, y_k) = \int \cdots_D \int p(x_1, x_2, \ldots, x_n)\, dx_1\, dx_2 \cdots dx_n,$$

where the region of integration D is defined by the inequalities

$$f_i(x_1, x_2, \ldots, x_n) < y_i \qquad (i = 1, 2, \ldots, k).$$

In the case of discrete random variables, the solution, obviously, is given by means of an n-fold sum which is also extended over the region D.

Consider now several important special cases.

The distribution function of a sum. Let us find the distribution function of the sum

$$\eta = \xi_1 + \xi_2 + \cdots + \xi_n,$$

if $p(x_1, x_2, \ldots, x_n)$ is the probability density function of the vector $(\xi_1, \xi_2, \ldots, \xi_n)$. The required function equals the probability that the point $(\xi_1, \xi_2, \ldots, \xi_n)$ falls in the half-space $\xi_1 + \xi_2 + \cdots + \xi_n < x$ and therefore

$$\Phi(x) = \int \cdots \int_{\Sigma x_i < x} p(x_1, x_2, \ldots, x_n)\, dx_1 dx_2 \cdots dx_n.$$

Consider the case $n = 2$ in more detail. The preceding formula now takes the form:

$$\Phi(x) = \int \int_{x_1 + x_2 < x} p(x_1, x_2)\, dx_1\, dx_2 = \int \int_{-\infty}^{x - x_1} p(x_1, x_2)\, dx_1\, dx_2. \quad (1)$$

If the variables ξ_1 and ξ_2 are independent, then $p(x_1, x_2) = p(x_1)$ $p(x_2)$ and equation (1) can be written in the form:

$$\Phi(x) = \int dx_1 \int_{-\infty}^{x - x_1} p_1(x_1) p_2(x_2)\, dx_2 = \int dx_1 \int_{-\infty}^{x} p_1(x_1) p_2(z - x_1)\, dz$$

$$= \int_{-\infty}^{x} dz \left\{ \int p_1(x_1) p_2(z - x_1)\, dx_1 \right\}. \quad (2)$$

In the general case formula (1) yields

$$\Phi(x) = \int_{-\infty}^{x} dx_1 \int p(z, x_1 - z)\, dz. \quad (3)$$

The last inequalities prove that if the multi-dimensional distribution of summands has a probability density function then their sum also has a probability density function. This density for two independent summands can be written in the form

$$p(x) = \int p_1(x-z)\,p_2(z)\,dz. \tag{4}$$

Consider some examples.

Example 1. Let ξ_1 and ξ_2 be independent and uniformly distributed in the interval (a, b). Find the density function for the sum $\xi_1 + \xi_2$.

The probability density functions of ξ_1 and ξ_2 are

$$p_1(x) = p_2(x) = \begin{cases} 0, & \text{if } x \leqslant a \text{ or } x > b, \\ \dfrac{1}{b-a}, & \text{if } a < x \leqslant b. \end{cases}$$

By formula (4), one finds that

$$p_\eta(x) = \int_a^b p_1(z)\,p_2(x-z)\,dz = \frac{1}{b-a}\int_a^b p_2(x-z)\,dz.$$

From the fact that for $x < 2a$

$$x - z < 2a - z < a,$$

and for $x > 2b$

$$x - z > 2b - z > b,$$

one concludes that for $x < 2a$ and $x > 2b$

$$p_\eta(x) = 0.$$

Now let $2a < y < 2b$. The function in the integral is different from 0 only for values of z satisfying the inequality

$$a < x - z < b$$

or, what is the same, the inequality

$$x - b < z < x - a.$$

Since $x > 2a$, it follows that $x - a > a$. Obviously, $x - a \leqslant b$ for $x < a + b$. Therefore, if $2a < x < a + b$, then

$$p_\eta(x) = \int_a^{x-a} \frac{dz}{(b-a)^2} = \frac{x-2a}{(b-a)^2}.$$

In an similar way, for $a + b < x \leqslant 2b$ one has

$$p_\eta(x) = \int_{x-b}^{b} \frac{dz}{(b-a)^2} = \frac{2b-x}{(b-a)^2}.$$

Collecting together all of the results obtained, one finds that

$$p_\eta(x) = \begin{cases} 0 & \text{for} \quad x \leqslant 2a \ x > 2b, \\ \dfrac{x-2a}{(b-a)^2} & \text{for} \quad 2a < x \leqslant a+b, \\ \dfrac{2b-x}{(b-a)^2} & \text{for} \quad a+b < x \leqslant 2b. \end{cases} \qquad (5)$$

The function $p_\eta(x)$ is called the Simpson distribution.

The computations in the last example can be substantially simplified if one makes use of geometrical reasoning. Let us, as usual, represent ξ_1 and ξ_2 as rectangular coordinates in the plane. Then the probability of the inequality $\xi_1 + \xi_2 < x$ for $2a < x \leq a + b$ equals the probability that a point lies in the doubly-shaded triangle in Figure 16. This probability can be easily calculated. It is equal to

$$F_\eta(x) = \frac{(x-2a)^2}{2(a-b)^2}.$$

For $a + b < x \leqslant 2b$, the probability of the inequality $\xi_1 + \xi_2 < x$ equals the probability that the point lies in the entire shaded region of the figure. This probability is given by

$$F_\eta(x) = 1 - \frac{(2b-x)^2}{2(b-a)^2}.$$

Differentiation with respect to x then yields formula (5).

In connection with the example considered, it is interesting to note the following.

Certain general questions in geometry led Nikolai Lobachevsky to the necessity of solving the following problem: Given a group of n

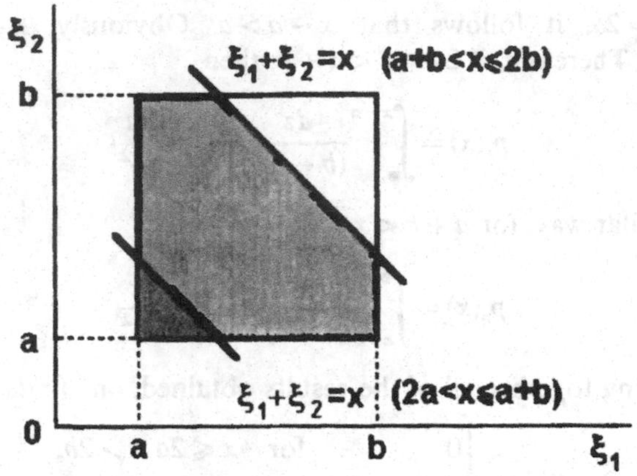

FIGURE 16. Geometrical Interpretation for Example 1.

mutually independent random variables $\xi_1, \xi_2, \ldots, \xi_n$, one needs to find the probability distribution of their arithmetic mean.

Lobachevsky solved this problem only for the case where all of the random variables (he considered errors of observation) were distributed uniformly in the interval $(-1, 1)$. It turns out in this case that the probability that the error in the arithmetic mean lies between -x and x equals

$$P_n(x) = \frac{1}{2^{n-1}} \times \sum (-1)^r \frac{[n - nx - 2r]^r}{(r!)^2 (n-r)!},$$

where the summation is extended over all the integers r from $r = 0$ to $r = [(n - nx)/2]$.

Example 2. A two-dimensional random variable (ξ_1, ξ_2) is normally distributed, i.e.,

$$p(x, y) = \frac{1}{2\pi\sigma_1\sigma_2\sqrt{1-r^2}}$$

$$\times \exp\left\{-\frac{1}{2(1-r^2)}\left(\frac{(x-a)^2}{\sigma_1^2} - 2r\frac{(x-a)(y-b)}{\sigma_1\sigma_2} + \frac{(y-b)^2}{\sigma_2^2}\right)\right\}.$$

Find the distribution function for the sum $\eta = \xi_1 + \xi_2$.

According to formula (3),

$$p_\eta(x)=\frac{1}{2\pi\sigma_1\sigma_2\sqrt{1-r^2}}$$

$$\int\exp\left\{-\frac{1}{2(1-r^2)}\left(\frac{(z-a)^2}{\sigma_1^2}-2r\frac{(z-a)(x-z-b)}{\sigma_1\sigma_2}+\frac{(x-z-b)^2}{\sigma_2^2}\right)\right\}dz.$$

For brevity, denote $x-a-b$ by v and $z-a$ by u. Then

$$p_\eta(x)=\frac{1}{2\pi\sigma_1\sigma_2\sqrt{1-r^2}}$$

$$\times\int\exp\left\{-\frac{1}{2(1-r^2)}\left(\frac{u^2}{\sigma_1^2}-2r\frac{u(v-u)}{\sigma_1\sigma_2}+\frac{(v-u)^2}{\sigma_2^2}\right)\right\}du.$$

Since

$$\frac{u^2}{\sigma_1^2}-2r\frac{u(v-u)}{\sigma_1\sigma_2}+\frac{(v-u)^2}{\sigma_2^2}=u^2\frac{\sigma_1^2+2r\sigma_1\sigma_2+\sigma_2^2}{\sigma_1^2\sigma_2^2}-2uv\frac{\sigma_1+r\sigma_2}{\sigma_1\sigma_2^2}+\frac{v^2}{\sigma_2}$$

$$=\left[u\frac{\sqrt{\sigma_1^2+2r\sigma_1\sigma_2+\sigma_2^2}}{\sigma_1\sigma_2}-\frac{v}{\sigma_2}\frac{\sigma_1+r\sigma_2}{\sqrt{\sigma_1^2+2r\sigma_1\sigma_2+\sigma_2^2}}\right]^2$$

$$+\frac{v^2}{\sigma_2^2}\left(1-\frac{(\sigma_1+r\sigma_2)^2}{\sigma_1^2+2r\sigma_1\sigma_2+\sigma_2^2}\right)=\left[u\frac{\sqrt{\sigma_1^2+2r\sigma_1\sigma_2+\sigma_2^2}}{\sigma_1\sigma_2}\right.$$

$$\left.-\frac{v}{\sigma_2}\frac{\sigma_1+r\sigma_2}{\sqrt{\sigma_1^2+2r\sigma_1\sigma_2+\sigma_2^2}}\right]^2+\frac{v^2(1-r^2)}{\sigma_1^2+2r\sigma_1\sigma_2+\sigma_2^2},$$

then after introducing the notation

$$t=\frac{1}{\sqrt{1-r^2}}\left[u\frac{\sqrt{\sigma_1^2+2r\sigma_1\sigma_2+\sigma_2^2}}{\sigma_1\sigma_2}-\frac{v}{\sigma_2}\frac{\sigma_1+r\sigma_2}{\sqrt{\sigma_1^2+2r\sigma_1\sigma_2+\sigma_2^2}}\right],$$

the expression for $p_\eta(x)$ can be reduced to the form

$$p_\eta(x)=\frac{\exp\{-v^2[2(\sigma_1^2+2r\sigma_1\sigma_2+\sigma_2^2)]\}}{2\pi\sqrt{\sigma_1^2+2r\sigma_1\sigma_2+\sigma_2^2}}\int e^{-t^2 2}dt.$$

Since

$$v = x - a - b \quad \text{and} \quad \int e^{-t^2 2} dt = \sqrt{2\pi},$$

then

$$p_\eta(x) = \frac{1}{\sqrt{2\pi(\sigma_1^2 + 2r\sigma_1\sigma_2 + \sigma_2^2)}} e^{-(x-a-b)^2/2(\sigma_1^2 + 2r\sigma_1\sigma_2 + \sigma_2^2)}.$$

In particular, if the random variables ξ_1 and ξ_2 are independent, then $r = 0$ and $p_\eta(x)$ takes the form

$$p_\eta(x) = \frac{1}{\sqrt{2\pi(\sigma_1^2 \sigma_2^2)}} e^{-(x-a-b)^2 \, 2(\sigma_1^2 + \sigma_2^2)}.$$

Thus one has obtained the following result: The sum of the random variables normally distributed is itself distributed according to the normal law.

It is interesting to note that for independent summands the inverse statement is also true (Cramer's Theorem): If the sum of two independent random variables is normally distributed then each of the summands is also distributed normally. The proof of this theorem requires more complex mathematical techniques and is omitted.

Example 3. The χ^2 distribution. Let $\xi_1, \xi_2, \ldots, \xi_n$ be independent random variables all distributed according to the same normal law, with parameters a and σ. The distribution function of the variable

$$\chi^2 = \frac{1}{\sigma^2} \sum_{k=1}^{n} (\xi_k - a)^2$$

is called the χ^2 distribution.

This distribution plays an important role in varied statistical problems.

Let us now compute the distribution function of the variable $\zeta = \chi/\sqrt{n}$. It will turn out to be independent of a and σ.

Obviously, for negative values of the argument, the distribution function $\Phi(y)$ of the random variable ζ equals 0. For positive values y the function $\Phi(y)$ equals the probability that the point $(\xi_1, \xi_2, \ldots, \xi_n)$

falls inside the following sphere:

$$\sum_{k=1}^{n}(x_k-a)^2=y^2\cdot n\cdot\sigma^2.$$

Thus

$$\Phi(y)=\underset{\Sigma x_i^2<y^2 n}{\int\cdots\int}\left(\frac{1}{\sqrt{2\pi}}\right)^n e^{-\Sigma_{i=1}^n(x_i^2/2)}dx_1\,dx_2\cdots dx_n.$$

To evaluate this integral, one passes to spherical coordinates, i.e., makes the substitution

$$x_1=\rho\cos\theta_1\,\cos\theta_2\cdots\cos\theta_{n-1},$$
$$x_2=\rho\cos\theta_1\,\cos\theta_2\cdots\sin\theta_{n-1},$$
$$\cdots\cdots\cdots\cdots\cdots\cdots\cdots\cdots$$
$$x_n=\rho\sin\theta_1.$$

This substitution yields

$$\Phi(y)=\int_{-\pi/2}^{\pi/2}\cdots\int_{-\pi/2}^{\pi/2}\int_0^{y\sqrt{n}}y\frac{1}{(\sqrt{2\pi})^n}e^{-\rho^2/2}\rho^{n-1}D(\theta_1\cdots\theta_{n-1})$$
$$\times\,d\rho\,d\theta_{n-1}\cdots d\theta_1=C_n\int_0^{y\sqrt{n}}e^{-\rho^2/2}\rho^{n-1}\,d\rho,$$

where the constant

$$C_n=\frac{1}{(\sqrt{2\pi})^n}\int_{-\pi/2}^{\pi/2}\cdots\int_{-\pi/2}^{\pi/2}D(\theta_1\cdots\theta_{n-1})d\theta_{n-1}\cdots d\theta_1$$

depends only on n.

This constant can be easily evaluated with the help of the equality

$$\Phi(+\infty)=1=C_n\int_0^{\infty}e^{-\rho^2/2}\rho^{n-1}d\rho=C_n\Gamma\left(\frac{n}{2}\right)\cdot 2^{n/2-1}.$$

From this, one finds:

$$\Phi(y)=\frac{1}{2^{n/2-1}\Gamma(n/2)}\int_0^{y\sqrt{n}}\rho^{n-1}e^{-\rho^2/2}\,d\rho.$$

The density function of the random variable ζ for $y \geqslant 0$ is

$$\varphi(y) = \frac{\sqrt{2n}}{\Gamma(n/2)}\left(\frac{y\sqrt{n}}{\sqrt{2}}\right)^{n-1} e^{-ny^2/2}. \tag{6}$$

Hence, for the particular case $n = 1$, one naturally obtains a density function which is twice as much as the density of the initial normal law:

$$\varphi(y) = \sqrt{2/\pi}\, e^{-y^2/2} \qquad (y \geqslant 0).$$

For $n = 3$, one obtains the well-known Maxwell distribution

$$\varphi(y) = \frac{3\sqrt{6}}{\sqrt{\pi}} y^2 e^{-3y^2/2}.$$

From formula (6), it is easy to derive the density function for the variable χ^2. For $x \leqslant 0$, *this density equals 0 and for* $x > 0$ equals

$$p_n(x) = \frac{x^{n/2-1} e^{-x/2}}{2^{n/2}\Gamma(n/2)}.$$

Table 11 represents main distributions tightly connected with χ^2 distribution. All of these distributions are widely used in practice.

Example 4. The distribution function of the quotient. Let the probability density function of the variable (ξ, η) be $p(x,y)$. Find the distribution function of the quotient $\zeta = \xi/\eta$.

Table 11

Random Variables	Density for $x > 0$
$\chi^2 = \frac{1}{\sigma^2}\sum_{k=1}^{n}(\xi_k - a)^2$	$\frac{x^{n/2-1}e^{-x/2}}{2^{n/2}\Gamma(n/2)}$
$\frac{1}{n}\chi^2 = \frac{1}{n\sigma^2}\sum_{k=1}^{n}(\xi_k - a)^2$	$\frac{(n/2)^{n/2}}{\Gamma(n/2)}x^{n/2-1}e^{-nx/2}$
$\chi = \sqrt{\frac{1}{\sigma^2}\sum_{k=1}^{n}(\xi_k - a)}$	$\frac{2}{2^{n/2}\Gamma(n/2)}x^{n-1}e^{-x^2/2}$
$\xi = \frac{x}{\sqrt{n}} = \sqrt{\frac{1}{n\sigma^2}\sum_{k=1}^{n}(\xi_k - a)^2}$	$\frac{\sqrt{2n}}{\Gamma(n/2)}\left(\frac{x\sqrt{n}}{\sqrt{2}}\right)^{n-1}e^{-nx^2/2}$

By definition,

$$F_\zeta(x) = \mathbf{P}\{\xi/\eta < x\}.$$

If ξ and η are considered as the rectangular coordinates of a point in the plane, then $F_\zeta(x)$ equals the probability that the point (ξ, η) falls in the region corresponding to points whose coordinates satisfy the inequality $\xi/\eta < x$. This region is shaded in Figure 17.

By the general formula, the required probability is

$$F_\zeta(x) = \int_0^\infty \int_{-\infty}^{zx} p(y, z)\,dy\,dz + \int_{-\infty}^0 \int_{zx}^{\infty} p(y, z)\,dy\,dz. \qquad (7)$$

From this it follows that if ξ and η are independent and $p_1(x)$ and $p_2(x)$ are their respective density functions, then

$$F_\zeta(x) = \int_0^\infty F_1(xz)\,p_2(z)\,dz + \int_{-\infty}^0 (1 - F_1(xz))\,p_2(z)\,dz. \qquad (7')$$

By differentiation of (7) one finds that

$$p_\zeta(x) = \int_0^\infty zp(zx, z)\,dz - \int_{-\infty}^0 zp(zx, z)\,dz. \qquad (8)$$

In particular, if ξ and η are independent, then

$$p_\zeta(x) = \int_0^\infty zp_1(zx)\,p_2(z)\,dz - \int_{-\infty}^0 zp_1(zx)\,p_2(z)\,dz. \qquad (8')$$

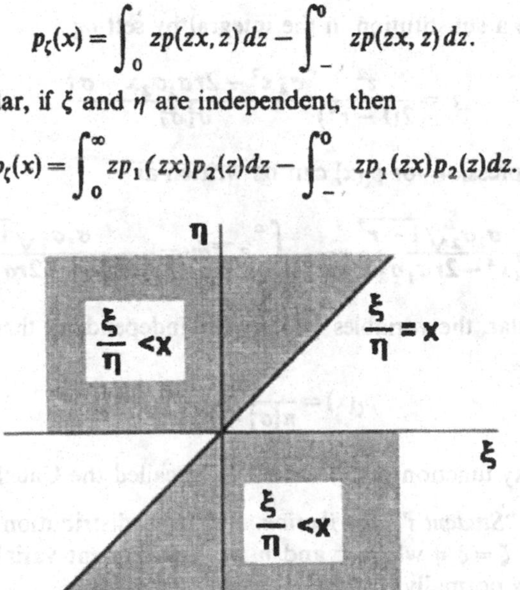

FIGURE 17 Geometrical Interpretation for Example 4.

Example 5. The random variable (ξ, η) is distributed according to the normal law, i.e.,

$$p(x, y) = \frac{1}{2\pi\sigma_1\sigma_2\sqrt{1-r^2}} \exp\left\{-\frac{1}{2(1-r^2)}\left[\frac{x^2}{\sigma_1^2} - 2r\frac{xy}{\sigma_1\sigma_2} + \frac{y^2}{\sigma_2^2}\right]\right\}.$$

Find the distribution function of quotient $\zeta = \xi/\eta$.

By formula (8)

$$p_\zeta(x) = \frac{1}{2\pi\sigma_1\sigma_2\sqrt{1-r^2}}$$

$$\times\left[\int_0^\infty - \int_{-\infty}^0\right] z \exp\left\{-\frac{z^2}{2(1-r^2)}\left[\frac{\sigma_2^2x^2 - 2r\sigma_1\sigma_2x + \sigma_1^2}{\sigma_1^2\sigma_2^2}\right]\right\} dz$$

$$= \frac{1}{\pi\sigma_1\sigma_2\sqrt{1-r^2}}\int_0^\infty z \exp\left\{-\frac{z^2}{2(1-r^2)}\cdot\frac{\sigma_2^2x^2 - 2r\sigma_1\sigma_2x + \sigma_1^2}{\sigma_1^2\sigma_2^2}\right\} dz.$$

Let us make a substitution in the integral by setting

$$u = \frac{z^2}{2(1-r^2)}\cdot\frac{\sigma_2^2x^2 - 2r\sigma_1\sigma_2x + \sigma_1^2}{\sigma_1^2\sigma_2^2}.$$

Then the expression for $p_\zeta(x)$ can be written as

$$p_\zeta(x) = \frac{\sigma_1\sigma_2\sqrt{1-r^2}}{\pi(\sigma_2^2x^2 - 2r\sigma_1\sigma_2x + \sigma_1^2)}\int_0^\infty e^{-u}\,du = \frac{\sigma_1\sigma_2\sqrt{1-r^2}}{\pi(\sigma_2^2x^2 - 2r\sigma_1\sigma_2x + \sigma_1^2)};$$

If, in particular, the variables ξ and η are independent, then

$$p_\zeta(x) = \frac{\sigma_1\sigma_2}{\pi(\sigma_1^2 + \sigma_2^2x^2)}.$$

The density function of the variable ζ is called the Cauchy law.

Example 6. "*Student's*" *distribution.* Find the distribution function of the quotient $\zeta = \xi/\eta$ where ξ and η are independent variables, and, in addition, ξ is normally distributed with the density

$$p_\xi(x) = \sqrt{\frac{n}{2\pi}}\, e^{-nx^2/2},$$

and $\eta = \chi/\sqrt{n}$ (see Example 3) with the density

$$p_\eta(x) = \frac{\sqrt{2n}}{\Gamma(n/2)}\left(\frac{y\sqrt{n}}{\sqrt{2}}\right)^{n-1} e^{-ny^2/2}.$$

By formula (8′)

$$p_\zeta(x) = \int_0^\infty z\sqrt{\frac{n}{2\pi}}\, e^{-nz^2x^2/2}\, \frac{\sqrt{2n}}{\Gamma(n/2)}\left(\frac{z\sqrt{n}}{\sqrt{2}}\right)^{n-1} e^{-nz^2/2}\, dz$$

$$= \frac{1}{\sqrt{\pi}\,\Gamma(n/2)}\int_0^\infty \left(\frac{z\sqrt{n}}{\sqrt{2}}\right)^{n-1} e^{-(nz^2/2)(x^2+1)}\, nz\, dz.$$

Making the substitution

$$u = \frac{nz^2}{2}(x^2+1),$$

one finds

$$p_\zeta(x) = \frac{(x^2+/1)^{-(n+1)/2}}{\sqrt{\pi}\,\Gamma(n/2)}\int_0^\infty u^{n-1/2} e^{-u}\, du = \frac{\Gamma\left(\frac{n+1}{2}\right)}{\sqrt{\pi}\,\Gamma(n/2)}(x^2+1)^{-n+1/2}.$$

The probability density function

$$p_\zeta(x) = \frac{\Gamma\left(\frac{n+1}{2}\right)}{\sqrt{\pi}\,\Gamma(n/2)}(1+x^2)^{-n+1/2}$$

is referred to as "Student's" law. ("Student" was the pseudonym of the English statistician W. L. Gosset, who first discovered this law by empirical means.)

For $n = 1$, Student's law transforms into Cauchy's law.

Example 7. Rotation of the coordinate axes. The distribution function of the two-dimensional random variable (ξ, η) is given. Find the distribution function for the random variables

$$\xi' = \xi\cos\alpha + \eta\sin\alpha,$$
$$\eta' = -\xi\sin\alpha + \eta\cos\alpha. \qquad (9)$$

Let $F(x, y)$ and $\Phi(x, y)$ denote the respective distribution functions of the variables (ξ, η) and (ξ', η'). If (ξ, η) and (ξ', η') are understood as the rectangular coordinates of a point in the plane, then it is easy to see that the coordinate system $\xi' 0 \eta'$ can be obtained from the coordinate system $\xi 0 \eta$ by rotating the latter through the angle a. Let us restrict ourselves to the case $0 < \alpha < (\pi/2)$. (Deriving analogous formulas for the remaining quantities of α is left to the reader.)

Let us denote the density of the random vector (ξ, η) by $p(x, y)$ and the density of the random vector (ξ', η') by $\pi(x, y)$. From (9) one finds

$$\xi' = \xi \cos \alpha - \eta \sin \alpha,$$

$$\eta' = \xi \sin \alpha + \eta \cos \alpha,$$

and therefore

$$\pi(x, y) = p(x \cos \alpha - y \sin \alpha,\ x \sin \alpha + y \cos \alpha). \tag{10}$$

This equality allows one to obtain the formula tying distribution functions of vectors (ξ, η) and (ξ', η'). When deriving the formula, one should keep in mind a geometrical interpretation.

Example 8. A two-dimensional random variable (ξ, η) is normally distributed:

$$p(x, y) = \frac{1}{2\pi\sigma_1\sigma_2\sqrt{1-r^2}} \exp\left\{-\frac{1}{2(1-r^2)}\left[\frac{x^2}{\sigma_1^2} - 2r\frac{xy}{\sigma_1\sigma_2} + \frac{y^2}{\sigma_2^2}\right]\right\}.$$

Find the density function of the random variables

$$\xi' = \xi \cos \alpha + \eta \sin \alpha,$$

$$\eta' = -\xi \sin \alpha + \eta \cos \alpha.$$

By equation (10), one has

$$\pi(x', y') = p(x' \cos \alpha - y' \sin \alpha,\ x' \sin \alpha + y' \cos \alpha)$$

$$= \frac{1}{2\pi\sigma_1\sigma_2\sqrt{1-r^2}} \exp\left\{-\frac{1}{2(1-r^2)}[Ax'^2 - 2Bx'y' + Cy'^2]\right\},$$

where

$$A = \frac{\cos^2\alpha}{\sigma_1^2} - 2r\frac{\cos\alpha\sin\alpha}{\sigma_1\sigma_2} + \frac{\sin^2\alpha}{\sigma_2^2},$$

$$B = \frac{\cos\alpha\sin\alpha}{\sigma_1^2} - r\frac{\sin^2\alpha - \cos^2\alpha}{\sigma_1\sigma_2} - \frac{\cos\alpha\sin\alpha}{\sigma_2^2},$$

$$C = \frac{\sin^2\alpha}{\sigma_1^2} + 2r\frac{\cos\alpha\sin\alpha}{\sigma_1\sigma_2} + \frac{\cos^2\alpha}{\sigma_2^2}.$$

From this formula it follows that under rotation of axes a normal two-dimensional distribution transforms into another normal two-dimensional distribution.

Note that if the angle α is selected in such a manner that

$$\operatorname{tg} 2\alpha = \frac{2r\sigma_1\sigma_2}{\sigma_1^2 - \sigma_2^2},$$

then $B = 0$ and

$$\pi(x', y') = \frac{1}{2\pi\sigma_1\sigma_2\sqrt{1-r^2}} e^{-(Ax'^2\, 2(1-r^2)) - (By'^2\, 2(1-r^2))}.$$

This equality suggests that any normally distributed two-dimensional random variable can be reduced by means of a rotation of axes to a system of two normally distributed independent random variables. This result can be extended to n-dimensional random variables.

It is possible to prove a stronger statement which exhaustively characterizes the normal distribution. Let there be a non-degenerate probability distribution in the plane, i.e., such a distribution which is not concentrated along a line. The necessary and sufficient condition for this distribution being normal consists in a possibility of choosing two distinct planar coordinate systems $\xi_1 O \xi_2$ and $\eta_1 O \eta_2$ such that the coordinates ξ_1 and ξ_2 (as well as η_1 and η_2), considered as random variables with the given probability distribution, are independent.

25 THE STIELTJES INTEGRAL

The further consideration essentially uses the concept of the Stieltjes integral. To assist the reading of the following sections, the definition and fundamental properties of the Stieltjes integral are given in a brief form without proofs.

Suppose that in the interval (a, b) there are defined a function $f(x)$ and a non-decreasing function of bounded variation $F(x)$. For definiteness, let us suppose that the function $F(x)$ is continuous from the left. If a and b are finite, one can subdivide the interval (a, b) into a finite number of subintervals (x_i, x_{i+1}) by means of the points x_i such that $a = x_0 < x_1 < x_2 < \cdots < x_n = b$. Let us now form the sum

$$\sum_{i=1}^{n} f(\tilde{x}_i)[F(x_i) - F(x_{i-1})],$$

where $\tilde{x}_i$ is an arbitrary value chosen in the subinterval (x_{i-1}, x_i), and begin to increase the number of points of subdivision simultaneously decreasing the length of the maximum subinterval to 0. If the sum written above tends to a definite limit

$$J = \lim_{n \to \infty} \sum_{i=1}^{n} (f(\tilde{x}_i)[F(x_i) - F(x_{i-1})], \tag{1}$$

then this limit is called the Stieltjes integral of the function $f(x)$ with respect to the integrating function $F(x)$. This integral is denoted by the symbol

$$J = \int_b^a f(x)\,dF(x). \tag{2}$$

The improper Stieltjes integral, i.e., one with the infinite interval of integration, is defined in the usual way. One considers the integral over some finite interval (a, b) and then, in an arbitrary way, allows the quantities a and b to approach $-\infty$ and $+\infty$, respectively. If the resulting limit

$$\lim_{\substack{a \to -\infty \\ b \to \infty}} \int_a^b f(x)\,dF(x),$$

exists, then this limit is called the Stieltjes integral of the function $f(x)$ with respect to the function $F(x)$ in the interval $(-\infty, +\infty)$ and is denoted by

$$\int f(x)\,dF(x).$$

It can be shown that if the function $f(x)$ is continuous and bounded, the limit of the sum (1) exists in the case of both finite and infinite limits of integration.

In certain cases, the Stieltjes integral also exists for unbounded functions $f(x)$. The consideration of such integrals is of considerable interest in the probability theory for calculating mathematical expectation, variance, moments, etc.

Note that throughout the sequel the integral of the function $f(x)$ is considered to exist if and only if the integral of $|f(x)|$ with respect to the same function $F(x)$ exists.

For the purposes of probability theory, it is important to extend the definition of the Stieltjes integral to the case where the function $f(x)$ has a finite or enumerable number of discontinuities. It can be shown that any bounded function that has a finite or enumerable number of points of discontinuing, in particular every function of bounded variation is integrable with respect to any function of bounded variation. For this, it is necessary to modify slightly the definition of a Stieltjes integral. In forming the limit (1), one must consider only those sequences of subdivisions of the interval of integration for which every point of discontinuity of the function $f(x)$ is a point of division in all the subdivisions, with the possible exception of a finite number of them.

It must be emphasized that in fixing the limits of integration it is important to specify whether or not either of the endpoints of the interval of integration is included. Indeed, from the definition of the Stieltjes integral, one obtains the following relation

$$\begin{aligned}\int_{a-0}^{b} f(x)\,dF(x) &= \lim_{n\to} \sum_{i=1}^{n} f(\tilde{x}_i)[F(x_i)-F(x_{i-1})]\\ &= \lim_{n\to} \sum_{i=2}^{n} f(\tilde{x}_i)[F(x_i)-F(x_{i-1})]\\ &\quad+ \lim_{x_1\to x_0=a} f(\tilde{x}_1)[F(x_1)-F(x_0)]\\ &= \int_{a+0}^{b} f(x)\,dF(x) + f(a)[F(a+0)-F(a)].\end{aligned}$$

Here the symbol $a-0$ signifies that a is included in the interval of integration, and the symbol $a+0$ signifies that a is excluded from this interval.

Thus, if $f(a) \neq 0$ and the function $F(x)$ has a jump at $x = a$, then

$$\int_{a-0}^{b} f(x)dF(x) - \int_{a+0}^{b} f(x)dF(x) = f(a)[F(a+0) - F(a-0)].$$

This shows that a Stieltjes integral which is taken over an interval that reduces to a single point can yield a non-zero result. Let us agree for concreteness that unless it is specifically stated to the contrary, the left-hand endpoint of the interval of integration is included in it and the right-hand endpoint is excluded. This condition allows one to write the following equation:

$$\int_{a}^{b} dF(x) = F(b) - F(a).$$

In fact, by definition

$$\int_{a}^{b} dF(x) = \lim_{n\to} \sum_{i=1}^{n} [F(x_i) - F(x_{i-1})] = \lim_{n\to} [F(x_n) - F(x_0)]$$

$$= F(b) - F(a)$$

(Let us recall that by assumption $F(x)$ is continuous on the left and therefore

$$F(b) = \lim_{S\to 0} F(b - \varepsilon).$$

In particular, if $F(x)$ is the distribution function of a random variable ξ then

$$\int_{a}^{b} dF(x) = F(b) - F(a) = \mathsf{P}\{a \leqslant \xi \leqslant b\},$$

$$\int_{-}^{b} dF(x) = F(b) = \mathsf{P}\{\xi < b\}.$$

If $F(x)$ has a derivative of which it is an integral, then from the formula of finite increments one can write

$$F(x_i) - F(x_{i-1}) = p(\tilde{x}_i)(x_i - x_{i-1}),$$

where

$$x_{i-1} < \tilde{x}_i < x_i$$

follows the equality

$$\int_a^b f(x)dF(x) = \lim_{n\to} \sum_{i=1}^n f(\tilde{x}_i)[F(x_i) - F(x_{i-1})]$$

$$= \lim_{n\to} \sum_{i=1}^n f(\tilde{x}_i)p(\tilde{x}_i)(x_i - x_{i-1}) = \int_a^b f(x)p(x)dx.$$

Thus in this case a Stieltjes integral is reduced to an ordinary integral.

If $F(x)$ has a jump at the point $x = c$, then by choosing the subdivisions so that for some value of the subscript $x_k < cx_{k+1}$ one has

$$\int_a^b f(x)dF(x) = \lim_{n\to} \sum_{i=1}^k f(\tilde{x}_i)[F(x_i) - F(x_{i-1})] + f(c)[F(x_{k+1}) - F(x_k)]$$

$$+ \lim_{n\to} \sum_{i=k+2}^n f(\tilde{x}_i)[F(x_i) - F(x_{i-1})]$$

$$= \int_a^c f(x)dF(x) + \int_{s+0}^b f(x)dF(x) + f(c)[F(c+0) - F(C-0)].$$

In particular, if the function $F(x)$ changes its values only at the points $c_1, c_2, \ldots, c_n, \ldots$ then

$$\int_a^b f(x)dF(x) = \sum_{n=1} f(c_n)[F(c_n+0) - F(c_n - 0)]$$

and the Stieltjes integral reduces to an infinite series.

Let us list the main properties of the Stieltjes integral which will be required later on. The reader can easily obtain the proofs of these properties by starting from the definition of a Stieltjes integral and using arguments similar to those which are used in the theory of ordinary integrals.

1. For $a < c_1 < c_2 < \cdots < c_n < b$, one has

$$\int_a^b f(x)dF(x) = \sum_{i=0}^n \int_{c_i}^{c_{i+1}} f(x)dF(x) \qquad [a = c_0, \quad b = c_{n+1}]$$

2. A constant factor may be removed from under the integral

$$\int_a^b cf(x)dF(x) = c\int_a^b f(x)dF(x).$$

3. The integral of a sum of a finite number of functions equals the sum of their integrals:

$$\int_a^b \sum_{i=1}^n f_i(x)dF(x) = \sum_{i=1}^n \int_a^b f_i(x)dF(x).$$

4. If $f(x) \geq 0$ and $b > a$, then

$$\int_a^b f(x)dF(x) \geqslant 0.$$

5. If $F_1(x)$ and $F_2(x)$ are monotonic functions of bounded variation, and c_1 and c_2 are arbitrary constants, then

$$\int_a^b f(x)d[c_1F_1(x) + c_2F_2(x)] = c_1\int_a^b f(x)dF_1(x) + c_2\int_a^b f(x)dF_2(x).$$

6. If

$$F(x) = \int_c^x g(u)\,dG(u)$$

where c is a constant, $g(u)$ is a continuous function and $G(u)$ is a non-decreasing function of bounded variation, then

$$\int_a^b f(x)dF(x) = \int_a^b f(x)g(x)dG(x).$$

Using the concept of a Stieltjes integral, one can write general formulas for the distribution functions of the sum of two independent random variables, $\xi_1 + \xi_2$

$$F(x) = \int F_1(x - z)\,dF_2(z) = \int F_2(x - z)dF_1(z)$$

as well as the distribution for the quotient of two random variables, ξ_1/ξ_2

$$F(x)=\int_0^{\infty} F_1(xz)dF_2(z)+\int_{-\infty}^{0} [1-F_1(xz)]dF_2(z)$$

under the assumption that $P\{\xi_2=0\}=0$.

26 EXERCISES

1. Prove that if $F(x)$ is a distribution function, then for any $h\neq 0$ the functions

$$\Phi(x)=\frac{1}{h}\int_x^{x+h} F(x)\,dx,\qquad \Psi(x)=\frac{1}{2h}\int_{x-h}^{x+h} F(x)dx$$

are also distribution functions.

2. The random variable ξ has $F(X)$ as its distribution function (or $p(x)$ as the density function). Find the distribution function (density function) for each of the random variables:

(a) $\eta=a\xi+b$, where a and b are real numbers;
(b) $\eta=\xi^{-1}$, where $P\{\xi=0\}=0$;
(c) $\eta=\tan\xi$;
(d) $\eta=\cos\xi$;
(e) $\eta=f(\xi)$, where $f(x)$ is a continuous monotonic function without intervals of constancy.

3. A line is drawn from the point $(0, a)$ making an angle φ with the y-axis. Find the distribution function for the abscissa of the point of intersection of this line with the x-axis if

(a) the angle φ is uniformly distributed in the interval $(0, \pi/2)$;
(b) the angle φ is uniformly distributed in the interval $(-\pi/2, \pi/2)$.

4. A point is chosen at random on the circle of radius R with center at the origin (in other words, the polar angle of the point chosen is uniformly distributed in the interval $(-\pi, \pi)$). Find the density function for

(a) the abscissa of the point selected;
(b) the length of the chord joining this point to the point $(-R, 0)$.

5. A point is chosen at random on the segment of the y-axis between the points $(0,0)$ and $(0, R)$ (i.e., the ordinate of this point is uniformly distributed in the interval $(0, R)$). Through this point we draw the chord of the circle $x^2+y^2=R^2$ that is perpendicular to the y-axis. Determine the distribution of the length of this chord.

6. The diameter of a circle is measured approximately. Assume that this quantity is uniformly distributed in the interval (a, b). Find the distribution of the area of the circle.

7. The density function of a random variable ξ is given by the expression

$$p(x)=\frac{a}{e^{-x}+e^{x}}.$$

Find

(a) the constant a;

(b) the probability that in two independent observations a random variable ξ will take on values less than 1.

8. The distribution function of a random vector (ξ, η) is given by

$$\text{a) } F(x, y)=F_1(x)\,F_2(y)+F_3(x);$$

$$\text{b) } F(x, y)=F_1(x)\,F_2(y)+F_3(x)+F_4(y).$$

Can the functions $F_3(x)$ and $F_4(x)$ be arbitrary? Are the components of the vector (ξ, η) dependent or independent?

9. Two points are chosen at random in the interval $(0, a)$ (i.e., the abscissa of each of them is uniformly distributed in the interval $(0, a)$). Determine the distribution function for the distance between these two points.

10. n points are chosen at random in the interval $(0, a)$. Assuming that the points are independent and each of them is distributed uniformly in interval $(0, a)$, find

(a) the density function for the abscissa of the k-th point from the left; (b) the joint density function of the abscissas of the k-th and m-th points from the left $(k<m)$.

11. n independent observations are performed over the values of a random variable ξ having a continuous distribution function. The results of the experiment are the following observed values: $x_1, x_2, \ldots, x_n$.

Determine the distribution function of each of the random variables:

(a) $\eta_n = \max(x_1, x_2, \dots, x_n)$;

(b) $\zeta_n = \min(x_1, x_2, \dots, x_n)$.

(c) the k-th largest value (the left among increasingly ordered values);

(d) the joint distribution of the k-th and m-th largest values.

12. The distribution function of the random vector $(\xi_1, \xi_2, \dots, \xi_n)$ is $F(x_1, x_2, \dots, x_n)$. The components of the vector observed as the result of a trial were as follows: $(z_1, z_2, \dots, z_n)$. Find the distribution function for values

(a) $\eta_n = \max(z_1, z_2, \dots, z_n)$;

(b) $\zeta_n = \min(z_1, z_2, \dots, z_n)$.

13. The random variable ξ has a continuous distribution function $F(x)$. How is the random variable $\eta = F(\xi)$ distributed?

14. The random variables ξ and η are independent; their density functions are given by the relations:

$$p_\xi(x) = p_\eta(x) = 0 \quad \text{for } x \leqslant 0,$$

$$p_\xi(x) = c_1 x^\alpha e^{-\beta x},\ p_\eta(x) = c_2 x^\gamma e^{-\beta x} \qquad \text{for } x > 0.$$

Find

(a) the constants c_1 and c_2;

(b) the density function of the sum $\xi + \eta$.

15. Find the distribution function for the sum of the independent variables ξ and η, the first of which is uniformly distributed in the interval $(-h, h)$, and the second of which has the distribution function $F(x)$.

16. The density function of the random vector (ξ, η, ζ) is given by

$$p(x, y, z) = \begin{cases} \dfrac{6}{(1 + x + y + z)^4} & \text{for } x > 0,\ y > 0,\ z > 0, \\ 0 & \text{otherwise} \end{cases}$$

Find the distribution of the random variable $\xi + \eta + \zeta$.

17. Determine the distribution function for the sum of the independent random variables ξ_1 and ξ_2 if their distributions are given by the conditions:

(a)

$$F_1(x) = F_2(x) = \frac{1}{2} + \frac{1}{\pi} \operatorname{arctg} x;$$

(b) uniformly distributed in the intervals $(-5, 1)$ and $(1, 5)$, respectively;

(c)

$$p_1(x) = p_2(x) = \frac{1}{2\alpha} e^{-|x|/\alpha}.$$

18. The density functions of the independent random variables ξ and η are

(a)

$$p_\xi(x) = p_\eta(x) = \begin{cases} 0 & \text{for } x \leqslant 0, \\ ae^{-ax} & \text{for } x > 0 \ (a > 0); \end{cases}$$

(b)

$$p_\xi(x) = p_\eta(x) = \begin{cases} 0 & \text{for } x \leqslant 0,\ x > a. \\ 1/a & \text{for } 0 < x \leqslant a. \end{cases}$$

Determine the density function of the variable $\zeta = \xi/\eta$.

19. Find the distribution function of the product of two independent random variables ξ and η if their distribution functions $F_1(x)$ and $F_2(x)$ are known.

20. The random variables ξ and η are independent and are distributed

(a) uniformly in the interval $(-a, a)$;

(b) normally, with parameters $a = 0$ and $e = 1$. Determine the distribution function of their product.

21. The lengths ξ and η of two sides of a triangle are independent random variables. On the basis of their distribution functions $F_\xi(x)$ and $F_\eta(x)$ find the distribution function of the third side if the angle between the two given sides has the constant value.

22. Prove that if the variables ξ and η are independent and their density functions are

$$p_\xi(x) = p_\eta(x) = \begin{cases} 0 & \text{for } x \leqslant 0, \\ e^{-x} & \text{for } x > 0, \end{cases}$$

then the variables $\xi + \eta$ and $\xi \ \eta$ are also independent.

23. Prove that if the variables ξ and η are independent and normally distributed with the parameters $a_1 = a_2 = 0$ and $e_1 = e_2 = e$, then the random variables $\zeta = \xi^2 + \eta^2$ and $k = \xi/\eta$ are also independent.

24. Prove that if the variables ξ and η are independent and distributed according to the χ law with parameters m and n, then the random variables $\zeta = \xi^2 + \eta^2$ and $k = \xi/\eta$ are independent.

25. The random variables $\xi_1, \xi_2, \ldots, \xi_n$ are independent and all have the same density function

$$p(x) = \frac{1}{\sigma\sqrt{2\pi}} e^{-(x-a)^2/2\sigma^2}$$

Find the two-dimensional density function of the random variables

$$\eta = \sum_{k=1}^{n} \xi_k$$

and

$$\zeta = \sum_{k=1}^{m} \xi_k$$

26. Prove that every distribution function has the following properties:

$$\lim_{x\to} x \int_x \frac{1}{z} dF(z) = 0, \quad \lim_{x\to +0} x \int_x \frac{1}{z} dF(z) = 0,$$

$$\lim_{x\to -} x \int_{-}^{x} \frac{1}{z} dF(z) = 0, \quad \lim_{x\to -0} x \int_{-}^{x} \frac{1}{z} dF(z) = 0.$$

27. Two series of independent trials are performed over the values of a random variable ξ having a continuous distribution function $F(x)$. As the results of experiments the observed values (increasingly ordered) are:

$$x_1 < x_2 < \cdots < x_M, \quad y_1 < y_2 < \cdots < y_N.$$

What is the probability of the inequality

$$y_\mu < x_{m+1} < y_{\mu+1},$$

where m and f are given quantities $(0 < m < M,\ 0 < f < N)$?

28. The random variable ξ has a continuous distribution function $F(x)$. n independent observations gave the values of $\xi: x_1 < x_2 < \cdots < x_n$

arranged in increasing order of magnitude. Find the density function of the variable

$$\eta = \frac{F(x_n) - F(x_2)}{F(x_n) - F(x_1)}.$$

29. Two random variables ξ and η are independent and identically distributed with the same density functions defined as

$$p_\xi(x) = p_\eta(x) = \frac{C}{1 + x^4}.$$

Find the constant C and show that the variable ξ/η distributed according to the Cauchy's law.

30. The random variables ξ and η are independent and their density functions equal, respectively,

$$p_\xi(x) = \frac{1}{\pi\sqrt{1 - x^2}} (|x| < 1);$$

$$p_\eta(x) = \begin{cases} 0 & \text{for } x \leqslant 0 \\ xe^{-x^2/2} & \text{for } x > 0. \end{cases}$$

Prove that the variable ξ/η is normally distributed.

31. Let ξ and η be independent and have the same probability densities

$$p_\xi(x) = p_\xi(x) = \begin{cases} 0 & \text{for } x \leqslant 0, \\ \lambda e^{-\lambda x} & \text{for } x > 0. \end{cases}$$

Prove that the quotient $\zeta = \xi/(\xi + \eta)$ is distributed uniformly on the interval $(0, 1)$.

32. The random variables ξ and η are independent and uniformly distributed on $(-1, 1)$. Calculate the probability that the roots of the equation $x^2 + \xi x + \eta = 0$ are real.

Chapter 6

NUMERICAL CHARACTERISTICS OF RANDOM VARIABLES

In the preceding chapter, it was shown that the most complete characterization of a random variable is provided by its distribution function. Indeed, the distribution function simultaneously indicates what values a random variable may assume and with what probabilities. However, in many cases, one needs to have considerably less knowledge about the random variable, restricted oneself to a summary description. In probability theory and its applications, some certain constants (numerical characteristics) play an important role. These constants are obtained from the probability functions of random variables in accordance with definite rules. Among these constants that serve to give a general quantitative characterization of random variables, the mathematical expectation, the variance, and the moments of various orders are of special importance.

27 MATHEMATICAL EXPECTATION

The discussion begins by considering the following illustrative example. Suppose that one needs 1 shot to fired from a certain gun to destroy some particular target with probability p_1, 2 shots for probability p_2, 3 shots for probability p_3, and so on. In addition, it is known that n shots definitely suffice to hit this target. Thus, it is known that

$$p_1 + p_2 + \cdots + p_n = 1.$$

The question is: How many shots, on the average, are needed to hit the target?

To answer this question, one argues as follows. Suppose that a sufficiently large number of shots are performed under the conditions specified above. Then, by Bernoulli's Theorem, one may assert that the relative frequency of the firings in which only a single shot is needed to hit the target approximately equals $p1$. Similarly, two shots are needed in approximately 100 p_2% of the firings, etc. Thus, "on the average" approximately

$$1 \cdot p_1 + 2 \cdot p_2 + \cdots + n \cdot p_n$$

shots are required to hit a single target.

Analogous questions concerning the calculation of the average value of a random variable arise in various applied problems. This is why in probability theory the special numerical characteristic called the mathematical expectation is introduced.
A definition of this concept is begun for discrete random variables, based on the example just considered.

Let

$$x_1, x_2, \ldots, x_n, \ldots$$

denote the possible values of a discrete random variable ξ, and

$$p_1, p_2, \ldots, p_n, \ldots$$

are their corresponding probabilities.

If the series

$$\sum_{n=1} x_n p_n$$

converges absolutely, then its sum is called the mathematical expectation of the random variable ξ and is denoted by $\mathbf{E}\xi$.

For continuous random variables, the following natural definition is proposed: If ξ is a continuous random variable with density function $p(x)$, then the mathematical expectation of the variable ξ is given by the integral

$$\mathbf{E}\xi = \int x p(x)\, dx \tag{1}$$

if the integral

$$\int |x|\, p(x)\, dx$$

exists.

The expectation of an arbitrary random variable ξ with a distribution function $F(x)$ is defined by the Stieltjes integral

$$\mathbf{E}\xi = \int x\,dF(x). \tag{2}$$

Using the definition of a Stieltjes integral, one can give a simple geometrical interpretation of the concept of mathematical expectation: it is equal to the difference between the area bounded by the y-axis, the line $y = 1$, and the curve $y = F(x)$ in the interval $(-\infty, 0)$. In Figure 18, the corresponding areas have been shaded and it is shown which sign (+ or −) should be taken when these areas are summed. Incidently, notice that the geometrical interpretation allows to write the mathematical expectation in the form

$$\mathbf{E}\xi = -\int_{-\infty}^{0} F(x)dx + \int_{0}^{\infty} (1 - F(x))dx. \tag{3}$$

This observation makes it possible in many cases to find the mathematical expectation almost without computation. Thus, the expectation of the random variable distributed according to the law given at the end of Section 19, is equal to 1/2.

Note that among the random variables considered earlier, the variable distributed according to Cauchy's law (Example 5, Section 21) has no mathematical expectation.

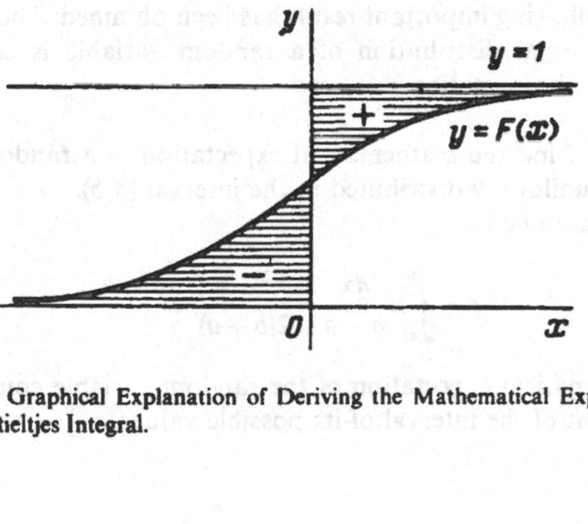

FIGURE 18. Graphical Explanation of Deriving the Mathematical Expectation with the Help of Stieltjes Integral.

Let us now consider some examples.

Example 1. One needs to find the mathematical expectation of a random variable ξ with a normal distribution normal

$$p(x) = \frac{1}{\sigma\sqrt{2\pi}} \exp\left(-\frac{(x-a)^2}{2\sigma^2}\right).$$

By formula (2), one finds that

$$\mathsf{E}\xi = \int x \frac{1}{\sigma\sqrt{2\pi}} \exp\left(-\frac{(x-a)^2}{2\sigma^2}\right) dx.$$

The substitution of variables $z = (x-a)/a$ reduces this integral to the form

$$\mathsf{E}\xi = \frac{1}{\sqrt{2\pi}} \int (\sigma z + a) e^{-z^2/2} dz = \frac{a}{\sqrt{2\pi}} \int z e^{-z^2/2} dz + \frac{a}{\sqrt{2\pi}} \int e^{-z^2/2} dz.$$

Since

$$\int e^{-z^2/2} dz = \sqrt{2\pi} \quad \text{and} \quad \int z e^{-z^2/2} dz = 0,$$

it follows that

$$\mathsf{E}\xi = a.$$

Thus the following important result has been obtained: The parameter a in the normal distribution of a random variable is equal to its mathematical expectation.

Example 2. Find the mathematical expectation of a random variable ξ which is uniformly distributed in the interval (a, b).

In this case one has

$$\mathsf{E}\xi = \int_a^b x \frac{dx}{b-a} = \frac{b^2 - a^2}{2(b-a)} = \frac{a+b}{2}.$$

The mathematical expectation of the random variable coincides with the midpoint of the interval of its possible values.

Example 3. Find the mathematical expectation of the random variable ξ with the Poisson distribution

$$\mathsf{P}\xi = k = \frac{a^k e^{-a}}{k!} \qquad (k = 0, 1, 2, \ldots).$$

It follows that

$$\mathsf{E}\xi = \sum_{k=0} k \cdot \frac{a^k e^{-a}}{k!} = \sum_{k=1} k \cdot \frac{a^k e^{-a}}{k!} = ae^{-a} \sum_{k=1} \frac{a^{k-1}}{(k-1)!} = ae^{-a} \sum_{k=0}^{\infty} \frac{a^k}{k!} = a.$$

If $F(x/B)$ is the conditional distribution function for a random variable ξ, then the integral

$$\mathsf{E}(\xi | B) = \int x \, dF(x|B) \tag{4}$$

is called the *conditional mathematical expectation of the random variable ξ with respect to the event B.*

Let $B_1, B_2, \ldots, B_n$ be a complete group of mutually exclusive events and $F(x/B_1), F(x/B_2), \ldots, F(x/B_n)$ be the conditional distribution functions of the variable ξ corresponding to these events. Let $F(x)$ denote the unconditional distribution function of the variable ξ. By the formula on total probability, one finds:

$$F(x) - \sum_{k=1}^{n} \mathsf{P}(B_k) F(x|B_k).$$

This equation together with (4) leads to the following formula:

$$\mathsf{E}\xi = \sum_{k=1}^{n} \mathsf{P}(B_k) \mathsf{E}(\xi | B_k).$$

Which can be rewritten in a different way:

$$\mathsf{E}\xi = \mathsf{E}\,\mathsf{E}(\xi | B_k). \tag{5}$$

In many cases, the later expression allows a considerable simplification of the computation of the mathematical expectation.

Example 4. A workman operates n similar machines which are set in a row (Figure 19). The distance between machines equals a. Assuming that with the probability equals $1/n$, that the workman will pay his attention to any machine, find the average path taken between machines.

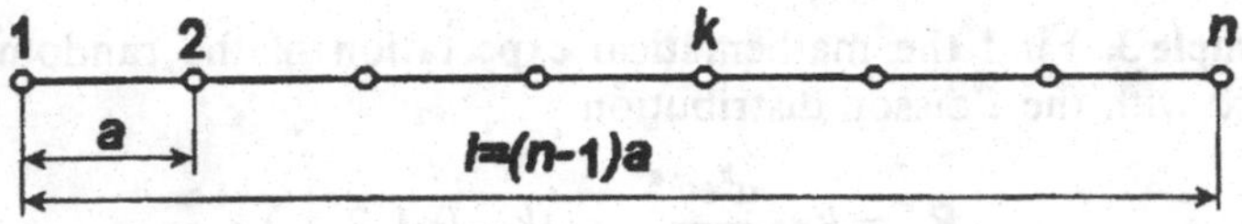

FIGURE 19. Disposition of Machines for Example 4.

Let us number the machines 1 to n, from left to right, and let B_k denote the event that the workman is at machine k. Let the next machine requiring the workman's attention be machine i. The path length λ in this case equals

$$\lambda_i^{(k)} = \begin{cases} (k-i)a & \text{for } k \geqslant i, \\ (i-k)a & \text{for } k < i. \end{cases}$$

By definition,

$$\begin{aligned} \mathbf{E}(\lambda|B_k) &= \frac{1}{n}\left(\sum_{i=1}^{k}(k-i)a + \sum_{i=k+1}^{n}(i-k)a\right) \\ &= \frac{a}{n}\left(\frac{k(k-1)}{2} + \frac{(n-k)(n-k+1)}{2}\right) \\ &= \frac{a}{2n}[2k^2 - 2(n+1)k + n(n+1)]. \end{aligned}$$

The probability that the workman is at machine k equals $1/n$, and therefore by formula (5) one finds

$$\mathbf{E}\lambda = \sum_{k=1}^{n} \frac{a}{2n^2}[2k^2 - 2(n+1)k + n(n+1)].$$

It is well-known that

$$\sum_{k=1}^{n} k^2 = \frac{n(n+1)(2n+1)}{6},$$

therefore

$$\mathbf{E}\lambda = \frac{a(n^2-1)}{3n} = \frac{1}{3}\left(1 + \frac{1}{n}\right),$$

where $l=(n-1)$ a is the distance between the first and last machines.

The mathematical expectation of an n-dimensional random variable $(\xi_1, \xi_2, \ldots, \xi_n)$ is defined as the set of n integrals[1]

$$a_k = \int\int\ldots\int x_k\, dF(x_1, \ldots, x_k, \ldots, x_n) = \int x\, dF_k(x) = \mathbf{E}\xi_k,$$

where $F_k(x)$ is the distribution function of the random variable ξ_k.

Example 5. The density function of a two-dimensional random variable (ξ_1, ξ_2) is given by the two-dimensional normal distribution

$$p(x_1, x_2) = \frac{1}{2\pi\sigma_1\sigma_2\sqrt{1-r^2}}\exp\left\{-\frac{1}{2(1-r^2)}\left[\frac{(x_1-a)^2}{\sigma_1^2} - \frac{2r(x_1-a)(x_2-b)}{\sigma_1\sigma_2} + \frac{(x_2-b)^2}{\sigma_2^2}\right]\right\}.$$

Find its mathematical expectation.

By definition

$$a_1 = \int\int x_1\, p(x_1, x_2)\, dx_1\, dx_2 = \int x_1\, p_1(x_1)\, dx_1$$

and

$$a_2 = \int\int x_2\, p(x_1, x_2)\, dx_1\, dx_2 = \int x_2\, p_2(x_2)\, dx_2.$$

In example 2 of Section 22 it was shown that

$$p_1(x_1) = \frac{1}{\sigma_1\sqrt{2\pi}}\exp\left\{-\frac{(x_1-a)^2}{2\sigma_1^2}\right\},$$

$$p_2(x_2) = \frac{1}{\sigma_2\sqrt{2\pi}}\exp\left\{-\frac{(x_2-b)^2}{2\sigma_2^2}\right\},$$

[1] A formal definition of an n-dimensional Stieltjes integral is not given here for two reasons: First, only discrete and continuous variables are actually under consideration and, second, in probability theory the abstract Lebesgue integral rather than the general Stieltjes integral is necessary. For more details see Chapter I of the monograph Limit Distributions for Sums of Independent Random Variables by Gnedenko and Kolmogorov (Reading, Mass., 1954).

and therefore, according to the results of Example 1 of the present sections one finds

$$a_1 = a, \quad a_2 = b.$$

Thus, the probabilistic meaning of the parameters a and b for the two-dimensional normal distribution is explained.

28 VARIANCE

The variance of a random variable ξ is defined as the mathematical expectation of the square of the deviation of ξ from $\mathbf{E}\xi$. The variance is denoted by $\mathbf{D}\xi$. Thus, by definition:

$$\mathbf{D}\xi = \mathbf{E}(\xi - \mathbf{E}\xi)^2 = \int_0^\infty x\,dF_\eta(x), \tag{1}$$

where $F_\eta(x)$ denotes the distribution function of the random variable $\eta = (\xi - \mathbf{E}\xi)^2$. Let us find the relation between functions $F_\eta(x)$ and $F_\xi(x)$. One has $F_\eta(x) = 0$ for $x \leqslant 0$ and for $x > 0$

$$\begin{aligned} F_\eta(x) &= \mathbf{P}\left\{\eta < x\right\} = \mathbf{P}\left\{(\xi - \mathbf{E}\xi)^2 < x\right\} \\ &= \mathbf{P}\left\{-\sqrt{x} < \xi - \mathbf{E}\xi < \sqrt{x}\right\} \\ &= \mathbf{P}\left\{\mathbf{E}\xi - \sqrt{x} < \xi < \mathbf{E}\xi + \sqrt{x}\right\} \\ &= F_\xi\left(\mathbf{E}\xi + \sqrt{x}\right) - F_\xi\left(\mathbf{E}\xi - \sqrt{x} + 0\right). \end{aligned}$$

Formula (1) can be rewritten as

$$\begin{aligned} \mathbf{D}\xi &= \int_0^\infty dx\left[F_\xi\left(\mathbf{E}\xi + \sqrt{x}\right) - F_\xi\left(\mathbf{E}\xi - \sqrt{x} + 0\right)\right] \\ &= \int_0^\infty x\,dF_\xi\left(\mathbf{E}\xi + \sqrt{x}\right) - \int_0^\infty x\,dF_\xi\left(\mathbf{E}\xi - \sqrt{x} + 0\right). \end{aligned}$$

Let us substitute in the first integral $z = E\xi + \sqrt{x}$ and in the second $z = E\xi - \sqrt{x}$. As a result, one has

$$\int_0^\infty x\,dF_\xi\left(\mathbf{E}\xi + \sqrt{x}\right) = \int_{\mathbf{E}\xi}^\infty (z - \mathbf{E}\xi)^2 dF_\xi(z),$$

$$\int_0^\infty x\,dF_\xi\left(\mathbf{E}\xi - \sqrt{x} + 0\right) = \int_{-\infty}^{\mathbf{E}\xi} (z - \mathbf{E}\xi)^2 dF_\xi(z).$$

Thus

$$\mathbf{D}\xi = \int (z - E\xi)^2 dF_\xi(z). \tag{2}$$

Since

$$(z - \mathbf{E}\xi)^2 = z^2 - 2z\mathbf{E}\xi + (\mathbf{E}\xi)^2 \quad \text{and} \quad \mathbf{E}\xi = \int z\,dF_\xi(z),$$

the formula (2) can be rewritten in another form

$$\mathbf{D}\xi = \int z^2 dF_\xi(z) - \left(\int z\,dF_\xi(z)\right)^2 = (\mathbf{E}\xi^2 - \mathbf{E}\xi)^2. \tag{3}$$

Because the variance is a non-negative quantity, from the last relation it follows that

$$\int z^2 dF_\xi(z) \geqslant \left(\int z\,dF_\xi(z)\right)^2.$$

This inequality is a special case of the familiar Bouniakowsky-Cauchy inequality.[2]

As was therefore mathematical expectation, the variance does not exist for all random variables. Thus, the Cauchy distribution considered earlier (Example 5, Section 21), does not have a finite variance.

Let us consider examples of the calculation of the variance.

---

[2]This inequality is also known as the Cauchy-Schwarz inequality.

Example 1. Find the variance of the random variable ξ distributed uniformly in the interval (a, b). In this case

$$\int x^2 dF_\xi(x) = \int_0^b \frac{x^2}{b-a} dx = \frac{b^3 - a^3}{3(b-a)} = \frac{b^2 + ab + a^2}{3}.$$

In the preceding section it was found that

$$\mathbf{E}\xi = \frac{a+b}{2}.$$

Thus

$$\mathbf{D}\xi = \frac{a^2 + ab + b^2}{3} - \left(\frac{a+b}{2}\right)^2 = \frac{(b-a)^2}{12}.$$

One sees that the variance depends only on the length of the interval (a, b) and is an increasing function of the length. The more spread out random values are, the larger their variance is. Thus, the variance plays the role of the measure of the dispersion of the values of the random variable about its mathematical expectation.

Example 2. Find the variance of a random variable ξ distributed normally

$$p(x) = \frac{1}{\sigma\sqrt{2\pi}} \exp\left\{-\frac{(x-a)^2}{2\sigma^2}\right\}.$$

One knows that $\mathbf{E}\xi = a$, therefore

$$\mathbf{D}\xi = \int (x-a)^2 p(x) dx = \frac{1}{\sigma\sqrt{2\pi}} \int (x-a)^2 e^{-(x-a)^2/2\sigma^2} dx.$$

Let us now make a substitution of variable in the integral $z = (x-a)/\sigma$ in the integral which leads to

$$\mathbf{D}\xi = \frac{\sigma^2}{\sqrt{2\pi}} \int z^2 e^{-z^2/2} dz.$$

Integrating by parts gives

$$\int z^2 e^{-z^2/2}\,dz = \int_{-}^{\infty} -ze^{-z^2/2} + \int e^{-z^2/2}dz = \sqrt{2\pi}.$$

Thus, one finally obtains:

$$\mathsf{D}\xi = \sigma^2.$$

Thus the probabilistic meaning of the second parameter in the normal distribution is also explained. So, the normal distribution is completely determined by the mathematical expectation and the variance. This fact is widely used in applied and theoretical investigations.

Note that in the case of a normally distributed random variable the variance also characterizes the measure of dispersion of a random variable. Although for any value of the variance a normally distributed random variable can assume all real values, the dispersion of its values will be smaller, as the variance is the smaller. Moreover, the closer these values are to the mathematical expectation, the larger are their probabilities. This fact was emphasized in the preceding chapter when the normal distribution was introduced.

Example 3. Find the variance of the random variable λ considered in Example 4 of Section 23.

Retaining the notation of Example 4, one finds that

$$\begin{aligned}\mathsf{E}(\lambda^2|B_k) &= \frac{1}{n}\left(\sum_{i=1}^{k}(k-i)^2a^2 + \sum_{i=k+1}^{n}(i-k)^2a^2\right)\\ &= \frac{a^2}{6n}[(k-1)\cdot k(2k-1)\\ &\quad + (n-k)(n-k+1)(2n-2k+1)]\\ &= \frac{a^2}{6}[6k^2 - 6(n+1)k + (2n+1)(n+1)]\end{aligned}$$

and therefore,

$$\mathsf{E}(\lambda^2) = \frac{1}{n}\sum_{k=1}^{n}\mathsf{E}(\lambda^2|B_k)$$

$$=\frac{a^2}{6n}[n(n+1)(2n+1)-3(n+1)^2 n$$

$$+n(n+1)(2n+1)]$$

$$=\frac{a^2}{6}(n^2-1).$$

From this it follows that

$$\mathbf{D}(\lambda)=\mathbf{E}(\lambda^2)-(\mathbf{E}\lambda)^2$$

$$=\frac{a^2}{6}(n^2-1)-\frac{a^2(n^2-1)^2}{9n^2}$$

$$=\frac{a^2(n^2-1)(n^2+2)}{18n^2}$$

$$=\frac{l^2}{18}\left(1+\frac{2}{n-1}+\frac{1}{n(n-1)}+\frac{2}{n^2(n-1)}\right).$$

The variance of an n-dimensional random variable $(\xi_1, \xi_2, \ldots, \xi_n)$ is defined as the collection of n^2 constants given by the formula

$$b_{jk}=\iint\cdots\int(x_j-\mathbf{E}\xi_j)(x_k-\mathbf{E}\xi_k)\,dF(x_1,x_2,\ldots,x_n) \tag{4}$$

$$(1\leqslant k\leqslant n,\ 1\leqslant j\leqslant n).$$

Since, for any real values of $t_j (1\leqslant j\leqslant n)$,

$$\int\cdots\iint\left\{\sum_{j=1}^{n}t_j(x_j-\mathbf{E}x_j)\right\}^2 dF(x_1,x_2,\ldots,x_n)=\sum_{j=1}^{n}\sum_{k=1}^{n}b_{jk}t_j t_k\geqslant 0,$$

as is known from the theory of quadratic forms, the quantities b_{jk} satisfy the inequalities

$$\begin{vmatrix} b_{11} & b_{12} & \cdots & b_{1k} \\ b_{21} & b_{22} & \cdots & b_{2k} \\ \cdots & \cdots & \cdots & \cdots \\ b_{k1} & b_{k2} & \cdots & b_{kk} \end{vmatrix}\geqslant 0 \quad \text{for} \quad k=1,2,\ldots,n.$$

It is obvious that

$$b_{kk} = \mathbf{D}\xi_k.$$

The quantities b_{jk} for $k \neq j$ are called that *mixed central moments of the 2nd order* (*covariance*) of the variables ξ_j and ξ_k. Obviously, $b_{jk} = b_{kj}$.

The following function of the moments of the 2nd order:

$$r_{ij} = \frac{b_{ij}}{\sqrt{b_{ii} b_{jj}}}$$

is called the correlation coefficient of the variables ξ_j and ξ_k. The correlation coefficient is a measure of linear relation between the variables ξ_j and ξ_k. Its value lies in the interval $[-1, 1]$ as it follows from the Bouniakowsky-Cauchy inequality. The correlation coefficient equals ± 1 only if ξ_j and ξ_k have a linear dependence.

Later it will be shown that for independent random variables the correlation coefficient equals 0.

Example 4. Find the variance of the two-dimensional random variable (ξ_1, ξ_2) distributed normally

$$p(x, y) = \frac{1}{2\pi\sigma_1\sigma_2\sqrt{1-r^2}} \times \exp\left\{-\frac{1}{2(1-r^2)}\left[\frac{(x-a)^2}{\sigma_1^2} - 2r\frac{(x-a)(y-b)}{\sigma_1\sigma_2} + \frac{(y-b)^2}{\sigma_2^2}\right]\right\}.$$

By formula (4), taking into account the results of Example 2 of the present section and Example 1 of Section 23, one finds:

$$\mathbf{D}\xi_1 = \sigma_1^2, \quad \mathbf{D}\xi_2 = \sigma_2^2.$$

Further,

$$b_{12} = b_{21} = \iint (x-a)(y-b)\,p(x, y)\,dx\,dy$$

$$= \frac{1}{2\pi\sigma_1\sigma_2\sqrt{1-r^2}} \int e^{-(y-b)^2/2\sigma_2^2} dy \times \int (x-a)(y-b) \times \exp\left\{-\frac{1}{2(1-r^2)}\left(\frac{x-a}{\sigma_1} - r\frac{y-b}{\sigma_2}\right)^2\right\} dx.$$

By substitution

$$z = \frac{1}{\sqrt{1-r^2}}\left(\frac{x-a}{\sigma_1} - r\frac{y-b}{\sigma_2}\right), \quad t = \frac{y-b}{\sigma_2}$$

the expression for b_{12} can be transformed to the form

$$b_{12} = b_{21} = \frac{1}{2\pi}\iint (\sigma_1\sigma_2\sqrt{1-r^2}\,tz + r\sigma_1\sigma_2 t^2)e^{-(t^2/2)-(z^2/2)}\,dz\,dt$$

$$= \frac{r\sigma_1\sigma_2}{2\pi}\int t^2 e^{-t^2/2}\,dt \int e^{-z^2/2}\,dz + \frac{\sigma_1\sigma_2\sqrt{1-r^2}}{2\pi}$$

$$\times \int te^{-t^2/2}\,dt \int ze^{-z^2/2}\,dz = r\sigma_1\sigma_2.$$

From this, one finds that

$$r = \frac{\iint (x-a)(y-b)p(x,y)\,dx\,dy}{\sigma_1\sigma_2} = \frac{\mathbf{E}(\xi_1 - \mathbf{E}\xi_1)(\xi_2 - \mathbf{E}\xi_2)}{\sqrt{\mathbf{D}\xi_1\,\mathbf{D}\xi_2}}.$$

Thus, the two-dimensional normal distribution is completely determined by the five quantities $\mathbf{E}\xi_1, \mathbf{E}\xi_2, \mathbf{D}\xi_1, \mathbf{D}\xi_2$, and r.

29 THEOREMS ON MATHEMATICAL EXPECTATION AND VARIANCE

THEOREM 1. *The mathematical expectation of a constant is that constant.*

PROOF: The constant C can be regarded as a discrete random variable which can take only one value C with probability 1. Therefore

$$\mathbf{E}C = C \cdot 1 = C.$$

THEOREM 2. *The mathematical expectation of the sum of two random variables equals the sum of their mathematical expectations:*

$$E(\xi + \eta) = E\xi + E\eta.$$

PROOF: First consider discrete random variables ξ and η. Let $a_1, a_2, \ldots, a_n, \ldots$ be the possible values of the variable ξ and $p_1, p_2, \ldots, p_n, \ldots$ be their probabilities; and let $b_1, b_2, \ldots, b_k, \ldots$ be the possible values of η and $q_1, q_2, \ldots, q_n, \ldots$ be their probabilities. The possible values of $\xi + \eta$ have the form $a_n + b_k$ $(k, n = 1, 2, \ldots)$. Let p_{nk} denote the probability that ξ assumes the value a_n and η assumes the value b_k. By definition of the mathematical expectation one can write

$$\mathbf{E}(\xi + \eta) = \sum_{n,k=1}^{\infty} (a_n + b_k) p_{nk}$$

$$= \sum_{n=1}^{\infty} \sum_{k=1}^{\infty} (a_n + b_k) p_{nk}$$

$$= \sum_{n=1}^{\infty} a_n \left(\sum_{k=1}^{\infty} p_{nk} \right) + \sum_{k=1}^{\infty} b_k \left(\sum_{n=1}^{\infty} p_{nk} \right).$$

Since, by the theorem on total probability,

$$\sum_{k=1}^{\infty} p_{nk} = p_n \quad \text{and} \quad \sum_{n=1}^{\infty} p_{nk} = q_k,$$

then

$$\sum_{n=1}^{\infty} a_n \sum_{k=1}^{\infty} p_{nk} = \sum_{n=1}^{\infty} a_n p_n = \mathbf{E}\xi$$

and

$$\sum_{k=1}^{\infty} b_k \left(\sum_{n=1}^{\infty} p_{nk} \right) = \sum_{k=1}^{\infty} b_k q_k = \mathbf{E}_\eta.$$

This completes the proof of the theorem for discrete summands. Similarly, if a two-dimensional density function $p(x, y)$ exists for the random variable (ξ, η), one finds by formula (3) of Section 23 that

$$\mathbf{E}\zeta = \mathbf{E}(\xi + \eta) = \int x \, dF_\zeta(x) = \int x \left(\int p(z, x - z) dz \right) dx$$

$$= \iint x p(z, x - z) \, dz \, dx = \iint (z + y) p(z, y) \, dz \, dy$$

$$= \iint zp(z,y)\,dz\,dy + \iint yp(z,y)\,dz\,dy$$

$$= \int zp_\xi(z)dz + \int yp_\eta(y)dy = \mathbf{E}\xi + \mathbf{E}\xi_\eta$$

Theorem 2 will be proved for the general case in Appendix 1.

COROLLARY 1. *The mathematical expectation of the sum of a finite number of random variables equals the sum of their mathematical expectations:*

$$\mathbf{E}(\xi_1 + \xi_2 + \cdots + \xi_n) = \mathbf{E}\xi_1 + \mathbf{E}\xi_2 + \cdots + \mathbf{E}\xi_n.$$

In fact, by virtue of the theorem just proved, one has

$$\mathbf{E}(\xi_1 + \xi_2 + \cdots + \xi_n) = \mathbf{E}\xi_1 + \mathbf{E}(\xi_2 + \xi_2 + \cdots + \xi_n)$$
$$= \mathbf{E}\xi_1 + \mathbf{E}\xi_2 + \mathbf{E}(\xi_3 + \cdots + \xi_n)$$
$$= \cdots = \mathbf{E}\xi_1 + \mathbf{E}\xi_2 + \cdots + \mathbf{E}\xi_n.$$

COROLLARY 2. *Consider the mathematical expectation of the sum*

$$\zeta_\mu = \xi_1 + \xi_2 + \cdots + \xi_\mu,$$

where μ is an integer random variable with finite mathematical expectation, random variables $\xi_1, \xi_2, \ldots$ do not depend on μ, and the set

$$\sum_{k=1} \mathbf{E}|\xi_k|\mathbf{P}\{\mu \geqslant k\}$$

converges. The mathematical expectation of the sum exists and is equal to

$$\mathbf{E}\zeta_\mu = \sum_{j=1} \mathbf{E}\xi_j \mathbf{P}\{\mu \geqslant j\}.$$

PROOF: Indeed, the conditional mathematical expectation of the sum, given that $\mu = k$, is equal to

$$\mathbf{E}\{\zeta_\mu | \mu = k\} = \mathbf{E}\xi_1 + \mathbf{E}\xi_2 + \cdots + \mathbf{E}\xi_k.$$

The unconditional mathematical expectation is

$$\mathbf{E}\zeta_\mu = \sum_{k=1} \mathbf{E}\{\zeta_\mu|\mu = k\}\cdot\mathbf{P}\{\mu = k\} = \sum_{k=1} \mathbf{P}\{\mu = k\} \sum_{j=1}^{k} \mathbf{E}\xi_j$$

$$= \sum_{j=1} \mathbf{E}\xi_j \sum_{k=j} \mathbf{P}\{\mu = k\} = \sum_{j=1} \mathbf{E}\xi_j \mathbf{P}\{\mu \geqslant j\}.$$

If the summands $\xi_1, \xi_2, \cdots$ are identically distributed, i.e., if $\mathbf{P}(\xi_1 < x\} = \mathbf{P}\{\xi_2 < x\} = \cdots = F(x)$, then

$$\mathbf{E}\zeta_\mu = \mathbf{E}\zeta_1 \cdot \mathbf{E}\mu.$$

Indeed,

$$\mathbf{E}\zeta_\mu = \sum_{k=1} \mathbf{P}\{\mu = k\} \sum_{j=1}^{k} \mathbf{E}\xi_j = \mathbf{E}\xi_1 \sum_{k=1} k\mathbf{P}\{\mu = k\} = \mathbf{E}\xi_1 \cdot \mathbf{E}\mu.$$

Example 1. The number of cosmic particles that fall on a given square is a random variable μ which is subject to the Poisson distribution with parameter a. Each particle has an energy ξ depending on chance. Find the mean energy $\mathscr{E}$ acquired by the square in one unit of time.

According to Corollary 2, one has:

$$\mathbf{E}\mathscr{E} = \mathbf{E}\xi \cdot \mathbf{E}\mu = a\mathbf{E}\xi.$$

Example 2. A particular target is fired at until the number of hits reaches n. Assuming that the shots are fired independently of each other and that all of probabilities of a hit in each trial equals p, find the mathematical expectation of the number of shells expended.

Let ξ_k denote the number of shots fired after hit $k-1$ until hit k. It is obvious that the number of shells required for the n hits is

$$\xi = \xi_1 + \xi_2 + \cdots + \xi_n$$

and therefore

$$\mathbf{E}\xi = \mathbf{E}\xi_1 + \mathbf{E}\xi_2 + \cdots + \mathbf{E}\xi_n.$$

But

$$\mathbf{E}\xi_1 = \mathbf{E}\xi_2 = \cdots = \mathbf{E}\xi_n$$

and

$$\mathbf{E}\xi_1 = \sum_{k=0}^{\infty} kq^{k-1}p = \frac{p}{(1-q)^2} = \frac{1}{p},$$

hence

$$\mathbf{E}\xi = n/p.$$

THEOREM 3. *The mathematical expectation of the product of two independent random variables ξ and η equals the product of their mathematical expectations.*

PROOF: Let considered random variables ξ and η be discrete. Let $a_1, a_2, \ldots, a_n, \ldots$ be the possible values of the variable ξ and $p_1, p_2, \ldots, p_n, \ldots$ be their probabilities; and let $b_1, b_2, \ldots, b_k, \ldots$ be the possible values of η and $q_1, q_2, \ldots, q_n, \ldots$ be their probabilities. The possible values of $\xi\eta$ have the form $a_n b_k$ $(k, n = 1, 2, \ldots)$. Let p_{nk} denote the probability that ξ assumes the value a_n and η assumes the value b_k. By definition of the mathematical expectation one can write

$$\mathbf{E}\xi_\eta = \sum_{k,n} a_k b_n p_k q_n = \sum_{k=1}^{\infty} \sum_{n=1}^{\infty} a_k b_n p_k q_n$$

$$= \left(\sum_{k=1}^{\infty} a_k p_k\right)\left(\sum_{n=1}^{\infty} b_n q_n\right) = \mathbf{E}\xi\mathbf{E}\eta.$$

The proof of the theorem for the case of continuous variables is only slightly more complicated. This proof is left to the reader.

COROLLARY 1. *A constant factor may be removed from under the operator of the mathematical expectation:*

$$\mathbf{E}C\xi = C\mathbf{E}\xi.$$

This statement is obvious because whatever ξ may be, the constant C and the variable ξ may be considered as independent variables.

THEOREM 4. *The variance of a constant is zero.*

PROOF: According to Theorem 1,

$$\mathbf{D}C = \mathbf{E}(C - \mathbf{E}C)^2 = \mathbf{E}(C - C)^2 = \mathbf{E}0 = 0.$$

THEOREM 5. *If c is constant, then*

$$\mathbf{D}\, c\xi = c^2 \mathbf{D}\xi.$$

PROOF: By virtue of the corollary to Theorem 3, one has

$$\mathbf{D}c\xi = \mathbf{E}[c\xi - \mathbf{E}c\xi]^2 = \mathbf{E}[c\xi - c\mathbf{E}\xi]^2$$

$$= \mathbf{E}c^2[\xi - \mathbf{E}\xi]^2 = c^2\mathbf{E}[\xi - \mathbf{E}\xi]^2 = c^2\mathbf{D}\xi.$$

THEOREM 6. *The variance of the sum of two independent random variables ξ and η equals the sum of their variances:*

$$\mathbf{D}(\xi + \eta) = \mathbf{D}\xi + \mathbf{D}\eta.$$

PROOF: Indeed,

$$\mathbf{D}(\xi + \eta) = \mathbf{E}[\xi + \eta - \mathbf{E}(\xi + \eta)]^2 = \mathbf{E}[(\xi - \mathbf{E}\xi) + (\eta - \mathbf{E}\eta)]^2$$

$$= \mathbf{D}\xi + \mathbf{D}\eta + 2\mathbf{E}(\xi - \mathbf{E}\xi)(\eta - \mathbf{E}\eta).$$

The variables ξ and η are independent, therefore the variables $\xi - \mathbf{E}\xi$ and $\eta - \mathbf{E}\eta$ are also independent. Hence

$$\mathbf{E}(\xi - \mathbf{E}\xi)(\eta - \mathbf{E}\eta) = \mathbf{E}(\xi - \mathbf{E}\xi) \cdot \mathbf{E}(\eta - \mathbf{E}\eta) = 0.$$

COROLLARY 1. *If $\xi_1, \xi_2, \ldots, \xi_n$ are random variables such that each of them is independent of the sum of the preceding ones, then*

$$\mathbf{D}(\xi_1 + \xi_2 + \cdots + \xi_n) = \mathbf{D}\xi_1 + \mathbf{D}\xi_2 + \cdots + \mathbf{D}\xi_n.$$

COROLLARY 2. *The variance of the sum of a finite number of pairwise independent random variables $\xi_1, \xi_2, \ldots, \xi_n$ equals the sum of their variances.*

PROOF: Indeed,

$$\mathbf{D}(\xi_1 + \xi_2 + \cdots + \xi_n) = \mathbf{E}\left(\sum_{k=1}^{n} (\xi_k - \mathbf{E}\xi_k)\right)^2$$

$$= \mathbf{E} \sum_{j=1}^{n} \sum_{k=1}^{n} (\xi_k - \mathbf{E}\xi_k)(\xi_j - \mathbf{E}\xi_j)$$

$$= \sum_{k=1}^{n} \sum_{j=1}^{n} \mathbf{E}(\xi_k - \mathbf{E}\xi_k)(\xi_j - \mathbf{E}\xi_j)$$

$$= \sum_{k=1}^{n} \mathbf{D}\xi_k + \sum_{k \neq j} \mathbf{E}(\xi_k - \mathbf{E}\xi_k)(\xi_j - \mathbf{E}\xi_j).$$

From the independence of any pair of quantities ξ_k and ξ_j $(j \neq k)$, it follows that

$$\mathbf{E}(\xi_k - \mathbf{E}\xi_k)(\xi_j - \mathbf{E}\xi_j) = 0.$$

This, obviously, completes the proof.

Example 1. The ratio

$$\frac{\xi - \mathbf{E}\xi}{\sqrt{\mathbf{D}\xi}}$$

is called the normalized deviation of the random variable ξ. Prove that

$$D\left(\frac{\xi - \mathbf{E}\xi}{\sqrt{D\xi}}\right) = 1.$$

Indeed, since ξ and $\mathbf{E}\xi$, regarded as random variables, are independent, by virtue of Theorems 5 and 6, one has

$$\mathbf{D}\left(\frac{\xi - \mathbf{E}\xi}{\sqrt{\mathbf{D}\xi}}\right) = \frac{\mathbf{D}\xi + \mathbf{D}(-\mathbf{E}\xi)}{\mathbf{D}\xi} = \frac{\mathbf{D}\xi}{\mathbf{D}\xi} = 1.$$

Example 2. If ξ and η are independent random variables, then

$$\mathbf{D}(\xi - \eta) = \mathbf{D}\xi + \mathbf{D}\eta.$$

By virtue of Theorems 5 and 6

$$\mathbf{D}(-\eta) = (-1)^2 \mathbf{D}\eta = \mathbf{D}\eta$$

and

$$\mathbf{D}(\xi - \eta) = \mathbf{D}\xi + \mathbf{D}\eta.$$

Example 3. Theorems 2 and 6 allow one to compute the mathematical expectation and the variance of μ, the number of occurrences of an event A in n independent trials.

Let p_k be the probability of occurrence of the event A in trial k and μ_k be the number of times the event A occurs in this trial. Obviously, that μ_k is a random variable taking the values 0 or 1 with the probabilities $q_k = 1 - p_k$ and p_k, respectively.

The variable μ can be thus represented in the form of a sum

$$\mu = \mu_1 + \mu_2 + \cdots + \mu_n .$$

Since

$$\mathbf{E}\mu_k = 0 \cdot q_k + 1 \cdot p_k = p_k$$

and

$$\mathbf{D}\mu_k = \mathbf{E}\mu_k^2 - (\mathbf{E}\mu_k)^2 = 0 \cdot q_k + 1 \cdot p_k - p_k^2 = p_k(1 - p_k) = p_k q_k ,$$

one can conclude from the theorems proved above that

$$\mathbf{E}\mu = p_1 + p_2 + \cdots + p_n$$

and

$$\mathbf{D}\mu = p_1 q_1 + p_2 q_2 + \cdots + p_n q_n .$$

For the case of Bernoulli trials, $p_k = p$ and therefore

$$\mathbf{E}\mu = np \quad \text{and} \quad \mathbf{D}\mu = npq .$$

Note that from this it follows that

$$\mathbf{E}\frac{\mu}{n} = p; \quad \mathbf{D}\frac{\mu}{n} = \frac{pq}{n}.$$

30 MOMENTS

The mathematical expectation of the random variable $(\xi - a)^k$ is called the moment of order k of the random variable ξ:

$$\nu_k(a) = \mathbf{E}(\xi - a)^k . \tag{1}$$

If $a = 0$, the moment is called the initial moment of order k. It is clear that the first initial moment is the mathematical expectation of the variable ξ.

If $a = \mathbf{E}\xi$, the moment is called the central moment of order k. It is easy to see that the central moment of first order vanishes and that the central moment of second order is the variance.

Initial moments will be denoted by ν_k with the subscript indicating the order of the moment.

A simple relation exists between central and initial. In fact,

$$\mu_n = \mathbf{E}(\xi - \mathbf{E}\xi)^n = \sum_{k=0}^{n} C_n^k(-\mathbf{E}\xi)^{n-k}\mathbf{E}\xi^k = \sum_{k=0}^{n} C_n^k(-\mathbf{E}\xi)^{n-k}\nu_k. \quad (2)$$

Since $\nu_1 = \mathbf{E}\xi$, one has

$$\mu_n = \sum_{k=2}^{n} (-1)^{n-k} C_n^k \nu_k \nu_1^{n-k} + (-1)^{n-1}(n-1)(\nu_1)^n. \quad (3)$$

Let us write out the relation between the moments for the first four values of n:

$$\begin{aligned} \mu_0 &= 1, \\ \mu_1 &= 0, \\ \mu_2 &= \nu_2 - \nu_1^2, \\ \mu_3 &= \nu_3 - 3\nu_2\nu_1 + 2\nu_1^2, \\ \mu_4 &= \nu_4 - 4\nu_3\nu_1 + 6\nu_2\nu_1^2 - 3\nu_1^4. \end{aligned} \quad (3')$$

These first few moments play an especially important role in mathematical statistics.

The quantity

$$m_k = \mathbf{E}|\xi - a|^k \quad (4)$$

is referred to as the absolute moment of order k.

According to the definition of the mathematical expectation, the quantity of $\mathbf{E}(\xi - a)^k$ has to be calculated by formula

$$\nu_k(a) = \int x\, dG(x), \quad (1')$$

where $G(x)$ is the distribution function of random variable $(\xi - a)^k$. However, one prefers to use another formula,

$$\nu_k(a) = \int (x - a)^k\, dF(x), \quad (5)$$

where $F(x)$ is the distribution function of random variable ξ. Formulas (1') and (5) do not contradict to each other if the following equality holds

$$\int x\,dG(x) = \int (x-a)^k\,dF(x).$$

Let us prove that this is true.

If k is an odd number, then quantity $(\xi - a)^k$ is a non-decreasing function in ξ, and therefore

$$G(x) = \mathbf{P}\left\{\left(\xi - a\right)^k < x\right\} = \mathbf{P}\left\{\xi - a < \sqrt[k]{x}\right\}$$

$$= \mathbf{P}\left\{\xi < a + \sqrt[k]{x}\right\} = F\left(a + \sqrt[k]{x}\right).$$

Thus for odd k one has

$$\mathbf{E}(\xi - a)^k = \int x\,dF\left(a + \sqrt[k]{x}\right).$$

It is easy to calculate that by the substitution

$$z = a + \sqrt[k]{x}$$

the integral obtained above can be reduced to (5).

If k is an even number, then the quantity $(\xi - a)^k$ is a non-negative value, and, consequently, $G(x) = 0$ for $x \leqslant 0$. For $x > 0$

$$G(x) = \mathbf{P}\left\{\left(\xi - a\right)^k < x\right\} = \mathbf{P}\left\{a - \sqrt[k]{x} < \xi < a + \sqrt[k]{x}\right\}$$

$$= F\left(a + \sqrt[k]{x}\right) - F\left(a - \sqrt[k]{x} + 0\right).$$

Thus for even k

$$\mathbf{E}(\xi - a)^k = \int_0^\infty x\,dF\left(a + \sqrt[k]{x}\right) - \int_0^\infty x\,dF\left(a - \sqrt[k]{x} + 0\right).$$

By substituting $z = a + \sqrt{x}$ in the first integral and $z = a - \sqrt{x}$ in the second one, we can reduce $\mathbf{E}(\xi - a)^k$ to the expression (5).

Thus, a particular case of the following theorem has been proven.

THEOREM 1. *If $F(x)$ is the distribution function of a random variable ξ, then*

$$\mathsf{E}f(\xi)=\int f(x)\,dF(x).$$

Since a random variable ξ has an mathematical expectation only if the integral defining the mathematical expectation converges absolutely, it is clear that the initial moment of order k of the variable ξ exists if and only if the integral

$$\int |x|^k\,dF_\xi(x)$$

converges. From this remark it follows that if the random variable ξ has a moment of order k, then it also has moments of all positive orders less than k. In fact, since for $r<k$ one has $|x|^k>|x|^r$ if $|x|>1$, then

$$\begin{aligned}\int |x|^r\,dF_\xi(x)&=\int_{|x|\leqslant 1}|x|^r\,dF_\xi(x)+\int_{|x|>1}|x|^r\,dF_\xi(x)\\&\leqslant\int_{|x|\leqslant 1}|x|^r\,dF_\xi(x)+\int_{|x|>1}|x|^k\,dF_\xi(x).\end{aligned}$$

The first integral on the right-hand side of this inequality is finite because the limits of integration are finite and the integrand is bounded. The second integral converges by assumption.

Example. Find the central and the absolute central moments of a random variable distributed normally:

$$p(x)=\frac{1}{\sigma\sqrt{2\pi}}\exp\left\{-\frac{(x-a)^2}{2\sigma^2}\right\}.$$

One has

$$\mu_k=\frac{1}{\sigma\sqrt{2\pi}}\int(x-a)^k\exp\left\{-\frac{(x-a)^2}{2\sigma^2}\right\}dx=\frac{\sigma^k}{\sqrt{2\pi}}\int x^k e^{-x^2/2}\,dx.$$

For odd k, $\mu_k = 0$ because the integrand is an odd function.
For even k,

$$\mu_k = m_k = \sqrt{\frac{2}{\pi}}\,\sigma^k \int_0^\infty x^k e^{-x^2/2}\,dx.$$

The substitution $x^2 = 2z$ reduces this integral to the form

$$\mu_k = m_k = \sqrt{\frac{2}{\pi}}\,\sigma^k 2^{(k-1)/2} \int_0^\infty z^{(k-1)/2} e^{-z}\,dz = \sqrt{\frac{2}{\pi}}\,\sigma^k 2^{(k-1)/2}\,\Gamma\left(\frac{k+1}{2}\right)$$

$$= \sigma^k (k-1)(k-3)\cdots 1 = \sigma^k \frac{k!}{2^{k/2}(k/2)!}.$$

For odd k, the absolute moment is

$$m_k = \sqrt{\frac{2}{\pi}}\,\sigma^k \int_0^\infty x^k e^{-x^2/2}\,dx = \sqrt{\frac{2}{\pi}}\,\sigma^k 2^{(k-1)/2}\,\Gamma\left(\frac{k+1}{2}\right)$$

$$= \sqrt{\frac{2}{\pi}}\,2^{(k-1)/2}\left(\frac{k-1}{2}\right)!\,\sigma^k.$$

The moments of a distribution cannot be arbitrary quantities. In fact, for any constants $t_0, t_1, \ldots, t_n$, the quadratic form

$$J_n = \int \left(\sum_{k=0}^{n} t_k (x-a)^k\right)^2 dF(x) = \sum_{j=1}^{n}\sum_{k=1}^{n} \nu_{k+j}(a) t_k t_j \geqslant 0$$

is non-negative, therefore the first moments $\nu_j(a)$ must satisfy the following inequalities:

$$\begin{vmatrix} \nu_0(a) & \nu_1(a) & \cdots & \nu_k(a) \\ \nu_1(a) & \nu_2(a) & \cdots & \nu_{k+1}(a) \\ \cdots & \cdots & \cdots & \cdots \\ \nu_k(a) & \nu_{k+1}(a) & \cdots & \nu_{2k}(a) \end{vmatrix} \geqslant 0 \qquad (k = 1, 2, \ldots, n).$$

The absolute moments also have to satisfy analogous inequalities.

One more theorem concerning absolute moments will be proved below.

THEOREM 2. *If the random variable ξ has an absolute moment of order k, then for arbitrary t and τ $(0<t<\tau<k)$*

$$\sqrt[t]{m_t} \leqslant \sqrt[\tau]{m_\tau} \leqslant \sqrt[k]{m_k},$$

where

$$m_t = \mathbf{M}\,|\xi - a|^t,$$

and a is any real value.

PROOF: Let us first prove the theorem for the case where t, τ and k are rational. Specifically, let

$$t = p/q, \quad \tau = s/q, \quad k = u/q,$$

where, by hypothesis, $p<s<u$.

Now, let r be some positive integer less than u. Consider the non-negative quadratic form

$$m_{(r-1)/q}\,u^2 + 2m_{r/q}\,uv + m_{(r+1)/q}\,v^2 = \int [u|x|^{(r-1)/2q} + v|x|^{(r+1)/2q}]^2\,dF(x).$$

The condition that it be non-negative is, as is well known, that the inequality

$$m_{r/q}^2 \leqslant m_{(r-1)/q} \cdot m_{(r+1)/q}$$

holds. This inequality can obviously be written in the following form:

$$m_{r/q}^{2r} \leqslant m_{(r-1)/q}^r \cdot m_{(r+1)/q}^r.$$

If one sequentially assigns to r the set of values from 1 to r, the sequence of inequalities will be obtained

$$m_{1/q}^2 \leqslant m_0 m_{2/q}, \quad m_{2/q}^{2.2} \leqslant m_{1/q}^2 m_{3/q}^2, \ldots\, m_{r/q}^{2r} \leqslant m_{(r-1)/q}^r m_{(r+1)/q}^r.$$

Note that always $m_0 = 1$. So, multiplying these inequalities and making some reductions one obtains the following inequality

$$m_{r/q}^{r+1} \leqslant m_{(r+1)/q}^r.$$

Hence

$$m_{r/q}^{1/r} \leqslant m_{(r+1)/q}^{1/(r+1)}$$

or, in another form,

$$m_{r/q}^{q/r} \leqslant m_{(r+1)/q}^{q/(r+1)}.$$

This inequality, obviously, proves the theorem for the case of rational t, τ and k.

Since the function m_t is continuous with respect to the argument t in the interval $0<t\leqslant k$, one can satisfy himself by means of a limit argument that the theorem remains valid for any values of t, τ, and k.

Note that the theorem just proved contains the following important property of moments:

$$m_1 \leqslant m_2^{1/2} \leqslant m_3^{1/3} \leqslant \cdots \leqslant m_k^{1/k} \leqslant m_{k+1}^{1/(k+1)} \leqslant \cdots$$

In the examples of the preceding sections, the first two moments of the random variable completely determined the distribution function if the form of the distribution was known in advance (this was for the normal, Poisson, uniform, and other distributions). In mathematical statistics, an important role is played by distribution laws that depend on more than two parameters. If it is known in advance that a random variable is subject to a distribution of a certain form and only the values of the parameters are unknown, then in the most important cases these unknown parameters can be determined in terms of the first few moments. However, if the form of the distribution function is unknown then, in the general case, knowledge of the first few moments or even of all the moments of integer order does not allow one to determine the desired distribution function. It is possible to construct examples in which distinct distributions have identical moments of all integral orders. In this connection, the following question arises (the Problem of Moments): Given the sequence of constants

$$c_0 = 1, c_1, c_2, c_3, \ldots$$

1) Under what conditions does there exist a distribution function $F(x)$ for which for all n the equality

$$c_n = \int x^n dF(x),$$

holds, and

2) When is this function unique?

At present there exists a complete solution to this problem, but we shall not consider it, since it is outside the scope of our book.

Among the other numerical characteristics, a most important role is played by the so-called semi-invariants (or cumulants). Their definition is postponed until Chapter 8. Here some facts are considered. The moment of a sum of independent random variables does not, in general, equal the sum of the moments of the summands. For the moment of sum of independent random variables ξ and η, the following equality is valid

$$\mathbf{E}(\xi+\eta)^n = \sum_{k=0}^{n} C_n^k \mathbf{E}\xi^k \mathbf{E}\eta^{n-k}.$$

The semi-invariants of different orders possess the property that the semi-invariant of a sum of independent variables equals the sum of the semi-invariants of the same order of the summands. It turns out that the semi-invariant of any order k is a rational function of the moments of orders less than or equal to k.

31 EXERCISES

1. A random variable ξ assumes non-negative integer values with the probabilities

(a)

$$P\{\xi=k\} = \frac{a^k}{(1+a)^{k+1}}$$

where $a>0$ is a constant. (This distribution is called the Pascal distribution).

(b)

$$p_k = P\{\xi=k\} = \left(\frac{\alpha\lambda}{1+\alpha\lambda}\right)^2 \cdot \frac{(1+\alpha)\cdots(1+(k-1)\alpha)}{k!} p_0$$

for all $k>0$ where $\alpha>0$, $\lambda>0$ and $p_0 = P\{\xi=0\} = (1+\alpha\lambda)^{-1}$. (This distribution is called the Polya distribution).

Find $\mathbf{E}\xi$ and $\mathbf{D}\xi$.

2. Let μ be the number of occurrences of an event A in n independent trials in each of which $P(A) = p$. Find
(a) $\mathbf{E}\mu^3$, (b) $\mathbf{E}\mu^4$, and (c) $\mathbf{E}|\mu - np|$.

3. The probability of occurrence of an event A in trial k equals p_i. Let μ be the number of occurrences of the event A in the first n independent trials. Find
(a) $\mathbf{E}\mu$, (b) $\mathbf{E}\mu$, (c)

$$\mathbf{E}\left(\mu - \sum_{i=1}^{n} p_i\right)^3,$$

and (d)

$$\mathbf{E}\left(\mu - \sum_{i=1}^{n} p_i\right)^4.$$

4. Prove that under the conditions of the preceding problem the maximum value of $\mathbf{D}\mu$ is attained for a given value of

$$a = \frac{1}{n}\sum_{i=1}^{n} p_i$$

under the condition $p_1 = p_2 = \cdots = p_n = a$.

5. Let μ be the number of occurrences of an event A in n independent trials in each of which $P(A) = p$. Further, let the variable η be equal to 0 or 1 depending on whether μ is even or odd. Determine $\mathbf{E}\eta$.

6. The density function of the random variable ξ is

$$p(x) = \frac{1}{2\alpha} e^{-|x-a|/\alpha}$$

(This distribution is called the Laplace distribution).
Find $\mathbf{E}\xi$ and $\mathbf{D}\xi$.

7. The density function of the magnitude of the velocity of a mole-cule is given by the Maxwell distribution

$$p(x) = \frac{4x^2}{\alpha^3\sqrt{\pi}} e^{-x^2/\alpha^2}$$

for $x>0$ and $p(x)=0$ for $x\leqslant 0$; $\alpha>0$ is a constant. Find the average speed and the average kinetic energy of a molecule (the mass of a molecule equals m), and the variances of the kinetic energy.

8. A molecule is in Brownian motion. The probability density of the distance x of a molecule from a reflecting barrier at time t if it was at a distance of t_0 from the barrier at time t_0 is given by the formula

$$p(x)=\begin{cases}\frac{1}{2\sqrt{\pi Dt}}\{e^{-(x+x_0)^2/4Dt}+e^{-(x-x_0)^2/4Dt}\} & \text{for} \quad x\geqslant 0,\\ 0 & \text{for} \quad x<0.\end{cases}$$

Find the mathematical expectation and variance of the magnitude of the displacement of the molecule during the time from t_0 to t. (D is a constant).

9. Prove that an arbitrary random variable ξ with possible values lying in the interval (a,b) satisfies the following inequalities:

$$a\leqslant \xi \leqslant b, \quad \mathbf{D}\xi\leqslant (b-a)^2/4.$$

10. Let $x_1, x_1, \ldots, x_n$ be the possible values of a random variable ξ. Prove that for $n\to\infty$

$$\text{a) } \mathbf{E}\xi^{n+1}/\mathbf{E}\xi^n \to \max_j x_j, \quad \text{b) } \sqrt[n]{\mathbf{E}\xi^n}\to \max_j x_j.$$

11. Let $F(x)$ be the distribution function of ξ. Prove that if $\mathbf{E}\xi$ exists then

$$\mathbf{E}\xi=\int_0^{\infty}[1-F(x)+F(-x)]dx$$

and that the condition

$$\lim_{x\to-\infty} xF(x)=\lim_{x\to\infty} x[1-F(x)]=0$$

is necessary and sufficient for the existence of $\mathbf{E}\xi$.

12. Two points are chosen at random in the interval (0, 1). Find the mathematical expectation and variance of the distance between them and the mathematical expectation of the n-th power of the distance between them.

13. A random variable ξ is distributed according to the lognormal distribution, i.e., for $x>0$ the density function of ξ is

$$p(x)=\frac{1}{x\sigma\sqrt{2\pi}}e^{-1/2\sigma^2(\ln x-a)^2}$$

with $p(x)=0$ for $x<0$. Find $\mathbf{E}\xi$ and $\mathbf{D}\xi$.

14. A random variable ξ is normally distributed with parameters a and σ. Find $\mathbf{E}|\xi-a|$.

15. The random variables $\xi_1,\xi_2,\dots,\xi_{n+m}$ $(n>m)$ are independent, identically distributed, and have finite variances. Determine the correlation coefficient of the two sums

$$s=\xi_1+\xi_2+\cdots+\xi_n,\quad \sigma=\xi_{m+1}+\xi_{m+2}+\cdots+\xi_{m+n}.$$

16. The random variables ξ and η are independent and normally distributed with the same parameters a and σ. Find the correlation coefficient of the variables $\alpha\xi+\beta\eta$ and $\alpha\xi-\beta\eta$, and also their joint distribution. (α and β are constants).

17. The random vector (ξ,η) is normally distributed. Its parameters are: $\mathbf{E}\xi=a$; $\mathbf{E}\eta=b$, $\mathbf{D}\xi=\sigma_1^2$, $\mathbf{D}\eta=\sigma_2^2$, and R is the correlation coefficient between ξ and η. Prove that $R=\cos q\pi$ where $q=\mathbf{P}\{(\xi-a)(\eta-b)<0\}$.

18. Let x_1 and x_2 be the results of two independent observations of a normally distributed variable ξ. Prove that $\mathbf{E}\max(x_1,x_2)=a+\sigma/\sqrt{\pi}$ where $a=\mathbf{E}\xi$ and $\sigma^2=\mathbf{D}\xi$.

19. A random vector (ξ,η) is normally distributed, with $\mathbf{E}\xi=\mathbf{E}\eta=0$, $\mathbf{D}\xi=\mathbf{D}\eta=1$, $\mathbf{E}\xi\eta=R$. Prove that

$$\mathbf{E}\max(\xi,\eta)=\sqrt{\frac{1-R}{\pi}}.$$

20. The unevenness in the length of cotton fiber is defined as the quantity

$$\lambda=\frac{a''-a'}{a},$$

where a is the mathematical expectation of the fiber length, a'' is the mathematical expectation of the length of those fibers whose length is larger than a, and a' is the mathematical expectation of those fibers whose length is less than a. Find the relation between the quantities

$$\text{a) } \lambda, a, \mathbf{E}|\xi-a|,\quad \text{b) } \lambda, a, \sigma.$$

if ξ is normally distributed.

21. The random variables $\xi_1, \xi_2, \ldots, \xi_n \ldots$ are independent and uniformly distributed in (0, 1). Let ν be the random variable equal to such value of k for which the sum

$$s_k = \xi_1 + \xi_2 + \cdots + \xi_k$$

initially exceeds 1. Prove that $\mathrm{E}\nu = e$.

22. Let ξ be a random variable with the density function

$$p_\xi(x) = \frac{1}{\pi} \cdot \frac{1}{1+x^2}.$$

Find $\mathrm{E}\min(|\xi|, 1)$.

Chapter 7

THE LAW OF LARGE NUMBERS

32 MASS PHENOMENA AND THE LAW OF LARGE NUMBERS

The vast experience accumulated by mankind shows that a phenomenon with probability very close to 1 is almost certain to take place. Similarly, an event with a very small probability of occurrence (in other words, this probability is very close to zero) happens very rarely. This fact plays a fundamental role in all practical inferences in probability theory. This empirical fact gives one the right to consider events with small probabilities as being practically impossible and those with probabilities very close to one as being practically certain. Nevertheless, there is no unique answer for a perfectly natural question: What should the probability of an event be if it is to be regarded as being practically impossible (or practically certain). This is quite understandable, since in any practical application it is necessary to take into account the importance of the kind of event with which we are dealing. Thus, for example, if the distance between two villages were measured and found to be 16,500 ft. and if the error in this measurement was greater than or equal to 50 ft., with a probability of 0.02, then one could ignore the possibility of such an error and consider the distance as actually being equal to 16,500 ft. In this example, an event which occurs with probability 0.02 is considered as having no practical importance, and it will not be taken into account for practical decisions. At the same time, there are other cases where one cannot disregard an event having a probability of 0.02, or even a much smaller probability. Imagine that one designs a large hydroelectric plant which requires an enormous expenditure of material and manpower. If it were ascertained that the probability of a catastrophic flood in the area of the plant location equaled 0.02, then this probability would be considered large, and would have to be taken into account in the planning of the project. Thus only practical considerations can suggest the criteria according to which various events are to be considered as being practically impossible or practically certain.

At the same time, it is necessary to emphasize that any event that has a positive probability, no matter how small, can occur. If the number of trials where such a rare event can occur is very large, then the probability of the event occurring at least once can come as close to one as desired. One should always keep this fact in mind. However, if the probability of some event is very small, it will nevertheless doubtlessly occur in some particular single trial. Thus, if a dealer insists that each of four players whom he evenly serves 36 cards will receive all cards of one suit only at the first deal, then it is natural to suspect that the dealer is unfair. For instance, he might have information that the cards were arranged in some specific order known to him. This belief is based on the fact that the probability of such a distribution when the cards are properly shuffled equals $(9!)^4 4!/36! < 1.1 \times 10^{-18}$, i.e., it is extremely small. Nevertheless, the fact that such a distribution of cards has occurred is a matter of record. This example illustrates the distinct difference between the concept of practical impossibility and, so to speak, categorical impossibility.

The discussion above clearly shows that events with probabilities close to 0, or to 1, are very important in practical applications as well as in general theory. It is clear that one of the fundamental problems of probability theory is the formulation of laws which predict probability close to 1. An important role must be played here by those laws which arise as the result of imposing a large number of independent or weakly dependent random factors. The law of large numbers is one such proposition in probability theory and, possibly, one of the most important.

It is natural that the law of large numbers should be understood as the aggregate of statements which declare that some event that depends on a set of random factors whose number increases without limit and each of which has only a negligible effect upon it, will occur with a probability as close to one as desired. This general description of theorems of the type of the law of large numbers can be formulated in a somewhat more definite way. Let a sequence of random variables

$$\xi_1, \xi_2, \ldots, \xi_n, \ldots \tag{1}$$

be given.

Consider the random variables ζ_n which are certain specified symmetric functions of the first n variables of the sequence (1):

$$\zeta_n = f_n(\xi_1, \xi_2, \ldots, \xi_n).$$

If there exists a sequence of constants $a_1, a_2, \ldots, a_n$ such that for any $\varepsilon > 0$

$$\lim_{n\to\infty} \mathbf{P}\{|\zeta_n - a_n| < \varepsilon\} = 1, \tag{2}$$

then the sequence (1) obeys the law of large numbers relative to the given functions f_n.

However, usually one understands the concept of the law of large numbers much more specifically. Namely, the consideration is limited by the case where f_n is the arithmetic mean of the variables $\xi_1, \xi_2, \ldots, \xi_n$.

If each a_n in relation (2) has the same value a, then one says that the sequence of random variables ζ_n converges in probability to a. In these terms, (2) means that $\zeta_n - a_n$ converges in probability to 0.

Observing a single phenomenon, one sees all its individual peculiarities, which veil the essence of the laws that hold when similar phenomena are observed a large number of times. It was observed a long time ago that particular factors which are not concerned with the essential nature of a process annihilate each other when the average of a large number of observations is considered.

This empirical result was noted more and more frequently, as a rule, without any attempt to find a theoretical explanation for it. Many authors did not even demand such an explanation because in their day the existence of laws governing both natural and social phenomena was regarded as nothing other than the manifestation of the laws of divine order.

Certain contemporary authors understate the content of the law of large numbers and even misinterpret its methodological significance by reducing it to an experimental observation of some dependence. Actually, the enduring scientific value of the investigations of Chebyshev, Markov, and others who investigated the law of large numbers is not in the simple fact that they observed and explained the empirical stability of the mean but in their determination of general sufficient conditions for the statistical stability of the mean.

Let us show how the law of large numbers operates via the following illustrative example. From the standpoint of modern physics, a gas is composed of an immense number of molecules in constant chaotic motion. Nobody can predict either the velocity of each individual molecule moving, or its position at any given moment of time. However, under specific conditions one can compute the average proportion of molecules that have a given velocity or the average proportion of them that are to be found in a given volume. As a matter of fact, this is exactly what the physicist wishes to know because the basic characteristics of a gas are its pressure, temperature, viscosity, etc. They are determined not by the sophisticated actions of a single molecule but by the behavior of the entire collection of all the molecules. Thus, the pressure of a gas is determined by the overall effect of the molecules hitting a unit area in a unit of time. The number of impinging molecules and their velocities vary in a way depending on chance. However, by the law of large numbers (in Chebyshev's form), the pressure must be almost constant. This effect of "equalizing" due to the law of large numbers is observed in physical phenomena with exceptional exactness. One can recall that under ordinary conditions, even very precise measurements do not allow one to observe visible deviations from Pascal's law of hydrostatic pressure. These deviations, called fluctuations of pressure, were indeed successfully observed after scientists learned how to isolate a comparatively small number of molecules. As a result, the influence of the individual molecule still remains strong since it is not completely neutralized.

33 CHEBYSHEV'S FORM OF THE LAW OF LARGE NUMBERS

Let us now formulate and prove the theorems of Chebyshev, Markov, and others. The method employed here is due to Chebyshev.

CHEBYSHEV'S INEQUALITY. *For every random variable ξ having a finite variance and for every* $\varepsilon > 0$, the inequality

$$\mathbf{P}\{|\xi - \mathbf{E}\xi| \geqslant \varepsilon\} \leqslant \frac{\mathbf{D}\xi}{\varepsilon^2} \tag{1}$$

holds.

PROOF: If $F(x)$ denotes the distribution function of the random variable ξ, then, obviously,

$$\mathbf{P}\{|\xi - \mathbf{E}\xi| \geqslant \varepsilon\} = \int_{|x-\mathbf{E}\xi|\geqslant\varepsilon} dF(x).$$

Since in the region of integration $|x - \mathbf{E}\xi|\varepsilon \geqslant 1$, it follows that

$$\int_{|x-\mathbf{E}\xi|\geqslant\varepsilon} dF(x) \leqslant \frac{1}{\varepsilon^2} \int_{|x-\mathbf{E}\xi|\geqslant\varepsilon} (x - \mathbf{E}\xi)^2\, dF(x).$$

The inequality will be only strengthened if the integration is be extended to all values of x:

$$\int_{|x-\mathbf{E}\xi|\geqslant\varepsilon} dF(x) \leqslant \frac{1}{\varepsilon^2} \int (x - \mathbf{E}\xi)^2\, dF(x) = \frac{\mathbf{D}\xi}{\varepsilon^2}.$$

Thus Chebyshev's inequality is proved.

CHEBYSHEV'S THEOREM. If $\xi_1, \xi_2, \ldots, \xi_n, \ldots$ *is a sequence of pairwise independent random variables possessing finite variances which are bounded by the same constant*

$$\mathbf{D}\xi_1 \leqslant C,\ \mathbf{D}\xi_2 \leqslant C, \ldots, \mathbf{D}\xi_n \leqslant C, \ldots$$

then for any constant $\varepsilon > 0$

$$\lim_{n\to\infty} \mathbf{P}\left\{\left|\frac{1}{n}\sum_{k=1}^{n} \xi_k - \frac{1}{n}\sum_{k=1}^{n} \mathbf{E}\xi_k\right| < \varepsilon\right\} = 1. \tag{2}$$

PROOF: By the assumptions of the theorem, one knows that

$$\mathbf{D}\left(\frac{1}{n}\sum_{k=1}^{n} \xi_k\right) = \frac{1}{n^2}\sum_{k=1}^{n} \mathbf{D}\,\xi_k$$

and therefore

$$\mathbf{D}\left(\frac{1}{n}\sum_{k=1}^{n} \xi_k\right) \leqslant \frac{C}{n}.$$

According to Chebyshev's inequality,

$$\mathbf{P}\left\{\left|\frac{1}{n}\sum_{k=1}^{n} \xi_k - \frac{1}{n}\sum_{k=1}^{n} \mathbf{E}\xi_k\right| < \varepsilon\right\} \geqslant 1 - \frac{\mathbf{D}\left(\frac{1}{n}\sum_{k=1}^{n} \xi_k\right)}{\varepsilon^2} \geqslant 1 - \frac{C}{n\varepsilon^2}.$$

Limit for $n \to \infty$ produces

$$\lim_{n\to\infty} \mathbf{P}\left\{\left|\frac{1}{n}\sum_{k=1}^{n}\xi_k - \frac{1}{n}\sum_{k=1}^{n}\mathbf{E}\xi_k\right| < \varepsilon\right\} \geqslant 1.$$

But since a probability cannot exceed 1, one concludes that the theorem is proven.

Note some important special cases of Chebyshev's Theorem.

1. BERNOULLI'S THEOREM. *Let μ be the number of occurrences of an event A in n independent trials and p be the probability of occurrence of the event A in each trial. Then for any $\varepsilon > 0$*

$$\lim_{n\to\infty} \mathbf{P}\left\{\left|\frac{\mu}{n} - p\right| < \varepsilon\right\} = 1. \tag{3}$$

PROOF: Indeed, by introducing the random variables μ_k equal to the number of occurrences of the event A in trial k, one has:

$$\mu = \mu_1 + \mu_2 + \cdots + \mu_n.$$

Since

$$\mathbf{E}\mu_k = p, \quad \mathbf{D}\mu_k = pq \leqslant 1/4,$$

it is clear that Bernoulli's Theorem is the simplest special case of Chebyshev's Theorem.

Since in practice one often needs to find. The approximate values of unknown probabilities from experiments, a large number of trials have been conducted for verifying the agreement between Bernoulli's Theorem and experiments. Researchers considered events whose probabilities could for one reason or another be regarded as known and for which one could easily conduct trials and ensure their independence and the constancy of the probabilities in each trial. All such experiments yielded excellent agreement with the theory. It is very interesting to cite the outcomes of some of these easily reproducible experiments.

In Example 5, Section 3, a deck of 36 playing cards was divided at random into two equal parts. This experiment was repeated one hundred times. We were interested in the following event: each part of 18 cards consists of nine red and nine black cards. In that case for $n = 100$ the difference between frequency and probability was sufficiently significant (about 0.02). By Laplace's Theorem the probability to observe. This or a larger deviation equals

$$\mathbf{P}\left\{\left|\frac{\mu}{n}-p\right|\geqslant 0.02\right\}=\mathbf{P}\left\{\left|\frac{\mu-np}{\sqrt{npq}}\right|\geqslant 0.02\sqrt{\frac{n}{pq}}\right\}\approx 1-2\Phi\left(0.02\sqrt{\frac{n}{pq}}\right)$$

$$=1-2\Phi\left(0.02\sqrt{\frac{100}{0.26\cdot 0.74}}\right)=1-2\Phi\,(0.455)\approx 0.65$$

Thus if the experiment described above is repeated a large number of times, an error not less than that mentioned above will be obtained in approximately 2/3 of the entire number of experiments.

The eighteenth century French naturalist, Buffon, tossed a coin 4,040 times, obtaining heads 2,048 times. The relative frequency of occurrence of heads in Buffon's experiment was approximately 0.507. The English statistician, Carl Pearson, tossed a coin 12,000 times and obtained heads 6,019 times. The relative frequency of heads in Pearson's experiment was 0.5016. On another occasion, he tossed a coin 24,000 times, with heads turning up 12,012 times; here, the relative frequency of heads turned out to be 0.5005. In all of these experiments the relative frequency deviated very little from the probability 0.5.

2. POISSON'S THEOREM. *If in a sequence of independent trials the probability of occurrence of an event A in trial k is p_k, then*

$$\lim_{n\to\infty}\mathbf{P}\left\{\left|\frac{\mu}{n}-\frac{p_1+p_2+\cdots+p_n}{n}\right|<\varepsilon\right\}=1,$$

where, as usual, μ denotes the number of times the event A occurs in the first n trials.

Introduce the random variables μ_k equal to the number of events A in the k-th trial and denote that

$$\mathbf{E}\mu_k=p_k,\quad \mathbf{D}\mu_k=p_kq_k\leqslant 1/4,$$

one can see that Poisson's Theorem is a special case of Chebyshev's Theorem.

3. *If the sequence of pairwise independent random variables $\xi_1, \xi_2, \ldots, \xi_n, \ldots$* is such that

$$\mathbf{E}\xi_1=\mathbf{E}\xi_2=\cdots=\mathbf{E}\xi_n=\cdots=a$$

and

$$\mathbf{D}\xi_1 \leqslant C, \quad \mathbf{D}\xi_2 \leqslant C, \ldots, \mathbf{D}\xi_n \leqslant C, \ldots,$$

then for any constant $\varepsilon > 0$

$$\lim_{n \to \infty} \mathbf{P}\left\{\left|\frac{1}{n}\sum_{k=1}^{n} \xi_k - a\right| < \varepsilon\right\} = 1.$$

This special case of Chebyshev's Theorem serves as a basis for the rule of the arithmetic mean, which is regularly used in the theory of measurement. Suppose that some physical quantity a is being measured. Repeating the measurement n times under identical conditions, the observer obtains the results $x_1, x_2, \ldots, x_n$ which do not completely coincide. As an approximate value for a, it is natural to take the arithmetic mean of the results of measurements:

$$a \sim \frac{x_1 + x_2 + \cdots + x_n}{n}.$$

Assume that the measurements are free of any systematic error, i.e., if

$$\mathbf{E}x_1 = \mathbf{E}x_2 = \cdots = \mathbf{E}x_n = a,$$

and if there is no uncertainty about the observed values themselves. Then according to the law of large numbers, one can obtain a value which is arbitrarily close to a with the probability that arbitrarily close to 1 by making the number of measurements, n, sufficiently large. It is necessary to emphasize that if a tool for measurement cannot give the accuracy better than some given δ (for instance, it is an increment of a discrete scale of the tool), i.e., each measurement is obtained with uncertainty $\pm\delta$, then the result will have the same uncertainty. This note makes clear that it is impossible to obtain a high level of accuracy by means of increasing the number of experiments if the measurement tool possesses some fixed uncertainty. One should avoid this misunderstanding in practical work.

Let us now formulate Markov's Theorem keeping in mind that its proof is an obvious corollary of Chebyshev's Theorem.

MARKOV'S THEOREM. *If the sequence of random variables* $\xi_1, \xi_2, \ldots, \xi_n, \ldots$ *is such that*

$$\frac{1}{n^2}\mathbf{D}\left(\sum_{k=1}^{n} \xi_k\right) \to 0, \tag{4}$$

for $n \to \infty$, then for any constant $\varepsilon > 0$

$$\lim_{n \to \infty} \mathbf{P}\left\{\left|\frac{1}{n}\sum_{k=1}^{n} \xi_k - \frac{1}{n}\sum_{k=1}^{n} \mathbf{E}\xi_k\right| < \varepsilon\right\} = 1.$$

If the random variables $\xi_1, \xi_2, \ldots, \xi_n, \ldots$ are pairwise independent, then Markov's condition takes on the following form:

$$\frac{1}{n^2}\sum_{k=1}^{n} \mathbf{D}\xi_k \to 0$$

as $n \to \infty$.

From this, it is clear that Chebyshev's Theorem is a special case of Markov's Theorem.

34 A NECESSARY AND SUFFICIENT CONDITION FOR THE LAW OF LARGE NUMBERS

It has already been indicated that the law of large numbers is one of the fundamental statements of probability theory. Therefore it is understandable why so much effort has been spent for establishing the most general conditions upon the variables $\zeta_1, \zeta_2, \ldots, \zeta_n, \ldots$ under which the law of large numbers holds.

The history of the problem is as follows. At the end of the Seventeenth and the beginning of the eighteenth centuries, Jacob Bernoulli proved the theorem now bearing his name. The theorem of Bernoulli was first published posthumously in 1713 in the tractate "Ars conjectandi" (The Art of Suggestions). Later on, at the beginning of the nineteenth century, Poisson proved an analogous theorem under more general conditions. Until the middle of the nineteenth century no further progress was made. In 1866 the great Russian mathematician Chebyshev discovered the method which was presented in the preceding section. Later, Markov observed that the reasoning of Chebyshev allowed a more general result to be obtained (see Section 27).

For a long time, further efforts did not yield any major advances, and only in 1926 did Kolmogorov derive conditions that were necessary and sufficient for a sequence of mutually independent random variables $\zeta_1, \zeta_2, \ldots, \zeta_n, \ldots$ to obey the law of large numbers. In 1923, Khinchine showed that if the random variables ζ_n were not only independent but also identically distributed then the existence of the mathematical expectation $\mathsf{E}\xi_n$ was a sufficient condition for the law of large numbers to apply.

In recent years, many works were devoted to determining those conditions which it is necessary to impose on dependent variables in order that they satisfy the law of large numbers. Markov's Theorem belongs to this class of propositions.

Using Chebyshev's method one can easily obtain a condition analogous to Markov's one which is not only sufficient but also necessary for the law of large numbers to hold for a sequence of arbitrary random variables.

THEOREM. *In order for the sequence $\xi_1, \xi_2, \ldots, \xi_n, \ldots$ of random variables (arbitrarily dependent) to satisfy the relation*

$$\lim_{n\to\infty} \mathsf{P}\left\{\left|\frac{1}{n}\sum_{k=1}^{n} \xi_n - \frac{1}{n}\sum_{k=1}^{n} \mathsf{E}\xi_k\right| < \varepsilon\right\} = 1, \tag{1}$$

for any positive ε, *it is necessary and sufficient that*

$$\mathsf{E}\frac{\left(\sum_{k=1}^{n} (\xi_k - \mathsf{E}\xi_k)\right)^2}{n^2 + \left(\sum_{k=1}^{n} (\xi_k - \mathsf{E}\xi_k)\right)^2} \to 0 \tag{2}$$

as $n \to \infty$.

PROOF: Suppose first that (2) is satisfied. Then one can show that (1) is also satisfied. Let $\Phi(x)$ denote the distribution function of the variable

$$\eta_n = \frac{1}{n}\sum_{k=1}^{n} (\xi_k - \mathsf{E}\xi_k).$$

It is easy to verify the following chain of relations

$$\begin{aligned}\mathsf{P}\left\{\left|\frac{1}{n}\sum_{k=1}^{n} (\xi_k - \mathsf{E}\xi_k)\right| \geqslant \varepsilon\right\} &= \mathsf{P}\{|\eta_n| \geqslant \varepsilon\} \\ &= \int_{|x|\geqslant\varepsilon} d\Phi_n(x) \leqslant \frac{1+\varepsilon^2}{\varepsilon^2}\int_{|x|\geqslant\varepsilon} \frac{x^2}{1+x^2}\, d\Phi_n(x) \\ &\leqslant \frac{1+\varepsilon^2}{\varepsilon^2}\int \frac{x^2}{1+x^2}\, d\Phi_n(x) = \frac{1+\varepsilon^2}{\varepsilon^2}\mathsf{E}\frac{\eta_n^2}{1+\eta_n^2}\end{aligned}$$

This inequality proves that the condition of the theorem is sufficient[1].

Let us now show that condition (2) is sufficient. It is easy to see that

$$\begin{aligned} \mathsf{P}\{|\eta_n| \geqslant \varepsilon\} &= \int_{|x| \geqslant \varepsilon} d\Phi_n(x) \geqslant \int_{|x| \geqslant \varepsilon} \frac{x^2}{1+x^2} d\Phi_n(x) \\ &= \int \frac{x^2}{1+x^2} d\Phi_n(x) - \int_{|x|<\varepsilon} \frac{x^2}{1+x^2} d\Phi_n(x) \\ &\geqslant \int \frac{x^2}{1+x^2} d\Phi_n(x) - \varepsilon^2 = \mathsf{E} \frac{\eta_n^2}{1+\eta_n^2} - \varepsilon^2. \end{aligned} \tag{3}$$

Thus

$$0 \leqslant \mathsf{E} \frac{\mu_n^2}{1+\mu_n^2} \leqslant \varepsilon^2 + \mathsf{P}\{|\mu_n| \geqslant \varepsilon\}.$$

Choosing ε sufficiently small and after this n sufficiently large, one can make the right-hand side of the last inequality arbitrarily small.

Note that all of the theorems that were proved in the preceding section follow in a simple way from the general theorem just proved. In fact, since the inequality

$$\frac{\eta_n^2}{1+\eta_n^2} \leqslant \eta_n^2 = \left[\frac{1}{n} \sum_{k=1}^{n} (\xi_k - \mathsf{E}\,\xi_k)\right]^2,$$

holds for any n and any ξ_k, the inequality

$$\mathsf{E} \frac{\eta_n^2}{1+\eta_n^2} \leqslant \frac{1}{n^2} \mathsf{D} \sum_{k=1}^{n} \xi_k$$

is provided if the variance exists.

Thus, if Markov's condition is satisfied, then so is condition (2), and therefore the sequence $\xi_1, \xi_2, \ldots, \xi_n, \ldots$ is subject to the law of large numbers.

Nevertheless one should note that in more complicated situations, such as when the random variables ξ_k are not assumed to have finite

[1]One can write the last equality on the basis of the definition

$$E\{f(\xi)\} = \int f(x)\, dF_\xi(x)$$

(see Theorem 1, Section 22).—*B. V.*

variances, the theorem just proved is of very little use in verifying the correctness of application of the law of large numbers, because condition (2) relates not to the individual variables but to their sum. Apparently, however, without making any assumptions about the random variables ξ_k or about some existing relation between them, one cannot expect to find necessary and sufficient conditions (especially conditions which are convenient for application).

If one assumes that random variables $\xi_1, \xi_2, \ldots, \xi_n, \ldots$ are mutually independent then it is possible to prove that the condition (2) is equivalent to the following:

$$\sum_{k=1}^{n} \mathsf{E} \frac{\zeta_k^2}{n^2 + \zeta_k^2} \to 0$$

for $n \to \infty$ where the notation

$$\zeta_k = \xi_k - \mathsf{E}\,\xi_k$$

is used.

One principal difficulty is faced in the practical application of the theorems just proved: can we accept that a phenomenon under investigation is taking place under the influence of independent causes? Does not the very concept of independence contradict the basic idea that all phenomena of the external world are interrelated? Whenever one undertakes a mathematical study of some natural phenomenon, technical operation, or social process, it must be based on a profound study of the essence of the phenomenon under investigation, of its qualitative characteristics. Any change in the external conditions under which our phenomenon occurs must be taken into account, and the mathematical model with all of the premises underlying its application must be modified as soon as it is discovered that the conditions of the phenomenon's existence have changed.

As a first approximation to reality, one may suppose that the causes of the phenomenon under investigation are independent and base any conclusions on this assumption. How successful such a description of a phenomenon has been and how successful our selection of the mathematical model for studying it, can only be judged by the agreement between the theory that is set up and practice. If the theoretical results should differ significantly from the experiment's, then one must re-examine the premises. In particular, if the question concerns the applicability of the law of large numbers, then perhaps one will have to reject the assumption that the causes are independent and go over to the assumption that they are at least weakly dependent.

But it has already been stated that enormous experience with using the laws of large numbers has shown that the condition of independence is satisfactory in many important practical problems in natural science and engineering.

35 THE STRONG LAW OF LARGE NUMBERS

One frequently makes unjustified inference from Bernoulli's Theorem that the relative frequency of an event A tends to the probability of the event A when the number of trials is increased indefinitely. The fact is that Bernoulli's Theorem merely establishes that when the number of trials n is made sufficiently large, the probability of the one single inequality

$$\left|\frac{\mu}{n}-p\right|<\varepsilon$$

becomes larger than $1-\eta$ for arbitrary $\eta>0$. In 1909, the French mathematician Emaile Borel proved a deeper statement. He found that for any $\varepsilon>0$ and $\eta>0$ one can show such n_0 that for any s the probability of simultaneous satisfaction of inequalities

$$\left|\frac{\mu}{n}-p\right|<\varepsilon$$

for all n such that $n_0 \leqslant n \leqslant n_0+s$ is larger than $1-\eta$.

This theorem will be proved on the basis of Kolmogorov's Theorem on the strong law of large numbers.

KOLMOGOROV'S INEQUALITY. *If mutually independent random variables $\xi_1, \xi_2, \ldots, \xi_n$ have finite variances then the probability of simultaneous existence of inequalities*

$$\left|\sum_{s=1}^{k}(\xi_s-\mathsf{E}\,\xi_s)\right|<\varepsilon \qquad (k=1,2,\ldots,n)$$

is not less than

$$1-\frac{1}{\varepsilon^2}\sum_{k=1}^{n}\mathbf{D}\xi_k.$$

PROOF: Introduce the notation

$$\eta_k=\xi_k-\mathbf{E}\xi_k,\quad S_k=\sum_{j=1}^{k}\eta_j.$$

Let E_k denote an event that

$$|S_j|<\varepsilon \text{ for } j\leqslant k-1 \text{ and } |S_k|\geqslant\varepsilon;$$

E_0 denote an event that $|S_j|<\varepsilon$ for $j\leqslant n$.

Since the event that at least for one $k\,(1\leqslant k\leqslant n)$ the inequality

$$|S_k|\geqslant\varepsilon \qquad (k=1,2,\ldots,n)$$

holds, or in other words, that $\max_{1\leqslant k\leqslant n}|S_k|\geqslant\varepsilon$ is equivalent to the event $\Sigma_{k=1}^{n}E_k$ then it follows that by mutual exclusiveness of the events E_k the equality

$$\mathbf{P}\left\{\max_{1\leqslant k\leqslant n}|S_k|\geqslant\varepsilon\right\}=\sum_{k=1}^{n}\mathbf{P}(E_k).$$

also holds.

By equation (5) of Section 23

$$\mathbf{D}S_n=\sum_{k=0}^{n}\mathbf{P}(E_k)\cdot\mathbf{E}(S_n^2\,|\,E_k)\geqslant\sum_{k=1}^{n}\mathbf{P}(E_k)\cdot\mathbf{E}(S_n^2|E_k).$$

It is evident that

$$\begin{aligned}\mathbf{E}(S_n^2|E_k)&=\mathbf{E}\{S_k^2+2\sum_{j>k}S_k\eta_j+\sum_{j>k}\eta_j^2+2\sum_{j>h>k}\eta_j\eta_h|E_k\}\\&\geqslant\mathbf{E}\{S_k^2+2\sum_{j>k}S_k\eta_j+2\sum_{j>h>k}\eta_j\eta_h|E_k\}.\end{aligned}$$

Since the occurrence of the event E_k imposes a restriction on only the first k of the random variables ξ_i, and the following ones, under this condition, remain independent of one another and of S_k, it follows that

$$\mathbf{E}(S_k\eta_j|E_k)=\mathbf{E}(S_k|E_k)\cdot\mathbf{E}(\eta_j|E_k)=0$$

and

$$\mathbf{E}(\eta_j \eta_h | E_k) = 0 \qquad (h \neq j, h > k, j > k \geq 1).$$

Besides, by (1) the inequality

$$\mathbf{E}(S_k^2 | E_k) \geq \varepsilon^2 \qquad (k \geq 1)$$

holds. Therefore it is possible to write that

$$\mathbf{D}S_n \geq \varepsilon^2 \sum_{k=1}^{n} \mathbf{P}\{E_k\}.$$

Finally,

$$\sum_{k=1}^{n} \mathbf{P}\{E_k\} = \mathbf{P}\left\{\max_{1 \leq k \leq n} |S_k| \geq \varepsilon\right\} \leq \frac{1}{\varepsilon^2} \mathbf{D}S_n.$$

Thus the Kolmogorov's inequality has been proved.

Let us say that the sequence of random variables $\xi_1, \xi_2, \xi_3, \ldots$ follows the law of large numbers if for any $\varepsilon > 0$ and $\eta > 0$ it is possible to find n_0 such that for any s and for all of n satisfying the condition $n_0 \leq n \leq n_0 + s$ the probability of the inequality

$$\max_{n_0 \leq n \leq n_0 + s} \left| \frac{1}{n} \sum_{k=1}^{n} \xi_k - \frac{1}{n} \sum_{k=1}^{n} \mathbf{E}\xi_k \right| < \varepsilon$$

is larger than $1 - \eta$.

KOLMOGOROV'S THEOREM. *If the sequence of random variables $\xi_1, \xi_2, \xi_3, \ldots$ satisfies the condition*

$$\sum_{n=1} \frac{\mathbf{D}\xi_n}{n^2} < +\infty,$$

then it follows the law of large numbers.

PROOF: Introduce the notation

$$\zeta_n = \xi_n - \mathbf{E}\xi_n, \quad S_n = \sum_{k=1}^{n} \zeta_k, \quad v_n = \frac{1}{n} S_n.$$

Consider the difference

$$P_m = \mathbf{P}\{\max |v_n| \geq \varepsilon, 2^m \leq n < 2^{m+1}\}.$$

Since

$$P_m \leqslant \mathbf{P}\{\max|v_n| \geqslant \varepsilon,\ 1 \leqslant n < 2^{m+1}\},$$

then by Kolmogorov's Inequality

$$P_m \leqslant \frac{1}{(2^m\varepsilon)^2} \sum_{j<2^{m+1}} \mathbf{D}\xi_j.$$

Since, further,

$$\mathbf{P}\left\{\max|v_n| \geqslant \varepsilon \text{ for } n > v\right\} \leqslant \sum_{m=\rho} P_m,$$

(where ρ is defined from the inequality $2^\rho \leqslant v < 2^{\rho+1}$) then it is clear that

$$\mathbf{P}\{\max|v_n| \geqslant \varepsilon \quad \text{for} \quad n > v\} \leqslant \frac{1}{\varepsilon^2} \sum_{m=\rho} \frac{1}{2^{2m}} \sum_{j<2^{m+1}} \mathbf{D}\xi_j$$

After changing the order of summation in the right-hand side of the last inequality, one has

$$\sum_{m=\rho}^{\infty} \frac{1}{2^{2m}} \sum_{j<2^{m+1}} \mathbf{D}\xi_j = \sum_{j=1} \mathbf{D}\xi_j\left(\sum_j \frac{1}{2^{2m}}\right),$$

where the sum Σ_j is expanded onto all $m \geqslant \rho$ for which $2^{m+1} > j$.

For $j \leqslant 2^{\rho+1}$, the coefficient of $\mathbf{D}\xi_j$ equals

$$\sum_{m \geqslant \rho} \frac{1}{4^m} = \frac{3}{4^{\rho-1}}.$$

For

$$2^{m_0+1} > j \geqslant 2^{m_0} \geqslant 2^{\rho+1}$$

this coefficient equals

$$\frac{3}{4^{m_0-1}} \leqslant \frac{3\cdot 16}{2^{2(m_0+1)}} \leqslant \frac{3\cdot 16}{j^2}.$$

Thus

$$\sum_{m=\rho}\frac{1}{2^{2m}}\sum_{j<2^{m+1}}\mathbf{D}\xi_j \leqslant \frac{3}{4^{\rho-1}}\sum_{j=1}^{2^{\rho+1}}\mathbf{D}\xi_j + 3\cdot 16\sum_{j=2^{\rho+1}+1}^{\infty}\frac{\mathbf{D}\xi_j}{j^2}$$

$$\leqslant \frac{3}{4^{\rho-1}}\sum_{j=1}^{\rho}\mathbf{D}\xi_j + 3\cdot 4\sum_{j=\rho+1}^{2^{\rho+1}}\frac{\mathbf{D}\xi_j}{2^{2(\rho-1)}} + 3\cdot 16\sum_{j=2^{\rho+1}+1}^{\infty}\frac{\mathbf{D}\xi_j}{j^2}$$

$$\leqslant \frac{3}{4^{\rho-1}}\sum_{j=1}^{\rho}\mathbf{D}\xi_j + 3\cdot 4\sum_{j=\rho+1}^{2^{\rho+1}}\frac{\mathbf{D}\xi_j}{j^2} + 3\cdot 16\sum_{j=2^{\rho+1}+1}^{\infty}\frac{\mathbf{D}\xi_j}{j}.$$

Because of convergence of the set $(\Sigma\mathbf{D}\xi_n)/n^2$, one can make the following two statements:

(1) The two sums of the latter inequality above, for large enough ρ, can be made as small as necessary.

(2) There exists such a constant C that $\mathbf{D}\xi_n < Cn^2$. It follows that

$$\frac{3}{4^{\rho+1}}\sum_{j=1}^{\rho}\mathbf{D}\xi_j \leqslant \frac{3\cdot C\rho^3}{4^{\rho-1}}$$

i.e., the first sum can also be made, for large enough ρ, as small as necessarry.

From the statements above it follows that for large enough n_0 the probability

$$\mathbf{P}\{\max|v_n| \geqslant \varepsilon \text{ for } n > v\}$$

can be set as small as necessary.

COROLLARY. *If variances of random variances ξ_k are bounded by the same constant C then the sequence of mutually independent random variables $\xi_1, \xi_2, \xi_3, \ldots$ follows the law of large numbers.*

A final result related to the strict law of large numbers has been also obtained by Kolmogorov for the sum of independent random variables identically distributed.

THEOREM. *The existence of the mathematical expectation is a necessary and sufficient condition of the applicability of the law of large numbers to s sequence of mutually independent random variables identically distributed.*

This theorem can be derived from the Kolmogorov theorem just proved.

Indeed, from the existence of the mathematical expectation of the random variable it follows that the integral $\int|x|dF(x)$ is bounded where $F(x)$ is the distribution function of the random variables ξ_n. Therefore

$$\begin{aligned}\sum_{n=1} \mathbf{P}\{|\xi|>n\} &= \sum_{n=1}^{\infty}\sum_{k\geqslant n} \mathbf{P}\{k<|\xi|\leqslant k+1\}\\ &= \sum_{n=1} k\,\mathbf{P}\{k<|\xi|\leqslant k+1\} \leqslant \sum_{k=0}\int_{k<|x|\leqslant k+1}|x|\,dF(x) \quad (1)\\ &< \int|x|dF(x)<\infty.\end{aligned}$$

Introduce the following random variables

$$\xi_n^* = \begin{cases}\xi_n & \text{for } |\xi_n|\leqslant n,\\ 0 & \text{for } |\xi_n|>n.\end{cases}$$

Then one gets

$$\mathbf{D}\xi_n^* \leqslant \mathbf{E}\,\xi_n^{*2} = \int_{-n}^{+n} x^2\,dF(x) \leqslant \sum_{k=0}^{n}(k+1)^2\,\mathbf{P}\{k<|\xi|\leqslant k+1\}$$

and

$$\begin{aligned}\sum_{n=1}\frac{\mathbf{D}\xi_n^*}{n^2} &\leqslant \sum_{n=1}^{\infty}\sum_{k=0}^{n}\mathbf{P}\{k<|\xi|\leqslant k+1\}\\ &\leqslant \sum_{k=0}^{\infty}\mathbf{P}\{k<|\xi|\leqslant k+1\}(k+1)^2\sum_{n\geqslant k}\frac{1}{n^2}\end{aligned}$$

Since

$$\sum_{n\geqslant k}\frac{1}{n^2}<\frac{1}{k^2}+\frac{1}{k}<\frac{2}{k},$$

by virtue of (1) one can find

$$\sum_{n=1}\frac{\mathbf{D}\xi_n^*}{n^2}<\infty,$$

where ξ_n^* satisfies the law of large numbers. Furthermore

$$\mathbf{P}\{\xi_n\neq\xi_n^* \text{ for some } n\geqslant N\} \leqslant \sum_{n\geqslant N}\mathbf{P}\{\xi_n\neq\xi_n^*\} = \sum_{n\geqslant N}\mathbf{P}\{|\xi_n|>n\}<\frac{\varepsilon}{4} \quad (2)$$

for $N \geqslant N_0(\varepsilon)$. Let us choose ν_0 so large that for $\nu \geqslant \nu_0(\varepsilon, \eta)$ the following inequalities hold:

$$\mathbf{P}\left\{\left|\frac{\sum_{k=1}^{N_0}(\xi_k - \mathbf{E}\xi_k)}{\nu}\right| \geqslant \frac{\eta}{3}\right\} \leqslant \frac{\varepsilon}{4} \tag{3}$$

$$\mathbf{P}\left\{\left|\frac{\sum_{k=1}^{N_0}(\xi_k^* - \mathbf{E}\xi_k^*)}{\nu}\right| \geqslant \frac{\eta}{3}\right\} \leqslant \frac{\varepsilon}{4}. \tag{4}$$

Finally, since ξ_n^* satisfies the law of large numbers then

$$\mathbf{P}\left\{\max|v_n^*| \geqslant \frac{\eta}{3}; n \geqslant \nu\right\} \leqslant \frac{\varepsilon}{4} \quad \text{for} \quad \nu \geqslant \nu_1(\varepsilon, \eta), \tag{5}$$

where

$$v_n^* = \frac{1}{n}\sum_{k=1}^{n}(\xi_n^* - \mathbf{E}\xi_n^*).$$

From (2), (3) and (4) follows

$$\mathbf{P}\{\max|v_n| \geqslant \eta;\ n \geqslant \nu\} \leqslant \varepsilon$$

for $\nu \geqslant \max(\nu_0, \nu_1, N_0)$, i.e., the random variables ξ_n also satisfy the law of large numbers.

The principal role of the law of large number in probability theory and its applications can hardly be exaggerated. Indeed, imagine for a moment that for a sum of identically distributed random variables with the bounded mathematical expectation the law of large numbers does not hold. Then with the probability arbitrarily close to 1 one can declare that cases will be repeated for which the arithmetical mean is far from the mathematical expectation. And this would take place even if the observations were done without uncertainty (the value of γ discussed in Section 28 equals 0). Would it be possible to say under these circumstances that the arithmetical mean can be taken as an approximate value of the measured parameter? Hardly ever.

36 GLIVENKO'S THEOREM

Let us now consider the Glivenko's Theorem which soon after being proved was called the main theorem of mathematical statistics. The topic is an estimation of the unknown distribution function of a random variable ξ by results of observations. Let the distribution function of this random variable be $F(x)$ and the results of a sequence of independent trials under identical circumstances be

$$x_1, x_2, \dots, x_n. \tag{1}$$

The sequence of the trial's results is ordered. If these values are numbered in correspondence to their increasing, the sequence (1) can be represented as

$$x_1^* \leqslant x_2^* \leqslant \cdots \leqslant x_n^*.$$

Such an ordered sequence is called the *variation sequence*.

The function $F_n(x)$ determined as

$$F_n(x) = \begin{cases} 0 & \text{for } x \leqslant x_1^*, \\ k/n & \text{for } x_k^* < x \leqslant x_{k+1}^*, \\ 1 & \text{for } x > x_n^*. \end{cases}$$

is called the *empirical distribution function*.

It is clear that an empirical distribution function is monotonic, continuous, and has its points of discontinuity only for values of argument coinciding with the values of the variation sequence. The size of a jump (for all different values of the variation sequence) equals $1/n$. Let us emphasize that for each x the ordinate $F_n(x)$ is a random variable with possible meaning equal to one of the following: $0, 1/n, \dots, (n-1)/n$, $n/n = 1$. It is easy to see tat the probability of the equality $F_n(x) = k/n$ equals

$$\mathsf{P}\left\{F_n(x) = \frac{k}{n}\right\} = C_n^k \{F(x)\}^k \{1 - F(x)\}^{n-k}.$$

In the simplest particular case where a random variable ξ can take only a finite number of meanings $a_1, a_2, \dots, a_s$ the terms of the variance sequence will be only these values. If $m_1, m_2, \dots, m_s$ $(m_1 + m_2 + \cdots + m_s = n)$ denote the number of trials where $\xi = a_1$, $\xi = a_2, \dots, \xi = a_n$ then by the law of large numbers these frequencies (for large enough n) will represent approximate values of unknown probabilities $p_1 = \mathbf{P}\{\xi = a_1\}$, $p_2 = \mathbf{P}\{\xi = a_2\}, \ldots, p_n = \mathbf{P}\{\xi = a_n\}$. Moreover, in this case the strict law of large numbers also holds.

Let us before formulation and proof of the theorem introduce several additional statements.

Consider some sequence of random variables $\xi_1, \xi_2, \ldots, \xi_n, \ldots$ The event consisting in convergence of this sequence to some random variable ξ has some certain probability as will be shown in the proof of Lemma 1.

If this probability equals 1, one says that the sequence $\{\xi_n\}$ converges to ξ almost certainly.[2]

In another form this statement can be expressed as follows: The sequence of random variables $\xi_1, \xi_2, \ldots, \xi_n, \ldots$ converges almost certainly to the random variable ξ if with the probability 1 for any integer r there exists such a number n that for all of $k > 0$ the following inequality holds:

$$|\xi_{n+k} < \xi| < 1/r$$

Obviously, the inequality

$$\mathbf{P}\{\xi_n \to \xi\} = 1 \tag{1}$$

can be written in another form:

$$\mathbf{P}\{\xi_n \nrightarrow \xi\} = 0. \tag{2}$$

This expression means that if there can be found such integer r that for any n and at least one k then the probability of the inequality

$$|\xi_{n+k} - \xi| > 1/r$$

equals 0.

LEMMA 1. *If for any integer* $r > 0$

$$\sum_{n=1} \mathbf{P}\{|\xi_n - \xi| \geqslant 1/r\} < +\infty, \tag{3}$$

then (1) takes place, or which is equivalent, (2) takes place.

PROOF: Let E_n^r denote an event that the inequality

$$|\xi_n - \xi| \geqslant 1/r.$$

[2] This concept very close to the concept of convergence almost everywhere in the theory of functions. - *B.V.*

holds. Introduce now the notation

$$S_n^r = \sum_{k=1} E_{n+k}^r.$$

From the inequality

$$\mathbf{P}\{S_n^r\} \leqslant \sum_{k=1} \mathbf{P}\{E_{n+k}^r\} = \sum_{l=n+1} \mathbf{P}\{|\xi_l - \xi| \geqslant 1/r\},$$

using (3) one derives the equality

$$\lim_{n\to\infty} \mathbf{P}\{S_n^r\} = 0. \tag{4}$$

Now let

$$S^r = S_1^r S_2^r S_3^r \cdots$$

Since the event S_n^r follows from S^r then by (4) one obtains

$$\mathbf{P}\{S^r\} = 0. \tag{5}$$

Introduce

$$S = S^1 + S^2 + S^3 + \cdots$$

As one can easily see this event means that such r can be found for any $n (n = 1, 2, 3, \ldots)$ that at least for one k $[k = k(n)]$ the inequality

$$|\xi_{n+k} - \xi| \geqslant 1/r$$

holds. Since

$$\mathbf{P}\{S\} \leqslant \sum_{r=1} \mathbf{P}\{S^r\},$$

then by (5)

$$\mathbf{P}\{S\} = 0,$$

Q.E.D.

LEMMA 2 (BOREL'S THEOREM). *Let μ be the number of an event A occurring in n independent trials in each of which the event A might occur with the probability p. Then for $n \to \infty$*

$$\mathbf{P}\left\{\frac{\mu}{n} \to p\right\} = 1.$$

Note that Borel's Theorem represents the simplest particular case of Kolmogorov's Theorem of the strict law of large numbers. Here another formulation of this particular case is represented different from that in Section 30.

PROOF: Events $\frac{\mu}{n}-p\to 0$ and $(\frac{\mu}{n}-p)^4\to 0$ are equivalent, which is obvious. Introduce auxiliary random variables μ_i equal to the number of occurrence events A in trial i. One finds that

$$\mathbf{E}\left(\frac{\mu}{n}-p\right)^4=\frac{1}{n^4}\sum_{i=1}^{n}\sum_{j=1}^{n}\sum_{k=1}^{n}\sum_{l=1}^{n}\mathbf{E}(\mu_j-p)(\mu_i-p)(\mu_k-p)(\mu_l-p).$$

Elementary calculations show that

$$\mathbf{E}\left(\frac{\mu}{n}-p\right)^4=\frac{pq}{n^4}[n(p^3+q^3)+3pq(n^2-n)]<\frac{1}{4n^2}.$$

By Chebyshev's Lemma

$$\mathbf{P}\left\{\left(\frac{\mu}{n}-p\right)^4\geqslant\frac{1}{r}\right\}\leqslant r\mathbf{E}\left(\frac{\mu}{n}-p\right)^4<\frac{r}{4n^2}.$$

It follows that the sequence

$$\sum_{n=1}^{\infty}\mathbf{P}\left\{\left(\frac{\mu}{n}-p\right)^4\geqslant\frac{1}{r}\right\}.$$

converges. Application of the previous lemma proves the statement above.

LEMMA 3. *If an event E is equivalent to the simultaneous occurrence of the infinite number of events $E_1, E_2, \ldots$, i.e. $E=E_1E_2\cdots$ and each next event E_{n+1} includes a foregoing event E_n then*

$$\mathbf{P}\{E\}=\lim_{n\to\infty}\mathbf{P}\{E_n\}$$

PROOF: Indeed, the event E_1 can be represented in the two following ways as a sum of mutually exclusive events:

$$E_1=E_1\bar{E}_2+E_2\bar{E}_3+\cdots+E_{n-1}\bar{E}_n+E_n$$

and

$$E_1=E_1\bar{E}_2+E_2\bar{E}_3+\cdots+E_{n-1}\bar{E}_n+E_n\bar{E}_{n+1}+\cdots+E.$$

It follows that

$$\mathbf{P}\{E_1\}=\mathbf{P}\{E_1\bar{E}_2\}+\mathbf{P}\{E_2\bar{E}_3\}+\cdots+\mathbf{P}\{E_{n-1}\bar{E}_n\}+\mathbf{P}\{E_n\}$$

and

$$\mathsf{P}\{E_1\} = \mathsf{P}\{E_1 E_2\} + \mathsf{P}\{E_2 E_3\} + \cdots + \mathsf{P}\{E_{n-1} E_n\}$$
$$+ \mathsf{P}\{E_n E_{n+1}\} + \cdots + \mathsf{P}\{E\}.$$

Comparing the two latter equations, one obtains the relationship

$$\mathsf{P}\{E\} = \mathsf{P}\{E_n\} - \sum_{k=n} \mathsf{P}\{E_k E_{k+1}\}.$$

Since the extracted value in the right-hand side of the latter equation is the remainder of the converged sequence, one has

$$\mathsf{P}\{E\} = \lim_{n\to\infty} \mathsf{P}\{E_n\}.$$

LEMMA 4. *If each event of finite or infinite sequence $E_1, E_2, \ldots, E_n, \ldots$ has the probability of occurrence equal to 1, then the probability of their joint occurrence also equals 1.*

PROOF: Consider at first two events, E_1 and E_2 for which

$$\mathsf{P}\{E_1\} = \mathsf{P}\{E_2\} = 1.$$

Since

$$\mathsf{P}\{E_1 + E_2\} = \mathsf{P}\{E_1\} + \mathsf{P}\{E_2\} - \mathsf{P}\{E_1 E_2\}$$

and

$$\mathsf{P}\{E_1 + E_2\} = 1,$$

then

$$\mathsf{P}\{E_1 E_2\} = 1.$$

and $\mathsf{P}\{E_1 + E_2\} = 1$, then $\mathsf{P}\{E_1 E_2\} = 1$. By induction, one writes that for any n events for which

$$\mathsf{P}\{E_1\} = \mathsf{P}\{E_2\} = \cdots = \mathsf{P}\{E_n\} = 1,$$

the equality

$$\mathsf{P}\{E_1 E_2 \cdots E_n\} = 1$$

also holds.

Now consider an infinite sequence of events $E_1, E_2, \ldots, E_n \ldots$ for which

$$\mathbf{P}\{E_2\} = \cdots = \mathbf{P}\{E_n\} = \cdots = 1.$$

It is obvious that

$$E_1 E_2 E_3 \cdots = E_1(E_1 E_2)(E_1 E_2 E_3) \cdots$$

and each next multiplier in the right-hand side of the equality includes a foregoing one. This gives by the latter theorem the condition

$$\mathbf{P}\{E_1 E_2 E_3 \cdots\} = \lim_{n\to} \mathbf{P}\{E_1 E_2 \cdots E_n\}.$$

This equality proves the theorem.

GLIVENKO'S THEOREM. *Let $F(x)$ be the distribution function of a random variable ξ, and $F_n(x)$ be the empirical distribution function of n independent observations of the random variable ξ. Then for $n \to \infty$*

$$\mathbf{P}\left\{ \sup_{- \ <x<\ } |F_n(x) - F(x)| \to 0 \right\} = 1.$$

PROOF: The minimum of x satisfying the inequality

$$F(x-0) = F(x) \leqslant \frac{k}{r} \leqslant F(x+0) \qquad (k = 1, 2, \ldots, r)$$

denote by $x_{r,k}$. Let A denote the event that $\xi < x_{r,k}$. It is clear that

$$\mathbf{P}\{A\} = F(x_{r,k}).$$

Since the frequency of the event A occurrence equals $F_n(x_{r,k})$ then by Borel's Theorem (Lemma 2), one has

$$\mathbf{P}\{F_n(x_{r,k}) \underset{n\to}{\to} F(x_{r,k})\} = 1. \tag{6}$$

Now let E_k^r be the event that for $n \to \infty$

$$F_n(x_{r,k}) \to F(x_{r,k}) \qquad (k = 1, 2, \ldots, r)$$

and

$$E^r = E_1^r E_2^r \cdots E_r^r.$$

It is clear that the event E^r is equivalent to the event that for $n \to \infty$

$$\max_{1 \leqslant k \leqslant r} |F_n(x_{r,k}) - F(x_{r,k})| \to 0.$$

Since by (6)

$$\mathsf{P}\{E_1^r\} = \mathsf{P}\{E_2^r\} = \cdots = \mathsf{P}\{E_r^r\} = 1,$$

then by Lemma 4

$$\mathsf{P}\{E^r\} = 1.$$

Then let

$$E = E^1 E^2 E^3 \cdots$$

By Lemma 4

$$\mathsf{P}\{E\} = 1.$$

Finally, let S denote the event that for $n \to \infty$

$$\sup_{-\infty < x < \infty} |F_n(x) - F(x)| \to 0.$$

For any x such that $x_{r,k} < x < x_{r,k+1}$ the following two equalities

$$F_n(x_{r,k} + 0) \leqslant F_n(x) \leqslant F_n(x_{r,k+1})$$

and

$$F(x_{r,k} + 0) \leqslant F(x) \leqslant F(x_{r,k+1}),$$

hold and, moreover,

$$0 \leqslant F(x_{r,k+1}) - F(x_{r,k} + 0) \leqslant 1/r.$$

From here it follows that

$$F_n(x_{r,k} + 0) - F(x_{r,k+1}) \leqslant F_n(x) - F(x) \leqslant F_n(x_{r,k+1}) - F(x_{r,k} + 0),$$

i.e.,

$$|F_n(x) - F(x)| \leqslant \max_{1 \leqslant k \leqslant r} |F_n(x_{r,k}) - F(x_{r,k})| + 1/r$$

and, consequently,

$$\sup_{-\infty<x<\infty} |F_n(x)-F(x)| \leqslant \max_{1\leqslant k\leqslant r} |F_n(x_{r,k})-F(x_{r,k})| + 1/r.$$

Since r is arbitrary, from the latter inequality it follows that $E \subset S$. This, obviously, proves that

$$\mathbf{P}\left\{\sup_{-\infty<x<\infty} |F_n(x)-F(x)| \to 0\right\} = 1.$$

37 EXERCISES

1. Prove that if the random variable ξ such that $\mathbf{E}e^{a\xi}$ exists ($a>0$ is a constant), then

$$\mathbf{P}\{\xi \geqslant \varepsilon\} \leqslant \frac{\mathbf{E}e^{a\xi}}{e^{a\varepsilon}}.$$

2. Let $f(x)>0$ be a non-decreasing function. Prove that if $\mathbf{E}f(|\xi-\mathbf{E}\xi|)$ exists, then

$$\mathbf{P}\{|\xi-\mathbf{E}\xi| \geqslant \varepsilon\} \leqslant \frac{\mathbf{E}f(|\xi-\mathbf{E}\xi|)}{f(\varepsilon)}.$$

3. A sequence of independent and identically distributed random variables $\{\xi_j\}$ is defined by the equations

(a)

$$\mathbf{P}\{\xi_n = 2^{k-\ln k - 2\ln\ln k}\} = \frac{1}{2^k} \qquad (k=1,2,3,\ldots),$$

(b)

$$\mathbf{P}\{\xi_n = k\} = \frac{c}{k^2 \ln^2 k} \qquad \left(k \geqslant 2, c^{-1} = \sum_{k=2} \frac{1}{k^2 \ln^2 k}\right).$$

Prove that the law of large numbers is applicable to each of the indicated sequences.

4. Prove that the law of large numbers is applicable to the sequence of independent random variables $\{\xi_n\}$ for which

$$\mathbf{P}\{\xi_n = n^\alpha\} = \mathbf{P}\{\xi_n = -n^\alpha\} = 1/2,$$

if and only if $\alpha < 0.5$.

5. Prove that if the independent random variables $\{\xi_n\}$ are such that

$$\max_{1 \leqslant k \leqslant n} \int_{|x| > A} |x|\, dF_k(x) \to 0, \quad \text{when} \quad A \to \infty.$$

then the law of large numbers is applicable to the sequence $\{\xi_n\}$.

6. Using the result of the preceding example, prove that if for a sequence of independent random variables $\{\xi_n\}$ there exist numbers $\alpha > 1$ and β such that $\mathbf{E}|\xi|^\alpha \leqslant \beta$ then the law of large numbers holdsfor the sequence $\{\xi_k\}$ (Markov's Theorem).

7. Given the sequence of random variables $\{\xi_k\}$ for which $\mathbf{D}\xi_n \leqslant C$ and $R_{ij \to 0}$ as $|i-j| \to \infty$. (R_{ij} is the correlation coefficient between ξ_i and ξ_j). Prove that the law of large numbers holds for the given sequence (Bernshtein's Theorem).

Chapter 8

CHARACTERISTIC FUNCTIONS

It was shown in the preceding chapters that the probability theory extensively uses the methods and analytical tools of various branches of mathematical analysis. A simple solution to many problems in probability theory, particularly those connected with sums of independent random variables, can be obtained by means of characteristic functions, the theory of which is developed in mathematical analysis, where it is more familiar as Fourier transformations. The present chapter presents the fundamental properties of characteristic functions.

38 THE DEFINITION AND SIMPLEST PROPERTIES OF CHARACTERISTIC FUNCTIONS

The mathematical expectation of the random variable $e^{it\xi}$ is called the characteristic function[1] of the random variable ξ. If $F(x)$ is the distribution function of the variable ξ, then by Theorem 1 of Section 22 its characteristic function is

$$f(t) = \int e^{itx}\, dF(x). \tag{1}$$

Let us denote a distribution function and the characteristic function that corresponds to it by a capital letter and the corresponding lower case letter, respectively.

From the fact that $|e^{itx}| = 1$ for all real values of t, it follows that the integral (1) exists for all distribution functions. Hence, a characteristic function can be defined for every random variable.

[1] t is a real parameter. The mathematical expectation of a complex random variable $\xi + i\eta$ is defined as $\mathrm{E}\xi + i\mathrm{E}\eta$. It is easy to verify that Theorems 1, 2, and 3 of Section 22 are valid in this case. – *B. G.*

THEOREM 1. *A characteristic function is uniformly continuous over the whole real line and satisfies the following relations:*

$$f(0) = 1, \quad |f(t)| \leq 1 \quad (-\infty < t < \infty). \tag{2}$$

PROOF: The relations (2) follow immediately from the definition of a characteristic function. Indeed, by (1) one has

$$f(0) = \int 1 \cdot dF(x) = 1$$

and

$$|f(t)| = \left|\int e^{itx}\,dF(x)\right| \leq \int |e^{itx}|\,dF(x) = \int dF(x) = 1.$$

It remains to prove the uniform continuity of the function $f(t)$. For this purpose, consider the difference

$$f(t+h) - f(t) = \int e^{itx}(e^{ixh} - 1)\,dF(x)$$

and estimate its absolute value. One has:

$$|f(t+h) - f(t)| \leq \int |e^{ixh} - 1|\,dF(x).$$

Let $\varepsilon > 0$ be an arbitrary quantity. Choose A sufficiently large that

$$\int_{|x|>A} dF(x) < \frac{\varepsilon}{4},$$

and a corresponding h sufficiently small that for $|x| < A$,

$$|e^{ixh} - 1| < \varepsilon/2.$$

Then

$$|f(t+h) - f(t)| \leq \int_{-A}^{A} |e^{ixh} - 1|\,dF(x) + 2\int_{|x|>A} dF(x) \leq \varepsilon.$$

This inequality proves the theorem.

THEOREM 2. *If $\eta = a\xi + b$, where a and b are constants, then*

$$f_\eta(t) = f_\xi(at)e^{ibt},$$

where $f_\eta(t)$ and $f_\xi(t)$ are the characteristic functions of the variables η and ξ.

PROOF: In fact,

$$f_\eta(t) = \mathbf{E}e^{it\eta} = \mathbf{E}e^{it(a\xi+b)} = e^{itb}\mathbf{E}e^{ita\xi} = e^{itb}f_\xi(at).$$

THEOREM 3. *The characteristic function of the sum of two independent random variables equals the product of their characteristic functions.*

PROOF: Let η and ξ be independent random variables and let $\zeta = \eta + \xi$. Then, obviously, $e^{it\xi}$ and $e^{it\eta}$ are also independent random variables. From this it follows that

$$\mathbf{E}e^{it\zeta} = \mathbf{E}e^{it(\xi+\eta)} = \mathbf{E}(e^{it\xi}e^{it\eta}) = \mathbf{E}e^{it\xi}\mathbf{E}e^{it\eta}.$$

This proves the theorem.

COROLLARY. *If*

$$\xi = \xi_1 + \xi_2 + \cdots + \xi_n,$$

where each term is independent of the sum of the preceding ones, then the characteristic function of the variable ξ equals the product of the characteristic functions of the summands.

The application of characteristic functions is essentially based on the property formulated in Theorem 3. As was shown in Section 21, this leads to a very complex operation of the distribution functions of the summands. In terms of characteristic functions, this complex operation is replaced by a simple multiplication of characteristic functions.

THEOREM 4. *If a random variable ξ has an absolute moment of order n, then its characteristic function is differentiable n times and for $k \leqslant n$*

$$f^{(k)}(0) = ki^k\,\mathbf{E}\,\xi^k. \tag{3}$$

PROOF: Formal differentiation of the characteristic function k times ($k \leqslant n$) results in the expression

$$f^{(k)}(t) = i^k \int x^k e^{itx}\,dF(x). \tag{4}$$

But

$$\left|\int x^k e^{itx}\,dF(x)\right| \leqslant \int |x|^k\,dF(x)$$

and, by the hypothesis of the theorem, it is therefore bounded. From this it follows that the integral (4) exists and hence that the differentiation is valid. By setting $t=0$ in (4), one finds:

$$f^{(k)}(0)=i^k\int x^k dF(x).$$

The mathematical expectation and variance can be very simply expressed in terms of the derivatives of the logarithm of the characteristic function. Let us set

$$\psi(t)=\ln f(t).$$

Then

$$\psi'(t)=\frac{f'(t)}{f(t)}$$

and

$$\psi''(t)=\frac{f''(t)\cdot f(t)-|f'(t)|^2}{f^2(t)}.$$

Taking into account equation (3) and the fact that $f(0)=1$, one finds:

$$\psi'(0)=f'(0)=i\,\mathsf{E}\xi$$

and

$$\psi''(0)=f''(0)-[f'(0)]^2=i^2\mathsf{E}\xi^2-[i\,\mathsf{E}\xi]^2=-\mathsf{D}\xi.$$

Hence

and

$$\left.\begin{aligned}\mathsf{E}\xi&=\frac{1}{i}\psi'(0)\\ \mathsf{D}\xi&=-\psi''(0)\end{aligned}\right\}. \tag{5}$$

The k-th derivative of the logarithm of the characteristic function at the point 0, multiplied by i^k, is called the semi-invariant of the k-th order of the random variable.

As follows immediately from Theorem 3, the semi-invariant of a sum of independent random variables equals the sum of the semi-invariants of the individual summands.

It was just shown that the first two semi-invariants are the mathematical expectation and the variance, i.e., the first-order moment and a certain rational function of the moments of first and second order. By carrying out the computation, one can easily satisfy oneself that a semi-invariant of any order k is a rational function of the first k moments. For illustrative purposes, let us consider the semi-invariants of the third and fourth orders:

$$i^3\psi'''(0) = -\{\mathsf{E}\xi^3 - 3\mathsf{E}\xi^3\,\mathsf{E}\xi + 2[\mathsf{E}\xi]^3\},$$

$$i^4\psi^{\mathrm{IV}}(0) = \mathsf{E}\xi^4 - 4\mathsf{E}\xi^3\mathsf{E}\xi - 3[\mathsf{E}\xi^2]^2 + 12\mathsf{E}\xi^2[\mathsf{E}\xi]^2 - 6[\mathsf{E}\xi]^4.$$

Let us now discuss several examples of characteristic functions.

Example 1. A random variable ξ is normally distributed with a mathematical expectation a and variance v^2. The characteristic function of the variable ξ is

$$\varphi(t) = \frac{1}{\sigma\sqrt{2\pi}}\int e^{itx - [(x-a)^2/2\sigma^2]}dx.$$

As a result of the substitution

$$z = \frac{x-a}{\sigma} - it\sigma$$

$\varphi(t)$ reduces to the form

$$\varphi(t) = e^{iat - \sigma^2t^2/2}\frac{1}{\sqrt{2\pi}}\int_{-\infty - it\sigma}^{\infty - it\sigma} e^{-z^2/2}\,dz.$$

Using the well-known fact that for any real value of α

$$\int_{-\infty - i\alpha}^{\infty - i\alpha} e^{-z^2/2}dz = \sqrt{2\pi},$$

one finds:

$$\varphi(t) = e^{iat - \sigma^2t^2\;2}.$$

By use of Theorem 4, one can easily compute the central moments for a normal distribution and in this alternative way obtain the result of the example discussed in Section 26.

Example 2. Find the characteristic function of a random variable ξ distributed according to Poisson's law.

By hypothesis, the variable ξ assumes only integer values with probabilities

$$\mathbf{P}\{\xi = k\} = \frac{\lambda^k e^{-\lambda}}{k!} \qquad (k = 0, 1, 2, \ldots),$$

where $\lambda > 0$ is a constant.

The characteristic function of the variable ξ is

$$f(t) = \mathbf{E}e^{it\xi} = \sum_{k=0} e^{itk}\mathbf{P}\{\xi = k\} = \sum_{k=0} e^{itk}\frac{\lambda^k}{k!}e^{-\lambda}$$

$$= e^{-\lambda}\sum_{k=0}\frac{(\lambda e^{it})^k}{k!} = e^{-\lambda + \lambda e^{it}} = e^{\lambda(e^{it} - 1)}.$$

Hence, according to (5), one finds:

$$\mathbf{E}\xi = \frac{1}{i}\psi'(0) = \lambda; \quad \mathbf{D}\xi = -\psi''(0) = \lambda.$$

These relations we previously obtained directly (Section 23, Example 3).

Example 3. A random variable ξ is uniformly distributed in the interval $(-a, a)$. Its characteristic function is

$$f(x) = \int_{-a}^{a} e^{itx}\frac{dx}{2a} = \frac{\sin at}{at}.$$

Example 4. Find the characteristic function of the variable μ representing the number of times an event A occurs in n independent trials in each of which the probability of occurrence of the event A is p.

The quantity μ may be represented as the sum

$$\mu = \mu_1 + \mu_2 + \cdots + \mu_n$$

of n independent variables each of which assumes only the two values 0 and 1, with probability $q = 1 - p$ and p, respectively. The variable μ_k assumes the value 1 if the event A occurs in the k-th trial and the value 0 if A does not occur in the k-th trial.

The characteristic function of the variable μ_k is given by

$$f_k(t) = \mathbf{E}e^{it\mu_k} = e^{it\cdot 0}q + e^{it\cdot 1}p = q + pe^{it}.$$

According to Theorem 3, the characteristic function of the variable μ is given by

$$f(t) = \prod_{k=1}^{n} f_k(t) = (q + pe^{it})^n.$$

Let us now find the characteristic function of the normalized variable

$$\eta = \frac{\mu - np}{\sqrt{npq}}.$$

By Theorem 2, this characteristic function is

$$f_\eta(t) = e^{-it\sqrt{np/q}} f\left(\frac{t}{\sqrt{npq}}\right) = e^{-it\sqrt{np/q}}(q + pe^{it/\sqrt{npq}})^n$$

$$= (qe^{-it\sqrt{p/nq}} + pe^{-it\sqrt{q/np}})^n.$$

Example 5. A characteristic function satisfies the condition

$$f(-t) = \overline{f(t)}$$

Indeed,

$$f(-t) = \int e^{-itx} dF(x) = \overline{\int e^{itx} dF(x)} = \overline{f(t)}$$

39 THE INVERSION FORMULA AND THE UNIQUENESS THEOREM

It was shown that the characteristic function of a random variable ξ can always be found from its distribution function. It is important that the converse theorem also holds: A distribution function is uniquely determined by its characteristic function.

THEOREM 1. *Let $f(t)$ and $F(x)$ be the characteristic function and distribution function of a random variable ξ, respectively. If x_1 and x_2 are points of continuity of the function $F(x)$, then*

$$F(x_2) - F(x_1) = \frac{1}{2\pi} \lim_{c \to} \int_{-c}^{c} \frac{e^{-itx_1} - e^{-itx_2}}{it} f(t)\,dt. \tag{1}$$

PROOF: From the definition of a characteristic function, it follows that the integral

$$J_c = \frac{1}{2\pi} \int_{-c}^{c} \frac{e^{-itx_1} - e^{-itx_2}}{it} f(t)\,dt$$

equals

$$J_c=\frac{1}{2\pi}\int_{-c}^{c}\int\frac{1}{it}[e^{it(z-x_1)}-e^{it(z-x_2)}]dF(z)dt.$$

The order of integration in the last integral may be changed, since the integral with respect to z converges absolutely and the integral with respect to t has finite limits of integration. Thus

$$J_c=\frac{1}{2\pi}\int\left[\int_{-c}^{c}\frac{e^{it(z-x_1)}-e^{it(z-x_2)}}{it}dt\right]dF(z)$$
$$=\frac{1}{2\pi}\int\left[\int_{0}^{c}\frac{e^{it(z-x_1)}-e^{-it(z-x_1)}-e^{-it(z-x_2)}+e^{-it(z-x_2)}}{it}dt\right]dF(z)$$
$$=\frac{1}{\pi}\int_{-}\int_{0}^{c}\left[\frac{\sin t(z-x_1)}{t}-\frac{\sin t(z-x_2)}{t}\right]dt\,dF(z).$$

From mathematical analysis, one knows that as $c\to\infty$

$$\frac{1}{\pi}\int_{0}^{c}\frac{\sin\alpha t}{t}dt\to\begin{cases}1/2, & \text{if}\quad \alpha>0,\\ -1/2, & \text{if}\quad \alpha<0,\end{cases}\tag{2}$$

and that this convergence is uniform for all c with respect to α for every area $\alpha>\delta>0$ (respectively, $a<-\delta$), and also for $|\alpha|\leqslant\delta$

$$\left|\frac{1}{\pi}\int_{0}^{c}\frac{\sin\alpha t}{t}dt\right|<1.\tag{3}$$

For definiteness, let us suppose that $x_1<x_2$ and let us represent the integral J_c in the form of the following sum:

$$J_c=\int_{-\infty}^{x_1-\delta}+\int_{x_1-\delta}^{x_1+\delta}+\int_{x_1+\delta}^{x_2-\delta}+\int_{x_2-\delta}^{x_2+\delta}+\int_{x_2+\delta}\psi(c,z;x_1,x_2)dF(z),$$

where for brevity the notation

$$\psi(c,z;x_1,x_2)=\frac{1}{\pi}\int_{0}^{c}\left\{\frac{\sin t(z-x_1)}{t}-\frac{\sin t(z-x_2)}{t}\right\}dt$$

is introduced and $\delta>0$ is so chosen that $x_1+\delta<x_2-\delta$.

In the area $-\infty < z - x_1 - \delta$ the inequalities $z - x_1 < -\delta$ and $z - x_2 > -\delta$ hold. Therefore, on the basis of (2), we conclude that

$$\int_{-\infty}^{x_1-\delta} \psi(c, z; x_1, x_2)\, dF(z) \to 0$$

as $c \to \infty$. Analogously, for $x_2 + \delta < x < +\infty$ and $c \to \infty$

$$\int_{x_2+\delta}^{\infty} \psi(c, z; x_1, x_2)\, dF(z) \to 0.$$

Further, since the inequalities $z - x_1 > \delta$ and $z - x_2 < \delta$ hold in the area $x_1 + \delta < z < x_1 - \delta$, it follows from (2) that as $c \to \infty$

$$\int_{x_1+\delta}^{x_2-\delta} \psi(c, z; x_1, x_2)\, dF(z) \to \int_{x_1+\delta}^{x_2-\delta} dF(z) = F(x_2 - \delta) - F(x_1 + \delta).$$

Finally, in view of (3), one can apply the estimates

$$\left| \int_{x_1-\delta}^{x_1+\delta} \psi(c, z; x_1, x_2)\, dF(z) \right| < 2 \int_{x_1-\delta}^{x_1+\delta} dF(z) = 2\,[F(x_1 + \delta) - F(x_1 - \delta\,]$$

and

$$\left| \int_{x_2-\delta}^{x_2+\delta} \psi(c, z; x_1, x_2)\, dF(z) \right| < 2 \int_{x_2-\delta}^{x_2+\delta} dF(z) = 2[F(x_2 + \delta) - F(x_2 - \delta)]$$

Thus we find that for any $\delta > 0$

$$\overline{\lim_{c\to}}\, J_c = F(x_2 - \delta) - F(x_1 + \delta) + R_1(\delta, x_1, x_2)$$

and

$$\underline{\lim_{c\to}}\, J_c = F(x_2 - \delta) - F(x_1 + \delta) + R_2(\delta, x_1, x_2),$$

where

$$|R_i(\delta, x_1, x_2)| < 2\{F(x_1 + \delta) - F(x_1 - \delta) + F(x_2 + \delta) - F(x_2 - \delta)\} \quad (i = 1, 2).$$

Now let $\delta \to 0$. From this and from the fact that the function $F(x)$ is continuous at the points x_1 and x_2 follow the equalities

$$\lim_{\delta \to 0} F(x_1 + \delta) = \lim_{\delta \to 0} F(x_1 - \delta) = F(x_1)$$

and

$$\lim_{\delta \to 0} F(x_2 + \delta) = \lim_{\delta \to 0} F(x_2 - \delta) = F(x_2).$$

And since J_c does not depend on δ, one has:

$$\lim_{c \to \infty} J_c = F(x_2) - F(x_1).$$

The equation (1) is referred to as the inversion formula. Let us now use this formula to prove the following important statement.

THEOREM 2 (The Uniqueness Theorem). *A distribution function is uniquely determined by its characteristic function.*

PROOF: From Theorem 1, it immediately follows that the formula

$$F(x) = \frac{1}{2\pi} \lim_{y \to -\infty} \lim_{c \to \infty} \int_{-c}^{+c} \frac{e^{-ity} - e^{-itx}}{it} f(t)\, dt$$

is applicable at each point of continuity of $F(x)$ where the limit in y is evaluated with respect to any set of points which are points of continuity of the function $F(x)$.

As an application of the last theorem, let us prove the following statement.

Example 1. If the independent random variables ξ_1 and ξ_2 are normally distributed, then their sum $\xi = \xi_1 + \xi_2$ is also normally distributed.

Indeed, if

$$\mathbf{E}\xi_1 = a_1, \quad \mathbf{D}\xi_1 = \sigma_1^2, \quad \mathbf{E}\xi_2 = a_2, \quad \mathbf{D}\xi_2 = \sigma_2^2,$$

then the characteristic functions of the variables ξ_1 and ξ_2 are given by

$$f_1(t) = e^{ia_1 t - 1/2(\sigma_1^2 t^2)}, \quad f_1(t) = e^{ia_2 t - 1/2(\sigma_2^2 t^2)}.$$

By Theorem 3 of Section 32, the characteristic function $f(t)$ of their sum is

$$f(t) = f_1(t) \cdot f_2(t) = e^{it(a_1 + a_2) - 1/2(\sigma_1^2 + \sigma_2^2)t^2}.$$

This is, incidentally, the characteristic function for a normal law with a mathematical expectation $a = a_1 + a_2$ and variance $\sigma = \sigma_1^2 + \sigma_2^2$. On the basis of the Uniqueness Theorem one can conclude that the distribution function of the variable ξ is normal.

Example 2. The independent random variables ξ_1 and ξ_2 are distributed according to Poisson's law, where

$$\mathsf{P}\{\xi_1 = k\} = \frac{\lambda_1^k e^{-\lambda_1}}{k!}, \quad \mathsf{P}\{\xi_2 = k\} = \frac{\lambda_2^k e^{-\lambda_2}}{k!}.$$

Then the random variable $\xi = \xi_1 + \xi_2$ is distributed according to Poisson's law with the parameter $\lambda = \lambda_1 + \lambda_2$.

Indeed, in Example 2 of the preceding section it was found that the characteristic functions of the random variables ξ_1 and ξ_2 are

$$f_1(t) = e^{\lambda_1(e^{it} - 1)}, \quad f_2(t) = e^{\lambda_2(e^{it} - 1)}.$$

By Theorem 3 of the preceding section, the characteristic function of their sum $\lambda = \lambda_1 + \lambda_2$ is

$$f(t) = f_1(t) \cdot f_2(t) = e^{(\lambda_1 + \lambda_2)(e^{it} - 1)},$$

i.e., it is the characteristic function for part of Poisson's law. By the uniqueness theorem, the unique distribution which has $f(t)$ as its characteristic function is the Poisson distribution for which

$$\mathsf{P}\{\xi = k\} = \frac{(\lambda_1 + \lambda_2)^k e^{-(\lambda_1 + \lambda_2)}}{k!} \qquad (k \geqslant 0).$$

D. A. Raikov has proved the deeper converse theorem: If the sum of two independent random variables is distributed according to Poisson's law, then each summand is also distributed according to Poisson's law.

Example 3. A characteristic function is real if and only if its corresponding distribution function is symmetric, i.e., the distribution function satisfies the condition

$$F(x) = 1 - F(-x+0)$$

for all values of x.

If a distribution function is symmetric, then its characteristic function is real. This is proved by the simple computation:

$$\begin{aligned} f(t) &= \int e^{itx} dF(x) \\ &= \int_0^{\infty} e^{-itx} dF(-x+0) + \int_0^{\infty} e^{itx} dF(x) - F(+0) - F(-0) \\ &= \int_0^{\infty} (e^{-itx} + e^{itx})\, dF(x) - F(+0) - F(-0) \\ &= 2\int_0^{\infty} \cos tx\, dF(x) - F(+0) - F(-0) = \int \cos tx\, dF(x). \end{aligned}$$

(Let us recall here that the lower limit is included into the interval of integration and the upper limit is excluded from this interval).

To prove the converse, consider the random variable $\eta = -\xi$. The distribution function for the variable η is

$$G(x) = \mathbf{P}\{\eta < x\} = \mathbf{P}\{\xi > -x\} = 1 - F(-x+0).$$

The characteristic functions of the variables ξ and η are connected by the relation

$$g(t) = \mathbf{E}e^{it\eta} = \mathbf{E}e^{-it\xi} = \overline{\mathbf{E}e^{it\xi}} = \overline{f(t)}.$$

Since by hypothesis $f(t)$ is real, then

$$\overline{f}(t) = f(t)$$

and this implies that $g(t) = f(t)$.

From the uniqueness theorem, one now concludes that the distribution functions of the variables ξ and η coincide, i.e., that

$$F(x) = 1 - F(-x+0),$$

Q.E.D.

40 HELLY'S THEOREMS

Two theorems of a purely analytical nature, the first and second theorems of Helly, will be required in the future.

Let us say that a sequence of non-decreasing functions

$$F_1(x), F_2(x), \ldots, F_n(x), \ldots$$

converges weakly to a non-decreasing function $F(x)$ if for $n \to \infty$ it converges to $F(x)$ at each of its points of continuity.

Henceforth, it will always be assumed that the functions $F_n(x)$ satisfies the supplementary condition

$$F_n(-\infty) = 0,$$

with no further mention of this.

Note immediately that in order for a sequence of functions to be weakly convergent it is sufficient that it converge to the function $F(x)$ on some everywhere-dense set D. Indeed, let x be an arbitrary point, and x' and x' be any two points of the set D such that $x' \leqslant x \leqslant x''$. Then also

$$F_n(x') \leqslant F_n(x) \leqslant F_n(x'').$$

Therefore,

$$\lim_{n\to\infty} F_n(x') \leqslant \varliminf_{n\to\infty} F_n(x) \leqslant \varlimsup_{n\to\infty} F_n(x) \leqslant \lim_{n\to\infty} F_n(x'').$$

And since by assumption

$$\lim_{n\to\infty} F_n(x') = F(x') \quad \text{and} \quad \lim_{n\to\infty} F_n(x'') = F(x''),$$

it also follows that

$$F(x') \leqslant \varliminf_{n\to\infty} F_n(x) \leqslant \varlimsup_{n\to\infty} F_n(x) \leqslant F(x'').$$

But the middle terms in these inequalities are independent of x' and x'', therefore

$$F(x-0) \leqslant \varliminf_{n\to\infty} F_n(x) \leqslant \varlimsup_{n\to\infty} F_n(x) \leqslant F(x+0).$$

If the function $F(x)$ is continuous at the point x, then

$$F(x-0)=F(x)=F(x+0)$$

Consequently, at the points of continuity of the function $F(x)$ one has:

$$\lim_{n\to\infty} F_n(x)=F(x).$$

THE FIRST HELLY'S THEOREM. *Any sequence of non-decreasing functions*

$$F_1(x), F_2(x), \ldots, F_n(x), \ldots \tag{1}$$

which are uniformly bounded contains at least one subsequence

$$F_{n_1}(x), F_{n_2}(x), \ldots, F_{n_k}(x), \ldots,$$

which converges weakly to some non-decreasing function $F(x)$.

PROOF: Let D be some denumerable everywhere-dense set of points $x'_1, x'_2, \ldots, x'_1, \ldots$. Consider the values of the functions of the sequence (1) at the point x_1':

$$F_1(x'_1), F_2(x'_1), \ldots, F_n(x'_1), \ldots$$

Since by assumption the set consisting of these values is bounded, it contains at least one subsequence

$$F_{11}(x'_1), F_{12}(x'_1), \ldots, F_{1n}(x'_1), \ldots, \tag{2}$$

which converges to some limiting value, which will be denoted by $G(x'_1)$. Now consider the set of values

$$F_{11}(x'_2), F_{12}(x'_2), \ldots, F_{1n}(x'_2), \ldots$$

Since this set is also bounded, it contains a subsequence which converges to some limiting value $G(x'_2)$. Thus, one can select from sequence (2) a subsequence

$$F_{21}(x), F_{22}(x), \ldots, F_{2n}(x), \ldots, \tag{3}$$

such that

$$\lim_{n\to\infty} F_{2n}(x_1') = G(x_1')$$

and

$$\lim_{n\to\infty} F_{2n}(x_2') = G(x_2')$$

hold simultaneously.

Continue to separate out such sequences

$$F_{k1}(x), F_{k2}(x), \ldots, F_{kn}(x), \ldots, \tag{4}$$

for which the relations

$$\lim_{n\to\infty} F_{kn}(x_r') = G(x_r')$$

hold simultaneously for all $r \leq k$. Now we consider the diagonal sequence

$$F_{11}(x), F_{22}(x), \ldots, F_{nn}(x), \ldots \tag{5}$$

One can note that this entire sequence has been selected from the sequence (2) and therefore

$$\lim_{n\to\infty} F_{nn}(x_1') = G(x_1').$$

Further, since the entire diagonal sequence, with the exception of the first term only, has been selected from the sequence (3),

$$\lim_{n\to\infty} F_{nn}(x_2') = G(x_2').$$

In general, the entire diagonal sequence with the exception of the first $k-1$ terms has been selected from the sequence (4), therefore

$$\lim_{n\to\infty} F_{nn}(x_k') = G(x_k').$$

for each value of k. The above result may be formulated as follows: The sequence (1) contains at least one subsequence converging at all the points x'_k of the set D to some function $G(x)$ defined on this set. Additionally, since the functions $F_{nn}(x)$ are non-decreasing and uniformly bounded, it is evident that the function $G(x)$ will also be non-decreasing and bounded.

It is now clear that the function $G(x)$ which is defined on the set D may be continued in such a way that it will be defined over the entire interval $-\vdash < x < +\infty$ while remaining non-decreasing and bounded.

The sequence (5) converges to this function on the everywhere-dense set D. Consequently, as proved above, it converges weakly to this function, Q.E.D.

We note that the function obtained by continuing the function G may prove not to be continuous from the left. But one can alter its values at the points of discontinuity so as to reestablish this property. The subsequence F_{nn} will converge weakly to the function thus "corrected".

THE SECOND HELLY'S THEOREM. *Let $f(x)$ be a continuous function and let the sequence of non-decreasing uniformly bounded functions*

$$F_1(x), F_2(x), \ldots, F_n(x), \ldots$$

converge weakly to the function $F(x)$ on some finite interval $a \leqslant x \leqslant b$ where a and b are points of continuity of the function $F(x)$. Then

$$\lim_{n\to\infty} \int_a^b f(x)dF_n(x) = \int_a^b f(x)dF(x).$$

PROOF: From the continuity of the function $f(x)$, it follows that given any positive constant ε no matter how small, one can find a subdivision of the interval $a \leqslant x \leqslant b$ by means of points $x_0 = a$, $x_1, \ldots x_n = b$ into the subintervals (x_k, x_{k+1}) such that in each of them the inequality $|f(x)-f(x_k)<\varepsilon$ holds. Using this fact, one may introduce an auxiliary function $f_\varepsilon(x)$ which assumes only a finite number of values and is defined by means of the equations

$$f_\varepsilon(x) = f(x_k) \quad \text{for} \quad x_k \leqslant x \leqslant x_{k+1}.$$

It is clear that the inequality

$$|f(x) - f_\varepsilon(x)| < \varepsilon$$

holds for all values of x in the interval $a \leqslant x \leqslant b$. One may select the points of division $x_1, x_2, \ldots, x_{N-1}$ beforehand so that they are the points of continuity of the function $F(x)$. Since the sequence of functions $F_1(x), F_2(x), F_3(x), \ldots$ converges to the function $F(x)$ the inequality

$$|F(x_k) - F_n(x_k)| < \frac{\varepsilon}{MN} \tag{6}$$

will be satisfied at all the points of subdivision for n sufficiently large, where M is the maximum of the absolute value of $f(x)$ in the interval $a \leqslant x \leqslant b$.

No explanations are needed for the following inequality

$$\left|\int_a^b f(x)dF(x) - \int_a^b f(x)dF_n(x)\right| \leqslant \left|\int_a^b f(x)dF(x) - \int_a^b f_\varepsilon(x)dF(x)\right| + \left|\int_a^b f_\varepsilon(x)dF(x) - \int_a^b f_\varepsilon(x)dF_n(x)\right| + \left|\int_a^b f_\varepsilon(x)dF_n(x) - \int_a^b f_\varepsilon(x)dF_n(x)\right|.$$

It is not difficult to calculate that the first term on the right-hand side does not exceed $\varepsilon[F(b) - F(a)]$ and that the third term does not exceed $\varepsilon[F_n(b) - F_n(a)]$. As regards the second term, it equals

$$\left|\sum_{k=0}^{N-1} f(x_k)[F(x_{k+1}) - F(x_k)] - \sum_{k=0}^{N-1} f(x_k)[F_n(x_{k+1}) - F_n(x_k)]\right|$$
$$= \left|\sum_{k=1}^{N-1} f(x_k)[F(x_{k+1}) - F_n(x_{k+1})] - \sum_{k=0}^{N-1} f(x_k)[F(x_k) - F_n(x_k)]\right|$$

and therefore, for n sufficiently large, does not exceed 2ε as follows from the inequality (6). Since the functions $F_n(x)$ are uniformly bounded, the sum

$$\varepsilon[F(b) - F(a)] + \varepsilon[F_n(b) - F_n(a)] + 2\varepsilon$$

can be made arbitrarily small by choosing ε sufficiently small.

THE GENERALIZED SECOND THEOREM OF HELLY. *If the function $f(x)$ is continuous and bounded over the entire real line*

$-\infty < x < +\infty$, *the sequence of non-decreasing uniformly bounded functions*

$$F_1(x), F_2(x), \ldots, F_n(x), \ldots$$

converges weakly to the function $F(x)$, *and*

$$\lim_{n\to\infty} F_n(-\infty) = F(-\infty), \quad \lim_{n\to\infty} F_n(+\infty) = F(+\infty),$$

then

$$\lim_{n\to\infty} \int f(x)dF_n(x) = \int f(x)dF(x).$$

PROOF: Let $A < 0$ and $B > 0$. Set

$$J_1 = \left| \int_{-\infty}^{A} f(x)dF(x) - \int_{-\infty}^{A} f(x)dF_n(x) \right|,$$

$$J_2 = \left| \int_{A}^{B} f(x)dF(x) - \int_{A}^{B} f(x)dF_n(x) \right|,$$

$$J_3 = \left| \int_{B}^{\infty} f(x)dF(x) - \int_{B}^{\infty} f(x)dF_n(x) \right|.$$

It is clear that

$$\left| \int f(x)dF(x) = \int f(x)dF_n(x) \right| \leqslant J_1 + J_2 + J_3.$$

The quantities J_1 and J_3 can be made arbitrarily small by choosing A and B sufficiently large in absolute value and moreover such that A and B are points of continuity of the function $F(x)$ and by choosing n sufficiently large. In fact, if M is the upper bound of $|f(x)|$ for $-\infty < x < +\infty$, then

$$J_1 \leqslant M[F(A) + F_n(A)],$$

$$J_3 \leqslant M[F(+\infty) - F(B)] + M[F_n(+\infty) - F_n(B)].$$

But

$$\lim_{A\to-\infty} F(A) = 0, \quad \lim_{B\to-\infty} F(B) = F(+\infty).$$

And since by assumption,

$$\lim_{n\to\infty} F_n(A) = F(A), \quad \lim_{n\to\infty} F_n(B) = F(B),$$

the assertion concerning J_1 and J_3 is proved. The quantity J_2 can be made arbitrarily small for n sufficiently large by virtue of Helly's Theorem for a finite interval. The theorem is thus proved.

41 LIMIT THEOREMS FOR CHARACTERISTIC FUNCTIONS

Two limit theorems, direct and converse, are most important for applying characteristic functions in deriving asymptotic formulas in probability theory. These theorems establish that the correspondence existing between distribution functions and characteristic functions is not only one-to-one but also continuous.

THE DIRECT LIMIT THEOREM. *If the sequence of distribution functions*

$$F_1(x), F_2(x), \ldots, F_n(x), \ldots$$

converges weakly to the distribution function $F(x)$, *then the sequence of characteristic functions*

$$f_1(t), f_2(t), \ldots, f_n(t), \ldots$$

converges to the characteristic function $f(t)$. *The convergence is uniform with respect to* t *in every finite interval.*

PROOF: Since

$$f_n(t) = \int e^{itx} dF_n(x), \quad f(t) = \int e^{itx} dF(x)$$

and the function e^{itx} is continuous and bounded in all axes $-\infty < t < +\infty$ then by the Generalized Helly Theorem, as $n \to \infty$

$$f_n(t) \to f(t).$$

The assertion that this convergence is uniform with respect to t in every finite interval can be verified by using literally the same reasoning as in the proof of Helly's Second Theorem.

THE CONVERSE LIMIT THEOREM. *If the sequence of characteristic functions*

$$f_1(t), f_2(t), \dots, f_n(t), \dots \tag{1}$$

converges to a continues function $f(t)$, *then the sequence of distribution functions*

$$F_1(x), F_2(x), \dots, F_n(x), \dots \tag{2}$$

converges weakly to some distribution function $F(x)$. (*In virtue of the direct limit theorem,* $f(t) = \int e^{itx} dF(x)$.)

PROOF: By Helly's First Theorem, we can conclude that the sequence (2) certainly contains a subsequence

$$F_{n_1}(x), F_{n_2}(x), \dots, F_{n_k}(x), \dots, \tag{3}$$

which converges weakly to some non-decreasing function $F(x)$. It is clear that one can consider the function $F(x)$ as continuous from the left:

$$\lim_{x' \to x-0} F(x') = F(x).$$

Generally speaking, the function $F(x)$ need not be a distribution function, since for this the conditions $F(-\infty) = 0$ and $F(+\infty) = 1$ also have to be satisfied. In fact, the limit of the sequence of functions

$$F_n(x) = \begin{cases} 0 & \text{for} \quad x \leqslant -n, \\ 1/2 & \text{for} \quad -n < x \leqslant n, \\ 1 & \text{for} \quad x > n, \end{cases}$$

is the function $F(x) \equiv 1/2$ and therefore both $F(-\infty)$ and $F(+\infty)$ are also equal to 1/2. However, under the conditions of the theorem, as will be shown, there are $F(-\infty) = 0$ and $F(+\infty) = 1$ almost surely.

Indeed, if this were not so, then taking into account that for the limit function $F(x)$ the relations $F(-\infty) \geqslant 0$ and $F(+\infty) \leqslant 1$ have to hold, one would have

$$\delta = F(+\infty) - F(-\infty) < 1.$$

Let us now take some positive number ε less than $1 - \delta$. Since by the hypothesis of the theorem, the sequence of characteristic functions (1) converges to the function $f(t)$, it follows that $f(0)=1$. But since in addition to this the function $f(t)$ is continuous, it is possible to choose a positive number τ so small that the inequality

$$\frac{1}{2\tau}\left|\int_{-\tau}^{\tau} f(t)\,dt\right| > 1-\varepsilon/2 > \delta+\varepsilon/2. \tag{4}$$

holds. But at the same time, one may choose $X > 4/\tau\varepsilon$ and K so large that for $k > K$

$$\delta_k = F_{n_k}(X) - F_{n_k}(-X) < \delta + \varepsilon/4.$$

Since $f_{n_k}(t)$ is a characteristic function, it follows that

$$\int_{-\tau}^{\tau} f_{n_k}(t)\,dt = \int\left[\int_{-\tau}^{\tau} e^{itx}\,dt\right]dF_{n_k}(x).$$

The integral on the right-hand side of this equation may be estimated in the following manner. On the one hand, since $|e^{itx}|=1$ one has

$$\left|\int_{-\tau}^{\tau} e^{itx}\,dt\right| \leqslant 2\tau.$$

On the other hand,

$$\int_{-\tau}^{\tau} e^{itx}\,dt = \frac{2}{x}\sin\tau x,$$

and since $|\sin\tau x| \leqslant 1$, we have for $|x| > X$

$$\left|\int_{-\tau}^{\tau} e^{itx}\,dt\right| < \frac{2}{X}.$$

Hence, using the first estimate for $|x| \leqslant X$ and the second for $|x| > X$ one obtains:

$$\left|\int_{-\tau}^{\tau} f_{n_k}(t)\,dt\right| \leqslant \left|\int_{|x|\leqslant X}\left(\int_{-\tau}^{\tau} e^{itx}\,dt\right)dF_{n_k}(x)\right| + \left|\int_{|x|>X}\left(\int_{-\tau}^{\tau} e^{itx}\,dt\right)dF_{n_k}(x)\right| < 2\tau\delta_k + \frac{2}{X}$$

and therefore

$$\frac{1}{2\tau}\left|\int_{-\tau}^{\tau} f_{n_k}(t)dt\right| < \delta + \frac{\varepsilon}{2}.$$

This inequality continues to hold in the limit:

$$\frac{1}{2\tau}\left|\int_{-\tau}^{\tau} f(t)dt\right| \leqslant \delta + \frac{\varepsilon}{2},$$

and this, evidently, contradicts inequality (4).

Thus, the function $F(x)$ to which the sequence of functions $F_{n_k}(x)$ converges weakly is a distribution function and by the direct limit theorem, its characteristic function is $f(t)$. In order to complete the proof of the theorem, it remains to show that the sequence (2) itself converges weakly to the function $F(x)$. Suppose the contrary. Then one could find a subsequence of functions

$$F_{n'_1}(x), F_{n'_2}(x), \ldots, F_{n'_k}(x), \ldots, \tag{5}$$

converging weakly to some function $F^*(x)$ which would differ from $F(x)$ in at least one of its points of continuity. According to what already has been proved, $F^*(x)$ has to be a distribution function with the characteristic function $f(t)$. By the uniqueness theorem, one must have

$$F^*(x) = F(x).$$

This contradicts the assumption.

Note that the conditions of the theorem are fulfilled in each of the following two cases:

(1) The sequence of characteristic functions $f_n(t)$ converges uniformly to some function $f(t)$ in every finite interval of t.

(2) The sequence of characteristic functions $f_n(t)$ converges to a characteristic function $f(t)$.

Example. As an example of using the limit theorems let us consider the proof of the DeMoivre-Laplace Theorem.

In Example 4 of Section 32 the characteristic function

$$f_n(t) = (qe^{-it\sqrt{p/nq}} + pe^{it\sqrt{q/np}})^n$$

of the random variable

$$\eta = \frac{\mu - np}{\sqrt{npq}}$$

was found.

By employing the MacLaurin expansion, one finds:

$$qe^{-it\sqrt{p/nq}} + pe^{it\sqrt{q/np}} = 1 - \frac{t^2}{2n}(1 + R_n),$$

where

$$R_n = 2\sum_{k=3}^{\infty} \frac{1}{k!}\left(\frac{it}{\sqrt{n}}\right)^{k-2} \frac{pq^k + q(-p)^k}{\sqrt{(pq)^k}}.$$

Since $R_n \to 0$ as $n \to \infty$, then

$$f_n(t) = \left[1 - \frac{t^2}{2n}(1 + R_n)\right]^n \to e^{-t^2/2}.$$

From the converse limit theorem follows that for any x

$$\mathsf{P}\left\{\frac{\mu - np}{\sqrt{npq}} < x\right\} \to \frac{1}{\sqrt{2\pi}} \int_{-\infty}^{x} e^{-z^2/2} \, dz.$$

as $n \to \infty$.

From the continuity of the limit function one easily deduces that the convergence is uniform in x.

42 POSITIVE-SEMIDEFINITE FUNCTIONS

The purpose of the present section is to give an exhaustive description of the class of characteristic functions. The fundamental theorem which is quoted below was simultaneously proved by Khinchine and Bochner and first published by Bochner.

To formulate and prove this theorem, one has to introduce a new notion. Let us say that a continuous function $f(t)$ of the real argument t is positive-semidefinite in the interval $-\infty < t < +\infty$ if for any real numbers $t_1, t_2, \ldots, t_m$, complex numbers $\xi_1, \xi_2, \ldots, \xi_m$ and integer n

$$\sum_{k=1}^{n}\sum_{j=1}^{n} f(t_j-t_k)\xi_j\bar{\xi}_k \geqslant 0. \tag{1}$$

Let us enumerate several of the simplest properties of positive-semidefinite functions.

1. $f(0)\geqslant 0$. In fact, if to set $n=1$, $t_1=0$ and $\xi_1=1$, then from the condition for the positive-semidefiniteness of the function $f(t)$, one finds:

$$\sum_{k=1}^{n}\sum_{j=1}^{n} f(t_k-t_j)\xi_k\bar{\xi}_j = f(0)\geqslant 0.$$

2. For any real value of t

$$f(-t)=\overline{f(t)}.$$

In order to prove this, one takes in (1) the values $n=2$, $t_1=0$, $t_2=t$, and leave ξ_1 and ξ_2 arbitrary. We have by assumption

$$\begin{aligned}0\leqslant \sum_{k=1}^{2}\sum_{j=1}^{2} f(t_k-t_j)\xi_k\bar{\xi}_j &= f(0-0)\xi_1\bar{\xi}_1+f(0-t)\xi_1\bar{\xi}_2\\ &\quad+f(t-0)\xi_2\bar{\xi}_1+f(t-t)\xi_2\bar{\xi}_2\\ &=f(0)(|\xi_1|^2+|\xi_2|^2)+f(-t)\xi_1\bar{\xi}_2+f(t)\bar{\xi}_1\xi_2,\end{aligned}$$

and therefore, the quantity

$$f(-t)\xi_1\bar{\xi}_2+f(t)\bar{\xi}_1\xi_2$$

has to be real. Thus, if one set $f(-t)=\alpha_1+i\beta_1, f(t)=\alpha_2+i\beta_2$,

$$\xi_1\bar{\xi}_2=\gamma+i\delta, \qquad \text{and} \qquad \bar{\xi}_1\xi_2=\gamma-i\delta$$

then it must be that

$$\alpha_1\delta+\beta_1\gamma-\alpha_2\delta+\beta_2\gamma=0.$$

Since ξ_1 and ξ_2 and, consequently, γ and δ, are arbitrary, it follows that

$$\alpha_1-\alpha_2=0, \quad \beta_1+\beta_2=0.$$

This implies the above assertion.

3. For any real value of t

$$|f(t)| \leqslant f(0).$$

Let us suppose that in inequality (2) $\xi_1 = f(t)$ and $\xi_2 = -|f(t)|$. Then by the preceding result,

$$2f(0)|f(t)|^2 - |f(t)|^2|f(t)| - |f(t)|^2|f(t)| \geqslant 0.$$

It gives for $|f(t)| \neq 0$

$$f(0) \geqslant |f(t)|.$$

If $|f(t)| = 0$ then again by property 1

$$f(0) \geqslant |f(t)|.$$

Incidentally, it follows that if a positive-semidefinite function is such that $f(0) = 0$ then $f(t) \neq 0$.

THE BOCHNER-KHINCHINE THEOREM. *In order for a continuous function $f(t)$ which satisfies the condition $f(0) = 1$ to be a characteristic function, it is necessary and sufficient that it be positivesemidefinite.*

PROOF: In one direction, the theorem is trivial. In fact, if

$$f(t) = \int e^{ixt}\, dF(x),$$

where $F(x)$ is some distribution function, then if n is any integer, $t_1, t_2, \ldots, t_n$ are arbitrary real numbers, and $\xi_1, \xi_2, \ldots, \xi_n$ are arbitrary complex numbers then

$$\sum_{k=1}^{n} \sum_{j=1}^{n} f(t_k - t_j)\xi_k \bar{\xi}_j = \sum_{k=1}^{n} \sum_{j=1}^{n} \left\{ \int e^{ix(t_k - t_j)}\, dF(x) \right\} \xi_k \bar{\xi}_j$$

$$= \int \sum_{k=1}^{n} \sum_{j=1}^{n} e^{ix(t_k - t_j)} \xi_k \bar{\xi}_j\, dF(x)$$

$$= \int \left(\sum_{k=1}^{n} e^{it_k x} \xi_k \right) \left(\sum_{j=1}^{n} e^{-it_j x} \bar{\xi}_j \right) dF(x)$$

$$= \int \left| \sum_{k=1}^{n} e^{it_k x} \xi_k \right|^2 dF(x) \geqslant 0.$$

The proof of sufficiency requires a more complicated argument.

Consider a sequence of numbers $f(k/n)$ depending on an integer parameter n. Since $f(x)$ is positive-semidefinite, for any N one has

$$\mathcal{P}_N^{(n)}(x) = \frac{1}{N}\sum_{k=0}^{N-1}\sum_{j=0}^{N-1} f\left(\frac{k-j}{n}\right)e^{-i(k-j)x} \geqslant 0.$$

It is easy to count that in this sum there are $N-|r|$ summands for which the difference $k-j$ equals r. Then, evidently, the number r can change from $-N+1$ up to $N-1$. Thus the equality

$$\mathcal{P}_N^{(n)}(x) = \sum_{r=-N}^{N}\left(1-\frac{|r|}{N}\right)f\left(\frac{r}{n}\right)e^{-irx}.$$

holds. Multiply both sides of the above equality by e^{isx} and integrate by x from $-\pi$ to π:

$$\int_{-\pi}^{\pi} e^{isx}\mathcal{P}_N^{(n)}(x)\,dx = \sum_{r=-N}^{N}\left(1-\frac{|r|}{N}\right)f\left(\frac{r}{n}\right)\int_{-\pi}^{\pi} e^{-irx}e^{isx}\,dx.$$

It is known that

$$\int_{-\pi}^{\pi} e^{-i(r-s)x}\,dx = \begin{cases} 0 & \text{for } r\neq s. \\ 2\pi & \text{for } r=s. \end{cases}$$

Therefore

$$\left(1-\frac{|s|}{N}\right)f\left(\frac{s}{n}\right) = \frac{1}{2\pi}\int_{-\pi}^{\pi}\mathcal{P}_N^{(n)}(x)e^{isx}\,dx = \int_{-\pi}^{\pi} e^{isx}\,dF_N^{(n)}(x),$$

where

$$F_N^{(n)}(x) = \frac{1}{2\pi}\int_{-\pi}^{x}\mathcal{P}_N^{(n)}(x)\,dx$$

is a non-decreasing function with

$$F_N^{(n)}(\pi) = \frac{1}{2\pi}\int_{-\pi}^{\pi}\mathcal{P}_N^{(n)}(x)\,dx = f(0) = 1,$$

i.e., it is a distribution function.

By the First Helly Theorem we can find a sequence $n_k \to \infty$ as $k\to\infty$ for which functions $F_N^{(n)}(x)$ converge to the limit function which is denoted by $F^{(n)}(x)$ (n is fixed!). The function $F^{(n)}(x)$ is a distribution function because $F_N^{(n)}(-\pi-\varepsilon)=0$ and $F_N^{(n)}(\pi+\varepsilon)=1$ for any N and arbitrary $\varepsilon > 0$. It also means that for arbitrary $\varepsilon > 0$

$$F^{(n)}(-\pi-\varepsilon)=0, \quad F^{(n)}(\pi-\varepsilon)=1.$$

By the Second Helly Theorem

$$\lim_{k\to} \int_{-\pi}^{\pi} e^{isx}\,dF_{N_k}^{(n)}(x)=\int_{-\pi}^{\pi} e^{isx}\,dF^{(n)}(x).$$

Thus[2]

$$f\left(\frac{s}{n}\right)=\int_{-\pi}^{\pi} e^{isx}\,dF^{(n)}(x)$$

for all integer s ($s=0, \pm 1, \pm 2, \ldots$).

Let us now consider a sequence of characteristic functions $f_n(t)$ which are defined by means of the equality

$$f_n(t)=\int_{-\pi n}^{\pi n} e^{itx}\,dF_n(x),$$

where

$$F_n(x)=F^{(n)}\left(\frac{x}{n}\right).$$

It is easy to check that for all of integer k

$$f_n\left(\frac{k}{n}\right)=f\left(\frac{k}{n}\right). \tag{3}$$

---

[2]Note that, in passing, the following Herglots' Theorem was proven. If the sequence of numbers C_n ($n=0, \pm 1, \ldots,$) is such that for arbitrarily chosen complex numbers $\xi_1, \xi_2, \ldots, \xi_N$ chosen and arbitrary N

$$\sum_{k=1}^{N}\sum_{j=1}^{N} c_{k-j}\xi_k\xi_j \geq 0,$$

then the sequence of c_n can be written in the form

$$c_n=\int_{-\pi}^{\pi} e^{inx}\,d\sigma(x)$$

where $\sigma(x)$ is non-decreasing function with a bounded variation.

But for any t one can choose such a sequence $k = k(n, t)$[3] that

$$0 \leqslant t - \frac{k}{n} < \frac{1}{n}.$$

From continuity of the function $f(t)$ it follows that

$$f(t) = \lim_{n\to\infty} f\left(\frac{k}{n}\right) = \lim_{n\to\infty} f_n\left(\frac{k}{n}\right). \tag{4}$$

If one proves that for all real t

$$f(t) = \lim_{n\to\infty} f_n(t), \tag{5}$$

then the proof of the theorem will have been completed since $f(t)$ is a continuous function and therefore by the Converse Limit Theorem for characteristic functions, it is a characteristic function.

For this purpose, note that from (3) and (4) follows the equality

$$\lim_{n\to\infty} f_n(t) = \lim_{n\to\infty} \left\{\left[f_n(t) - f_n\left(\frac{k}{n}\right)\right] + f_n\left(\frac{k}{n}\right)\right\}$$

$$= f(t) + \lim_{n\to\infty} \left[f_n(t) - f_n\left(\frac{k}{n}\right)\right]. \tag{6}$$

Denote $\eta = t - (k/n)$. According to the choice of numbers k one has $0 \leqslant \eta < (1/n)$. By the definition of the function $f_n(t)$ one has

$$\left|f_n(t) - f_n\left(\frac{k}{n}\right)\right| = \left|\int_{-\pi n}^{\pi n} e^{i(k/n)x}(e^{i\eta x} - 1)\, dF_n(x)\right|$$

$$\leqslant \int_{-\pi n}^{\pi n} |e^{i\eta x} - 1|\, dF_n(x). \tag{7}$$

---

[3]Everywhere below k is understood as numbers $k(n, t)$.

Using the Cauchy-Bouniakowsky Inequality, one finds that

$$\int_{-\pi n}^{\pi n} |e^{i\theta x} - 1|\, dF_n(x) \leqslant \sqrt{\int_{-\pi n}^{\pi n} |e^{i\theta x} - 1|^2 dF_n(x)}$$

$$= \sqrt{\int_{-\pi n}^{\pi n} 2(1 - \cos\theta x)\, dF_n(x)}$$

$$= \sqrt{2(1 - Rf_n(\theta))}, \tag{8}$$

where a symbol $Rf_n(\eta)$ denotes real part of $f_n(\eta)$. Since $\cos z \leqslant \cos \alpha z$ for $0 \leqslant \alpha < 1$ and $-\pi \leqslant z < \pi$ then

$$1 - Rf_n(\theta) = \int_{-\pi n}^{\pi n} (1 - \cos\theta x)\, dF_n(x) = \int_{-\pi}^{\pi} (1 - \cos\theta \cdot nz)\, dF_n(zn)$$

$$\leqslant \int_{-\pi}^{\pi} (1 - \cos z)\, dF_n(zn).$$

And since

$$F_n(zn) = F^{(n)}(z),$$

it follows

$$1 - Rf_n(\theta) \leqslant \int_{-\pi}^{\pi} (1 - \cos z)\, dF^{(n)}(z) = 1 - R\int_{-\pi}^{\pi} e^{iz}\, dF^{(n)}(z).$$

By (3) one finds that

$$1 - Rf_n(\theta) \leqslant 1 - Rf\left(\frac{1}{n}\right). \tag{9}$$

The inequalities (7), (8), and (9) together allow us to write

$$\left|f_n(t) - f_n\left(\frac{k}{n}\right)\right| \leqslant \sqrt{2\left(1 - Rf\left(\frac{1}{n}\right)\right)}.$$

From the continuity of the function $f(t)$ it follows now that

$$\lim_{n\to\infty}\left[f_n(t) - f_n\left(\frac{k}{n}\right)\right] = 0.$$

Thus the relation (5), as it is clear now from (6), has been proved.

43 CHARACTERISTIC FUNCTIONS OF MULTI-DIMENSIONAL RANDOM VARIABLES

In the present section, the basic properties of characteristic functions of multi-dimensional random variables are presented without proof.

The characteristic function of an n-dimensional random variable $(\xi_1, \xi_2, \ldots, \xi_n)$ is defined as the mathematical expectation of the variable

$$e^{i(t_1\xi_1 + t_2\xi_2 + \cdots + t_n\xi_n)}$$

where $t_1, t_2, \ldots, t_n$ are real variables:

$$f(t_1, t_2, \ldots, t_n) = \mathsf{E} \exp\left(i \sum_{k=1}^{n} t_k \xi_k\right). \tag{1}$$

If $F(x_1, x_2, \ldots, x_n)$ is the distribution function of the variable $(\xi_1, \xi_2, \ldots, \xi_n)$ then as is known from our discussions[4]

$$f(t_1, t_2, \ldots, t_n) = \int \ldots \int \left(\exp i \sum_{k=1}^{n} t_k x_k\right) dF(x_1, \ldots, x_n). \tag{2}$$

Analogously to the one-dimensional case, the characteristic function of an n-dimensional random variable is uniformly continuous over the entire space $(-\infty < t_j < +\infty,\ 1 \leqslant j \leqslant n)$ and satisfies the following relations

$$f(0, 0, \ldots, 0) = 1,$$

$$|f(t_1, t_2, \ldots, t_n)| \leqslant 1 \, (-\infty < t_k < +\infty,\ k = 1, 2, \ldots),$$

$$f(-t_1, -t_2, \ldots, -t_n) = \overline{f(t_1, t_2, \ldots, t_n)}.$$

From the characteristic function $f(t_1, t_2, \ldots, t_n)$ of the random variable $(\xi_1, \xi_2, \ldots, \xi_n)$ it is easy to find the characteristic function of any k-dimensional $(k < n)$ variable $(\xi_{j1}, \xi_{j2}, \ldots, \xi_{jk})$ whose components are the variables ξ_s $(1 < s < n)$. For this, it is necessary to set all the arguments t_s for $s \neq j_r$ $(1 \leqslant r \leqslant k)$ equal to zero in expression (2). Thus, for

[4] See Theorem 1 of Section 24 and the remark concerning multi-dimensional Stieltjes integrals in Section 23.

example, the characteristic function of the variable ξ_1 equals

$$f_1(t_1)=f(t_1, 0,\dots,0).$$

If the components of the variable $(\xi_1, \xi_2,\dots,\xi_n)$ are independent random variables, then from the definition it follows that its characteristic function equals the product of the characteristic functions of the components

$$f(t_1, t_2,\dots, t_n)=f(t_1)\cdot f(t_2)\dots f(t_n).$$

Just as in the one-dimensional case, the moments of different orders can readily be found from the multi-dimensional characteristic functions.

Thus, for example,

$$\mathsf{E}\xi_1^{k_1}\xi_2^{k_2}\cdots\xi_n^{k_n}=\iint\dots\int x_1^{k_1}x_2^{k_2}\cdots x_n^{k_n}\,dF(x_1,x_2,\dots,x_n)$$

$$=(i)^{\Sigma_j k_j}\left[\frac{\partial^{k_1+k_2+\cdots+k_n}f(t_1,t_2,\dots,t_n)}{\partial_{t_1}^{k_1}\partial_{t_2}^{t_2}\cdots\partial_{t_n}^{k_n}}\right]_{t_1=t_2=\cdots=t_n=0}$$

To compute a characteristic function, it is useful to know the following theorem, which the reader will have no difficulty in proving.

THEOREM 1. *If $f(t_1, t_2,\dots, t_n)$ is the characteristic function of the variable $(\xi_1,\xi_2,\dots,\xi_n)$ then the characteristic function of the variable $(\sigma_1\xi_1+a_1, \sigma_2\xi_2+a_2\cdots\sigma_n\xi_n+a_n)$ where α_1 and σ_i $(1\leqslant i\leqslant n)$ are real constants is*

$$\exp\left(i\sum_{n=1}^{n}a_k t_k\right)\cdot f(\sigma_1 t_1,\sigma_2 t_2,\dots,\sigma_n t_n).$$

Example 1. Let us compute the characteristic function of a two-dimensional random variable which is distributed according to the normal law:

$$p(x, y)=\frac{1}{2\pi(1-r^2)}\exp\left\{-\frac{1}{2(1-r^2)}[x^2-2rxy+y^2]\right\}. \qquad (3)$$

By formula (2)

$$f(t_1, t_2) = \frac{1}{2\pi(1-r^2)} \iint e^{i(t_1 x + t_2 y)} p(x, y)\, dx\, dy.$$

By changing variables, one can reduce $f(t_1, t_2)$ to the form

$$f(t_1, t_2) = e^{-(1/2)(t_1^2 + 2rt_1 t_2 + t_2^2)} \frac{1}{2\pi} \iint e^{-(1/2)(u^2+v^2)}\, du\, dv$$

$$= e^{-(1/2)(t_1^2 + 2rt_1 t_2 + t_2^2)}.$$

Example 2. By application of Theorem 1, one can find the characteristic function of the variable (η_1, η_2) distributed according to the normal law

$$p(x, y) = \frac{1}{2\pi\sigma_1\sigma_2(1-r^2)} \times \exp\left\{-\frac{1}{2(1-r^2)}\left[\frac{(x-a)^2}{\sigma_1^2} - 2r\frac{(x-a)(y-b)}{\sigma_1^2\sigma_2^2} + \frac{(y-b)^2}{\sigma_2^2}\right]\right\}. \quad (4)$$

If one sets $\eta_1 = \sigma_1\xi_1 + a$ and $\eta_2 = \sigma_2\xi_2 + b$ then the variable (ξ_1, ξ_2) will be distributed according to the normal law (3). By Theorem 1 the characteristic function of the random variable (ξ_1, ξ_2) is

$$\varphi(t_1, t_2) = \exp\left[iat_1 + iat_2 - \frac{1}{2}(\sigma_1^2 t_1^2 + 2\sigma_1\sigma_2 r t_1 t_2 + \sigma_2^2 t_2^2)\right].$$

The following theorem follows from the definition of the characteristic function.

THEOREM 2. *If $f(t_1, t_2, \ldots, t_n)$ is the characteristic function of the variable $(\xi_1, \xi_2, \ldots, \xi_n)$ then the characteristic function of the sum $\xi_1 + \xi_2 + \cdots + \xi_n$ is*

$$f(t) = f(t, t, \ldots, t).$$

REMARK. Observe that

$$f(t) = f(tt_1, tt_2, \ldots, tt_n)$$

is the characteristic functions of the sum $t_1\xi_1 + t_1\xi_2 + \cdots + t_1\xi_n$.

Example 3. Let us use Theorem 2 to determine the distribution function of the sum $\eta_1+\eta_2$ if (η_1,η_2) is distributed according to the law (4).

By Theorem 2, the characteristic function of the sum $\eta_1+\eta_2$ is

$$f(t)=\exp\left[it(a+b)-\frac{t^2}{2}(\sigma_1^2+2r\sigma_1\sigma_2+\sigma_2^2)\right].$$

One knows (Example 1 of Section 32) that this is the characteristic function for a normal law with mathematical expectation $a+b$ and variance $\sigma_1^2+2r\sigma_1\sigma_2+\sigma_2^2$. This result was obtained earlier directly (Section 21, Example 2).

It is important to note that the following theorem also holds in the multi-dimensional case.

THEOREM 3. *A distribution function $F(x_1,x_2,\ldots,x_n)$ is uniquely determined by its characteristic function.*

The proof of this proposition is based on the inversion formula.

THEOREM 4. *If $f(t_1,t_2,\ldots,t_n)$ is the characteristic function and $F(x_1, x_2,\ldots,x_n)$ the distribution function of the random variable $(\xi_1,\xi_2,\ldots,\xi_n)$ then*

$$\mathsf{P}\{a_k\leqslant\xi_k<b_k,\ k=1,2,\ldots,n\}=\lim_{T\to\infty}\frac{1}{(2\pi)^n}\int_{-T}^{T}\int_{-T}^{T}\cdots\frac{e^{it_ka_k}-e^{it_kb_k}}{it_k}$$
$$\times f(t_1,\ldots,t_n)\,dt_1\,dt_2\cdots dt_n,$$

where a_k and b_k are any real numbers satisfying this single requirement: the probability of the point falling on the surface of the parallelepiped $a_k\leqslant\xi_k<b_k$ $(k=1,2,\ldots,n)$ equals 0.

Just as in the one-dimensional case, the direct limit theorem for characteristic functions and its converse both hold. We shall not stop to consider them here.

Example 4. An n-dimensional random variable $(\xi_1,\xi_2,\ldots,\xi_n)$ is said to have a non-degenerate (proper) n-dimensional normal distribution if its density function is given by

$$p(x_1,x_2,\ldots,x_n)=Ce^{-(1/2)Q(x_1,x_2,\ldots,x_n)},$$

where

$$Q(x_1, x_2, \ldots, x_n) = \sum_{i,j} b_{ij}(x_i - a_i)(x_j - a_j)$$

is a positive-definite quadratic form, and C, a_i, and b_{ij} are real constants.

Simple computations show[5] that

$$C = (\sqrt{2\pi})^{-n}\sqrt{D},$$

where

$$D = \begin{vmatrix} b_{11} & b_{12} & \cdots & b_{1n} \\ b_{21} & b_{22} & \cdots & b_{2n} \\ \cdots & \cdots & \cdots & \cdots \\ b_{n1} & b_{n2} & \cdots & b_{nn} \end{vmatrix}$$

Let D_{ij} denote the minor of D corresponding to the element b_{ij} then

$$\mathbf{E}\,\xi_j = a_j,\ \sigma_j^2 = \mathbf{D}\,\xi_j = \frac{D_{jj}}{D} \qquad (j = 1, 2, \ldots, n),$$

$$r_{ij} = \frac{\mathbf{E}(\xi_i - a_i)(\xi_j - a_j)}{\sigma_i\sigma_j} = \frac{D_{ij}}{\sqrt{D_{ii}D_{jj}}} \qquad (i, j = 1, 2, \ldots, n).$$

The determinant D and all its principal minors are positive.

By the usual operations, it is easily verified that the characteristic function of the variable $(\xi_1, \xi_2, \ldots, \xi_n)$ equals

$$(\xi_{i_1}, \xi_{i_2}, \ldots, \xi_{i_k}).$$

Thus, the n-dimensional normal distribution is completely determined by assignment of its mathematical expectation and variance.

From this expression for the characteristic function of an n-dimensional normal distribution, one sees that the distribution of the

[5]The usual procedure in such computations is to make a change of variables which reduces the form Q to a sum of squares and then to carry out the compurations in the new variables.

variable

$$f(t_1,t_2,\ldots,t_n)=e^{i\sum_{j=1}^{n}a_j t_j - -1 2\sum_{k=1}^{n}\sum_{j=1}^{n}\sigma_j\sigma_k\rho_{jk}t_j t_k}$$

for any $1 \leqslant i_1 < i_2 < \cdots < i_k \leqslant n$ is a k-dimensional normal distribution.

44 LAPLACE-STIELTJES TRANSFORM

For non-negative random variables the use of Laplace-Stieltjes transforms often appears more convenient than the use of characteristic functions. Let ξ be a non-negative random variable and $F(x)$ its distribution function. The integral

$$f(s)=\int_{-0}^{\infty}e^{-sx}dF(x) \tag{1}$$

is called Laplace-Stieltjes transform of $F(x)$. This transform possesses a set of useful properties repeating that of characteristic functions.

(1) Laplace-Stieltjes transform is an analytical function in the right hyper-plane. Its odd derivatives are negative and even ones are positive.

(2) $f(0)=1$.

(3) A distribution function is uniquely defined by its Laplace-Stieltjes transform.

(4) A sequence of functions converges at each point of continuity if and only if their Laplace-Stieltjes transforms converge uniformly on each finite interval of its argument.

(5) The Laplace-Stieltjes transform of a sum of random variables equals the product of Laplace-Stieltjes transforms of its summands.

(6)

$$f^{(n)}(0)=(-1)^n\int_{-0}^{\infty}x^n dF(x).$$

Below there are Laplace-Stieltjes transforms for some distributions.

(1) Threshold distribution: $F(x)=0$ for $x \leqslant a$ and $F(x)=1$ for $x>a$.

$$f(t)=e^{-as}$$

(2) Exponential distribution: $F(x)=1-e_{-\lambda x}$ for $x \geqslant 0$,

$$f(s)=\frac{\lambda}{\lambda+s}.$$

(3) Gamma distribution:

$$F'(x)=\frac{\beta^{\alpha}x^{\alpha-1}}{\Gamma(\alpha)}e^{-\beta x}$$

and α and β are positive constants,

$$f(s)=\frac{\beta^{\alpha}}{(\beta+s)^{\alpha}}.$$

(4) Poisson distribution:

$$p_k=\frac{\lambda^k e^{-\lambda}}{k!},$$

$k=0,1,2,\ldots.$

$$f(s)=e^{-\lambda(1-e^{-s})}.$$

(5) Geometrical distribution: $p_k=\mathbf{P}\{\xi=k\}=qp^k, \quad k=0,1,2,\ldots$

$$f(s)=\frac{q}{1-pe^{-s}}, \qquad q=1-p.$$

Let us illustrate the application of Laplace-Stieltjes transforms to an engineering problem.

In engineering, reliability is considered to be one of the most important technical properties of systems. It means, in general, that the system is able to perform its operations without failures. To increase reliability, one applies a wide spectrum of engineering measures including redundancy with renewal (this means repair or replacement).

Consider some unit A_1 which has a redundant support by unit A_2. This means that the redundant unit is switched onto the operational position instantaneously if the main unit has failed. The failed unit is immediately sent to repair. The question is: What could give such a redundant unit?

For solving the problem one needs to make some suggestions.

(1) Both units are identical.

(2) A unit being at the moment redundant is always in the idle state if another unit is in the operating state.

(3) Time of failure for each unit is a random variable ξ with the distribution function $F(x)$.

(4) Renewal of a failed unit is complete, i.e., after the renewal a unit performs as new.

(5) Repair time for each unit is a random variable η with the distribution function $G(x)$.

Let η denote time of failure of the system, i.e., of the pair of main and redundant units. The distribution function of this random variable is denoted by $\Phi(x)$. For convenience, one considers function

$$\bar{\Phi}(x) = 1 - \Phi(x) = P\{\zeta \geqslant x\}.$$

(In general, a notation of any distribution function with a bar will denote a complimentary function, i.e., $F(x) + F^{-}(x) = 1$.)

The event $\zeta \geqslant x$ can occur by two quite different ways:

(1) unit A_1 does not fail until a specified moment of time x;

(2) unit A_1 fails before x but the system consisting of unit A_1 (sent for repair) and unit A_2 (which became operating on the main position) will have successfully worked up to the moment x.

Let the probability that the system starting with failed unit A_1 which has not failed during time u be denoted by $\omega^{-}(u)$. Using this notation one can write

$$\bar{\Phi}(x) = \bar{F}(x) + \int_0^x \bar{\omega}(x-z)dF(z). \tag{2}$$

This equality is not sufficient to solve the problem since a new unknown probability ω is introduced. For this probability, one needs to write an additional equation. Consider the probability $\omega^{-}(x)$. The event of interest can occur again by two different ways:

(1) the system described (i.e., starting with the first failure of unit A_1) will have successfully operated up to the moment t;

(2) unit A_2 standing on the operation position has failed before x but unit A_1 has been repaired up to this moment, and the system as a whole from this moment will have successfully operated up to the moment t. This leads to the following additional equation

$$\bar{\omega}(x) = \bar{F}(x) + \int_0^x \bar{\omega}(x-z)\,G(z)\,dF(z). \tag{3}$$

For solution of the obtained system of integral equations one uses Laplace-Stieltjes transform. For these purposes let us introduce the following transforms

$$\varphi(s) = \int_0^\infty e^{-sx}\,d\Phi(x), \qquad f(s) = \int_0^\infty e^{-sx}\,dF(x),$$

$$\tilde{\omega}(s) = \int_0^\infty e^{-sx}\,d\omega(x), \qquad g(s) = \int_0^\infty e^{-sx}\,G(x)\,dF(x).$$

In the terms of these transforms equations (2) and (3) take the form

$$\varphi(s) = f(s)\,\tilde{\omega}(s), \qquad \tilde{\omega}(s) = f(s) - g(s) + \tilde{\omega}(s)g(s)$$

It follows that

$$\varphi(s) = f(s)\frac{f(s) - g(s)}{1 - g(s)}. \tag{4}$$

Formally speaking, the solution has been found because with the help of inverse transform the function $\Phi(x)$ can be restored. However let us continue to analyze solution (4) with the goal of finding the system's mean time of failure. From (4) follows

$$\frac{\varphi'(s)}{\varphi(s)} = \frac{f'(s)}{f(s)} + \frac{f'(s) - g'(s)}{f(s) - g(s)} + \frac{g'(s)}{1 - g(s)}.$$

It follows that for $s = 0$ one obtains the equality

$$\varphi'(0) = -f'(0)\left(1 + \frac{1}{1 - g(0)}\right).$$

But it is known that

$$T = \mathbf{E}\zeta = -\varphi'(0), \qquad a = \int_0^\infty x\,dF(x) = -f'(0)$$

and that

$$\alpha = 1 - g(0) = \int_0^\infty \bar{G}(x)\,dF(x)$$

is the probability that for a unit the random repair time is longer than the time of failure. Thus, finally one has

$$T = a\left(1 + \frac{1}{\alpha}\right). \tag{5}$$

From this formula one sees that the system's mean time of failure can be increased by two means:

(1) to increase the mean time of failure for each of the units;
(2) to decrease the mean repair time.

Let us assume that improving the process of the system renewal one approaches such a distribution function $G_n(x)$ of a random repair time that $\alpha_n \to 0$ (but never reaches 0).

Let us prove that if the mean time of failure is finite and equals a and at the same time $\alpha_1 \to 0$, then the following limit relation

$$\mathbf{P}\left\{\frac{\zeta_n}{T_n} > x\right\} \to e^{-x} \quad \text{for} \quad x \geqslant 0 \quad (n \to \infty).$$

holds. In other words, it will be proved that the asymptotic distribution of the system's time of failure of two units is exponential.

The problem is to show that Laplace-Stieltjes transform of the distribution of the value ζ_n/T_n converges to $1/(1+s)$. For this purpose let us transform the expression

$$\varphi_n\left(\frac{s}{T_n}\right) = f\left(\frac{s}{T_n}\right)\frac{f\left(\frac{s}{T_n}\right) - g_n\left(\frac{s}{T_n}\right)}{1 - g_n\left(\frac{s}{T_n}\right)}$$

as follows

$$\varphi_n\left(\frac{s}{T_n}\right)=f\left(\frac{s}{T_n}\right)\frac{1}{1+\dfrac{1-f\left(\dfrac{s}{T_n}\right)}{\dfrac{s}{T_n}}\cdot\dfrac{s}{T_n\left(f\left(\dfrac{s}{T_n}\right)-g\left(\dfrac{s}{T_n}\right)\right)}}$$

It is obvious that

$$\frac{1-f\left(\dfrac{s}{T_n}\right)}{\dfrac{s}{T_n}}=-\frac{f\left(\dfrac{s}{T_n}\right)-f(0)}{\dfrac{s}{T_n}}\to -f'(0)=a \qquad (n\to\infty).$$

It is clear that

$$T_n\left[f\left(\frac{s}{T_n}\right)-g\left(\frac{s}{T_n}\right)\right]\sim\frac{a}{\alpha_n}\left[f\left(\frac{s}{T_n}\right)-g\left(\frac{s}{T_n}\right)\right]$$

$$=a\left[1-1+\frac{f\left(\dfrac{s}{T_n}\right)-g\left(\dfrac{s}{T_n}\right)}{\alpha_n}\right].$$

Now one can write

$$1-\frac{f\left(\dfrac{s}{T_n}\right)-g\left(\dfrac{s}{T_n}\right)}{\alpha_n}=\frac{1}{\alpha_n}\int_0^\infty(1-e^{sx/T_n})(1-G_n(x))\,dF(x).$$

Let us evaluate the integral on the right-hand side of the latter relation. For this purpose, divide it by two parts

$$\left(\int_0^{\sqrt{T_n}}+\int_{\sqrt{T_n}}\right)(1-e^{-sx/T_n})(1-G_n(x))\,dF(x).$$

In the first summand one can use the inequality $1-e^{-x}\leqslant x$ and in the second one can use the inequality $1-e^{-x}\leqslant 1$. Finally, one obtains that

$$\int_0^{\sqrt{T_n}} (1 - e^{-sx/T_n})(1 - G_n(x))\, dF(x) \leqslant \frac{s\sqrt{T_n}}{T_n} \int_0^{\infty} (1 - G_n(x))\, dF(x) = o(\alpha_n),$$

$$\int_{\sqrt{T_n}}^{\infty} (1 - e^{-sx/T_n})(1 - G_n(x))\, dF(x) \leqslant \int_{\sqrt{T_n}}^{\infty} (1 - G_n(x))\, dF(x) = o(\alpha_n).$$

Finally, all of estimates above-obtained allow us to write

$$\varphi_n\left(\frac{s}{T_n}\right) = \frac{1}{1+s}(1 + o(1)),$$

Q.E.D.

45 EXERCISES

1. Prove that the functions

$$f_1(t) = \sum_{k=0}^{\infty} a_k \cos kt\; f_2(t) = \sum_{k=0}^{\infty} a_k e^{i\lambda_k t},$$

(where $a_k \geqslant 0$ and $\Sigma a_k = 1$) are characteristic functions. Find the probability distributions which correspond to them.

2. Find characteristic functions for the following probability densities

(a) $p(x) = \dfrac{a}{2} e^{-a|x|};$

(b) $p(x) = \dfrac{a}{\pi(a^2 + x^2)};$

(c) $p(x) = \begin{cases} 0 & \text{for } |x| \geqslant a, \\ \dfrac{a - |x|}{a^2} & \text{for } |x| < a; \end{cases}$

(d) $p(x) = \dfrac{2\sin^2(ax/2)}{\pi a x^2}.$

REMARK. The alert reader will notice that Examples (a) and (b), and also (c) and (d) are, so to speak, converses.

3. Prove that functions

$$\varphi_1(t)=\frac{1}{\operatorname{ch} t},\quad \varphi_2(t)=\frac{1}{\operatorname{sh} t},\quad \varphi_3(t)=\frac{1}{\operatorname{ch}^2 t}$$

are the characteristic functions corresponding to the density functions

$$p_1(x)=\frac{1}{2\operatorname{ch}(\frac{\pi x}{2})},\quad p_2(x)=\frac{\pi}{4\operatorname{ch}^2(\frac{\pi x}{2})},\quad p_3(x)=\frac{x}{2\operatorname{sh}(\frac{\pi x}{2})}.$$

4. Find the probability distribution of each of the random variables whose characteristic function equals

(a) $\cos t$, (b) $\cos^2 t$, (c) $\dfrac{a}{a+it}$, (d) $\dfrac{\sin at}{at}$.

5. Prove that the function defined by the relations

$$f(t)=f(-t),\quad f(t+2a)=f(t),$$

$$f(t)=\frac{a-t}{a}\quad \text{for}\quad 0\leqslant t\leqslant a,$$

is a characteristic function.

REMARK. The characteristic functions in Examples 2(d) and 5 have the following striking property:

$$f_2(t)=f_s(t)\quad \text{for}\quad |t|\leqslant a,$$

$$f_2(t)\neq f_s(t)\quad \text{for}\quad |t|>a\ t\neq \pm 2a,\ldots$$

Thus, there exist characteristic functions whose values coincide on an arbitrary large interval $(-a, a)$ and which are not identically equal. The first of these two examples of such characteristic functions was pointed out by the author. Later, M. Krein gave necessary and sufficient conditions for two characteristic functions that coincide in some interval $(-a, a)$ to be identical.

6. Prove that it is possible to find independent random variables ξ_1, ξ_2, and ξ_3 such that the probability distributions of ξ_2 and Ξ_3 are

different but that the distribution functions of the sums $\xi_1 + \Xi_2$ and $\xi_1 + \Xi_3$ are identical.
Hint: Use the results of Examples 2(a) and 5.

7. Prove that if $f(t)$ is a characteristic function then the function

$$f(t) = \begin{cases} f(t) & \text{for } |t| \leqslant a, \\ f(t+2a) & \text{for } -\infty < t \leqslant \infty \end{cases}$$

is also a characteristic function.
Hint: Use the Bochner-Khinchine Theorem.

8. Prove that if $f(t)$ is a characteristic function, then

$$\varphi(t) = e^{f(t)-1}$$

is also a characteristic function.

9. Prove that if $f(t)$ is a characteristic function, then

$$\varphi(t)\frac{1}{t}\int_0^t f(t)dt$$

is also a characteristic function.

10. Prove that the inequality

$$1 - f(2t) \leqslant 4(1 - f(t)),$$

holds for any real characteristic function $f(t)$. This means that the inequality

$$1 - |f(2t)|^2 \leqslant 4(1 - |f(t)|^2).$$

holds for any characteristic function.

11. Prove that the inequality

$$1 + f(2t) \geqslant 2[f(t)]^2$$

holds for any real characteristic function.

12. Prove that if $F(x)$ is a distribution function and $f(t)$ its corresponding characteristic function, then the relation

$$\lim_{T\to} \frac{1}{2T}\int_{-T}^{T} f(t)e^{-itx}\,dt = F(x+0) - F(x-0)$$

holds for any value of x.

13. Prove that if $F(x)$ is a distribution function and $f(t)$ its corresponding characteristic function, and if x_k are the abscissas of the jumps in the function $F(x)$, then

$$\lim_{T\to\infty}\frac{1}{2T}\int_{-T}^{T}|f(t)|^2\,dt=\sum_k[F(x_k+0)-F(x_k-0)].$$

14. Prove that if a random variable has a density function, then its characteristic function tends to 0 as $t\to\infty$.

15. A random variable ξ is distributed according to the Poisson distribution with the mean $\mathbf{E}\xi=\lambda$. Prove that for $\lambda\to\infty$, the distribution of the variable $(\xi-\lambda)/\sqrt{\lambda}$ tends to the normal distribution with the parameters $a=0$ and $\sigma=1$.

16. The random variable ξ has the density function

$$p(x)=\begin{cases}0 & \text{for } x\leqslant 0,\\ \dfrac{\beta^\alpha}{\Gamma(\alpha)}x^{\alpha-1}e^{-\beta x} & \text{for } x>0.\end{cases}$$

Prove that the distribution of the variable $(\beta\xi-\alpha)/\sqrt{\alpha}$ tends to the normal distribution with parameters $a=0$ and $\sigma=1$ as $\alpha\to\infty$.

REMARK. The results of Exercises 15 and 16 enable one to use normal distribution tables to compute the probabilities $\mathrm{E}\{a\leqslant\xi\leqslant b\}$ for large values of λ (or α). In particular, it turns out that the indicated limiting relation gives excellent accuracy for the χ^2-distribution when $n\geqslant 30$. This last observation is used constantly in statistics.

17. Prove that if $\varphi(t)$ is a characteristic function and $\Psi(t)$ is a function such that for some sequence $\{h_n\}$ (h_n ∞ as $n\to\infty$) the product $f_n(t)=\varphi(t)\Psi(h_n t)$ is also a characteristic function, then the function $\Psi(t)$ is a characteristic function.

Chapter 9

THE CLASSICAL LIMIT THEOREM

46 STATEMENT OF THE PROBLEM

The DeMoivre-Laplace Integral Limit Theorem, which has been proved in Chapter 3, has served as a basic source for a number of investigations of fundamental significance both for the theory of probability and for its various applications to the natural sciences, engineering and economics. In order to better understand all aspects of these investigations, let us slightly change the form of the presentation of the DeMoivre-Laplace Theorem. Specifically, if one denotes the number of times event A occurs in the trial h by μ_k then the number of occurrences of event A in n sequential trials equals $\Sigma_{k=1}^n \mu_k$. Furthermore, in Example 3 of Section 25 it was calculated that $\mathbf{E}\, \Sigma_{k=1}^n \mu_k = np$ and $\mathbf{D}\, \Sigma_{k=1}^n \mu_k = npq$.

Therefore, the DeMoivre-Laplace Theorem can be expressed in the following form: As $n \to \infty$,

$$\mathbf{P}\left\{a \leqslant \frac{\sum_{k=1}^{n} (\mu_k - \mathbf{E}\mu_k)}{\sqrt{\sum_{k=1}^{n} \mathbf{D}\mu_k}} < b\right\} \to \frac{1}{\sqrt{2\pi}} \int_a^b e^{-z^2/2}\, dz \tag{1}$$

This statement in a verbal form is: If there exist independent random variables each assuming only two values 0 and 1 with probabilities q and $p = 1 - q$, respectively $(0 < p < 1)$, then the probability that the sum of the deviations of these random variables from their mathematical expectations divided by the square root of the sum of their variances lies between a and b and approaches the integral

$$\frac{1}{\sqrt{2\pi}} \int_a^b e^{-z^2/2}\, dz.$$

uniformly in a and b as the number of summands increases indefinitely.

The following questions naturally arise: How closely is relation (1) connected with the special choice of the variables μ_k? Would it also hold if weaker restrictions were imposed on the distribution functions of the summands? The formulation of this problem, as well as its solution, is essentially due to P. L. Chebyshev and his pupils, A. A. Markov and A. M. Liapounov. Their investigations have shown that one should impose a very general restriction on the random variables concerning the fact that an individual summand has to have a negligible effect on the entire sum. The next section includes a precise formulation of this condition. The reasons for the great importance of these results in applications lie in the nature of mass phenomena. As mentioned before, the study of laws governing mass phenomena constitutes the subject matter of probability theory.

One of the most important mathematical models serving as a basis for applying the results of probability theory to natural science and engineering is the following. Suppose that a process takes place under the influence of a large number of independent random factors each of which might change the development of the phenomenon or process insignificantly. The researcher, being interested in studying the whole process rather than the effect of individual factors, merely looks at the summary effect of these factors. Let us consider two typical examples.

Example 1. The outcome of an experiment is inevitably affected by a great number of factors that produce errors in the measurement. They are the errors due to the state of the measuring device, to change of environments, human errors of the observer depending on his psychological or physical state, etc. Each of these factors may produce an insignificant error. In other words, the error of measurement actually observed is a random variable which is the sum of a great number of mutually independent or slightly dependent random variables of insignificant magnitude. And although these random variables, as well as their distribution functions, are not known, their effect on the outcome of the measurement is noticeable and must therefore be made the subject of study.

Example 2. Large batches of identical items are produced by mass production in many branches of industry. Let us focus our attention on some numerical characteristics of such a product. Inasmuch as the considered article is manufactured in accordance with industrial norms, there exists a certain standard numerical meaning of this characteristic. Actually, however, there is always some deviation from this specified value. In a well-organized manufacturing process, such deviations can only be caused by hazard factors, each of which has only an imperceptible effect. The cumulative effect, however, might bring an observable deviation from the norm.

Everybody could present a number of similar examples.

The problem thus arises of investigating the laws governing the behavior of the value of the sums of a larger number of independent random variables each of which has only a minor effect on the entire sum. Later on, this statement will be given a more precise meaning. In order to study such sums of a very large but finite number of terms, one instead considers a sequence of sums having an increasing number of terms and then takes into account corresponding limit distribution functions of the sums. This sort of passing from a finite formulation of a problem to limit is very typical in modern mathematics and in many branches of the natural sciences.

Thus, let us begin with a strong formulation of the problem. Let a sequence of mutually independent random variables

$$\xi_1, \xi_2, \ldots, \xi_n, \ldots,$$

be given. Every random variable considered has a mathematical expectation and variance. The following notation will be used:

$$a_k = \mathbf{E}\,\xi_k, \quad b_k^2 = \mathbf{D}\xi_k, \quad B_n^2 = \sum_{k=1}^{n} b_k^2 = \mathbf{D} \sum_{k=1}^{n} \xi_k.$$

The question arises: What conditions must be imposed on the random variables ξ_k so that the distribution functions of the sums

$$\frac{1}{B_n} \sum_{k=1}^{n} (\xi_k - a_k) \tag{2}$$

converge to the normal distribution law? In the following section it will be shown that the sufficient condition is presented by the Lindeberg Condition for any $\tau > 0$

$$\lim_{n \to \infty} \frac{1}{B_n^2} \sum_{k=1}^{n} \int_{|x-a_k| > \tau B_n} (x - a_k)^2 \, dF_k(x) = 0,$$

where $F_k(x)$ denotes the distribution function of random variable ξ_k.

Let us make clear what this condition means. Let A_k denote the event that

$$|\xi_k - a_k| > \tau B_n (k = 1, 2, \ldots, n)$$

Estimate the probability

$$\mathsf{P}\{\max_{1 \leqslant k \leqslant n} |\xi_k - a_k| > \tau B_n\}.$$

Since

$$\mathsf{P}\{\max_{1 \leqslant k \leqslant n} |\xi_k - a_k| > \tau B_n\} = \mathsf{P}\{A_1 + A_2 + \cdots + A_n\}$$

and

$$\mathsf{P}\{A_1 + A_2 + \cdots + A_n\} \leqslant \sum_{k=1}^{n} \mathsf{P}\{A_k\},$$

by noting that

$$\mathsf{P}\{A_k\} = \int_{|x - a_k| > \tau B_n} dF_k(x) \leqslant \frac{1}{(\tau B_n)^2} \int_{|x - a_k| > \tau B_n} (x - a_k)^2 \, dF_k(x)$$

one finds the inequality

$$\mathsf{P}\{\max_{1 \leqslant k \leqslant n} |\xi_k - a_k| \geqslant \tau B_n\} \leqslant \frac{1}{\tau^2 B_n^2} \sum_{k=1}^{n} \int_{|x - a_k| > \tau B_n} (x - a_k)^2 \, dF_k(x).$$

By virtue of the Lindeberg Condition, the last sum tends to 0 as $n \to \infty$ for any constant $\tau > 0$. Thus, the Lindeberg Condition represents the distinctive requirement that the terms $(\xi_k - a_k)/B_k$ in the sum (2) are to be uniformly small.

We note once more that Markov and Liapounov had already investigated quite thoroughly the meaning of the conditions sufficient for the distribution functions of the sums (2) to converge to the normal law.

47 LINDEBERG'S THEOREM

Let us begin with the proof of the sufficiency of the Lindeberg Condition.

THEOREM. *If a sequence of mutually independent random variables* $\xi_1, \xi_2, \ldots, \xi_n$ *satisfies the Lindeberg Condition*

$$\lim_{n\to\infty} \frac{1}{B_n^2} \sum_{k=1}^{n} \int_{|x-a_k|>\tau B_n} (x-a_k)^2\, dF_k(x) = 0, \tag{1}$$

for any constant $\tau > 0$ *then as* $n \to \infty$ *uniformly in* x

$$\mathsf{P}\left\{\frac{1}{B_n}\sum_{k=1}^{n} (\xi_k - a_k) < x\right\} \to \frac{1}{\sqrt{2\pi}} \int_{-\infty}^{x} e^{-z^2/2}\, dz.$$

PROOF: For brevity, let us introduce the notation

$$\xi_{nk} = \frac{\xi_k - a_k}{B_n}, \quad F_{nk}(x) = \mathsf{P}\{\xi_{nk} < x\}.$$

Obviously, we have

$$\mathsf{E}\,\xi_{nk} = 0, \quad \mathsf{D}\,\xi_{nk} = \frac{1}{B_n^2}\mathsf{D}\,\xi_k$$

and, consequently, it follows

$$\sum_{k=1}^{n} \mathsf{D}\,\xi_{nk} = 1. \tag{2}$$

It is easy to see that in this notation the Lindeberg's Condition takes the following form:

$$\lim_{n\to\infty} \sum_{k=1}^{n} \int_{|x|>\tau} x^2\, dF_{nk}(x) = 0. \tag{1'}$$

The characteristic function of the sum

$$\frac{1}{B_n}\sum_{k=1}^{n} (\xi_k - a_k) = \sum_{k=1}^{n} \xi_{nk}$$

equals

$$\varphi_n(t) = \prod_{k=1}^{n} f_{nk}(t).$$

The next step is to prove that

$$\lim_{n\to\infty} \varphi_n(t) = e^{-t^2/2}.$$

Let us, first of all, note that the factors $f_{nk}(t)$ tend to 1 as $n \to \infty$ uniformly in k $(1 \leqslant k \leqslant n)$. Indeed, taking into account that $\mathbf{E}\xi_{nk} = 0$, one finds that

$$f_{nk}(t) - 1 = \int (e^{itx} - 1 - itx)\, dF_{nk}(x).$$

Since, for any real α inequality[1]

$$|e^{i\alpha} - 1 - i\alpha| \leqslant \alpha^2/2, \tag{3}$$

one has

$$|f_{nk}(t) - 1| \leqslant \frac{t^2}{2} \int x^2\, dF_{nk}(x).$$

Let $\varepsilon > 0$ be an arbitrary number. Then it is clear that

$$\int x^2\, dF_{nk}(x) = \int_{|x| \leqslant \varepsilon} x^2\, dF_{nk}(x) + \int_{|x| > \varepsilon} x^2\, dF_{nk}(x) \leqslant \varepsilon^2 + \int_{|x| > \varepsilon} x^2\, dF_{nk}(x).$$

According to (1′), the last term can be made smaller than ε^2 for all sufficiently large values of n. Thus, for all sufficiently large n and for t in any finite interval $|t| \leqslant T$

$$|f_{nk}(t) - 1| \leqslant \varepsilon^2 T^2.$$

uniformly in k $(1 \leqslant k \leqslant n)$.

---

[1] This inequality and a series of similar ones can be derived, for example, as follows. From the fact that

$$|e^{i\alpha} - 1| = \left| \int_0^\alpha e^{ix}\, dx \right| \leqslant \alpha$$

where $\alpha > 0$ there follows the inequality

$$|e^{i\alpha} - 1 - i\alpha| = \left| \int_0^\alpha (e^{ix} - 1)\, dx \right| \leqslant \frac{\alpha^2}{2}.$$

It then follows from this inequality that

$$\left| e^{i\alpha} - 1 - i\alpha + \frac{\alpha^2}{2} \right| = \int_0^\alpha (e^{ix} - 1 - ix)\, dx| \leqslant \int_0^\alpha |e^{ix} - 1 - ix|\, dx \leqslant \int_0^\alpha \frac{x^2}{2}\, dx = \frac{\alpha^3}{6}, \tag{3'}$$

and so on.

From this one concludes that

$$\lim_{n\to\infty} f_{nk}(t) = 1 \tag{4}$$

uniformly in k $(1 \leqslant k \leqslant n)$ and that for all sufficiently large n, the inequality

$$|f_{nk}(t) - 1| < 1/2 \tag{5}$$

is satisfied for t lying in an arbitrary finite interval $|t| \leqslant T$. Therefore, in the interval $/t/ \leqslant T$ one can write the expansion (where ln represents the principal value of the logarithm)

$$\begin{aligned} \ln \varphi_n(t) &= \sum_{k=1}^{n} \ln f_{nk}(t) = \sum_{k=1}^{n} \ln [1 + (f_{nk}(t) - 1)] \\ &= \sum_{k=1}^{n} (f_{nk}(t) - 1) + R_n, \end{aligned} \tag{6}$$

where

$$R_n = \sum_{k=1}^{n} \sum_{s=2} \frac{(-1)^{s-1}}{s} (f_{nk}(t) - 1)^s.$$

In view of (5),

$$|R_n| \leqslant \sum_{k=1}^{n} \sum_{s=2} \frac{1}{2} |f_{nk}(t) - 1|^s = \frac{1}{2} \sum_{k=1}^{n} \frac{|t_{nk}(t) - 1|^2}{1 - |f_{nk}(t) - 1|} \leqslant \sum_{k=1}^{n} |f_{nk}(t) - 1|^2.$$

Since

$$\sum_{k=1}^{n} |f_{nk}(t) - 1| = \sum_{k=1}^{n} \left| \int (e^{itx} - 1 - itx) dF_{nk}(x) \right| \leqslant \frac{t^2}{2} \sum_{k=1}^{n} \int x^2 dF_{nk}(x) = \frac{t^2}{2},$$

it follows that

$$|R_n| \leqslant \frac{t^2}{2} \max_{1 \leqslant k \leqslant n} |f_{nk}(t) - 1|.$$

Thus it follows from (4) that, as $n \to \infty$,

$$R_n \to 0 \tag{7}$$

uniformly for t in an arbitrary finite interval $|t| \leqslant T$. But

$$\sum_{k=1}^{n} (f_{nk}(t) - 1) = -\frac{t^2}{2} + \rho_n, \tag{8}$$

where

$$\rho_n = \frac{t^2}{2} + \sum_{k=1}^{n} \int (e^{itx} - 1 - itx) \, dF_{nk}(x).$$

Let $\varepsilon > 0$ be an arbitrary number. Then by virtue of (2′)

$$\rho_n = \sum_{k=1}^{n} \int_{|x| \leqslant \varepsilon} \left(e^{itx} - 1 - itx - \frac{(itx)^2}{2} \right) dF_{nk}(x)$$

$$+ \sum_{k=1}^{n} \int_{|x| > \varepsilon} \left(\frac{t^2 x^2}{2} + e^{itx} - 1 - itx \right) dF_{nk}(x).$$

The inequalities (3) and (3′) enable one to obtain the following estimate:

$$|\rho_n| \leqslant \frac{|t|^3}{6} \sum_{k=1}^{n} \int_{|x| \leqslant \varepsilon} |x|^3 \, dF_{nk}(x) + t^2 \sum_{k=1}^{n} \int_{|x| > \varepsilon} x^2 \, dF_{nk}(x)$$

$$\leqslant \frac{|t|^3}{6} \cdot \varepsilon \sum_{k=1}^{n} \int_{|x| \leqslant \varepsilon} x^2 \, dF_{nk}(x) + t^2 \sum_{k=1}^{n} \int_{|x| > \varepsilon} x^2 \, dF_{nk}(x)$$

$$= \frac{|t|^3}{6} \cdot \varepsilon + t^2 \left(1 - \frac{|t|}{6} \varepsilon \right) \sum_{k=1}^{n} \int_{|x| > \varepsilon} x^2 \, dF_{nk}(x).$$

According to condition (1), the second term for any $\varepsilon > 0$ can be made less than any $\eta > 0$ by making n sufficiently large. And since $\varepsilon > 0$ is arbitrary, one can select it so small that for any T and any $\eta > 0$ the inequality

$$|\rho_n| < 2\eta \qquad (n \geqslant n_o(\varepsilon, \eta, T))$$

is satisfied for all t in the interval $|t| \leqslant T$. This inequality shows that

$$\lim_{n \to} \rho_n = 0 \tag{9}$$

uniformly in every finite interval $|t| \leqslant T$. Equations (6), (7), (8), and (9) allow one to determine that the relation

$$\lim_{n \to} \ln \varphi_n(t) = -t^2/2$$

uniformly in every finite interval $|t| \leqslant T$.

The theorem is proved.

COROLLARY. *If the independent random variables* $\xi_1, \xi_2, \dots, \xi_n, \dots$ *are identically distributed and have finite and non-zero variances, then*

$$P\left\{ \frac{1}{B_n} \sum_{k=1}^{n} (\xi_k - E\,\xi_k) < x \right\} \to \frac{1}{\sqrt{2\pi}} \int_{-}^{x} e^{-z^2/2}\,dz$$

as $n \to \infty$ *uniformly in x.*

PROOF: It suffices to verify that the Lindeberg Condition is satisfied under the given assumptions. Note that in this case

$$B_n = b\sqrt{n},$$

where b is the variance of the individual terms. By putting $E\xi_k = a$, one may write the following obvious equations:

$$\sum_{k=1}^{n} \frac{1}{B_n^2} \int_{|x-a|>\tau B_n} (x-a)^2\,dF_k(x) = \frac{1}{nb^2} n \int_{|x-a|>\tau B_n} (x-a)^2\,dF_1(x)$$

$$= \frac{1}{b^2} \int_{|x-a|>\tau B_n} (x-a)^2\,dF_1(x).$$

By the assumption that the variance is positive and finite, one concludes that the integral on the right-hand side of the last equation tends to 0 as $n \to \infty$.

LIAPOUNOV'S THEOREM. *If for a sequence of mutually independent random variables* $\xi_1, \xi_2, \dots, \xi_n, \dots$ *a positive number* δ *can be found such that as* $n \to \infty$

$$\frac{1}{B_n^{2+\delta}} \sum_{k=1}^{n} E|\xi_k - a_k|^{2+\delta} \to 0, \tag{10}$$

uniformly in x then for $n \to \infty$ also uniformly in x

$$\mathsf{P}\left\{\frac{1}{B_n}\sum_{k=1}^{n}(\xi_k - a_k) < x\right\} \to \frac{1}{\sqrt{2\pi}}\int_{-\infty}^{x} e^{-z^2/2}\,dz.$$

PROOF: Again, it suffices to verify that the Liapounov Condition (condition (10)) implies the Lindeberg Condition. But this is obvious from the following chain of inequalities:

$$\frac{1}{B_n^2}\sum_{k=1}^{n}\int_{|x-a_k|>\tau B_n}(x-a_k)^2\,dF_k(x)$$

$$\leqslant \frac{1}{B_n^2(\tau B_n)^{\delta}}\sum_{k=1}^{n}\int_{|x-a_k|>\tau B_n}|x-a_k|^{2+\delta}\,dF_k(x) \leqslant \frac{1}{\tau^{\delta}}\,\frac{\sum_{k=1}^{n}\int |x-a_k|^{2+\delta}\,dF_k(x)}{B_n^{2+\delta}}.$$

48 THE LOCAL LIMIT THEOREM

Let us now give the sufficient conditions for another classical limit theorem, the Local Limit Theorem. The discussion will be restricted by the case of mutually independent summands that have identical distributions.

It is said that a discrete random variable ξ has a lattice distribution if there exist numbers a and $h > 0$ such that all possible values of ξ can be represented in the form $a + kh$ where parameter k may take on any integer values $(-\infty < k < +\infty)$.

The Poisson and the Bernoulli distributions are among the examples of lattice distributions.

Let us now express the condition for a lattice distribution in terms of characteristic functions. For this purpose, the following lemma must be proved.

LEMMA. *A random variable ξ has a lattice distribution if and only if for some $t \neq 0$ the absolute value of its characteristic function equals* 1.

PROOF: Indeed, if ξ has a lattice distribution and p_k is the probability of the equality $\xi = a + kh$, then the characteristic function of the random variable ξ is

$$f(t) = \sum_{k=-\infty}^{\infty} p_k e^{it(a+kh)} = e^{iat}\sum_{k=-\infty}^{\infty} p_k e^{itkh}.$$

From this, one finds that

$$f\left(\frac{2\pi}{h}\right) = e^{2\pi i(a/h)} \sum_{k=-\infty}^{\infty} p_k e^{2\pi i k} = e^{2\pi i(a/h)}.$$

Thus it is shown that for every lattice distribution

$$\left|f\left(\frac{2\pi}{h}\right)\right| = 1.$$

Let us now assume that for some $t \neq 0$

$$|f(t_1)| = 1,$$

and show that ξ then has a lattice distribution. The last equation implies that for some η

$$f(t_1) = e^{i\theta}.$$

Thus

$$\int e^{it_1 x}\, dF(x) = e^{i\theta}$$

and therefore

$$\int e^{i(t_1 x - \theta)}\, dF(x) = 1.$$

From this, it follows that

$$\int \cos(t_1 x - \theta)\, dF(x) = 1.$$

In order for this relation to be possible, it is necessary that the function $F(x)$ may increase only at those values of x for which

$$\cos(t_1 x - \theta) = 1.$$

This means that all of the possible values of ξ must be of the form

$$x = \frac{\theta}{t_1} + k\frac{2\pi}{t_1},$$

Q.E.D.

The number h is called the distribution span.

The distribution span is maximal if for any choice of b ($-\infty < b < +\infty$) and $h_1 > h$, all of the possible values of ξ cannot be represented in the form $b + kh_1$.

To illustrate the distinction between the notions of distribution span and maximal distribution span, let us consider the following example. Suppose that ξ takes on merely odd integer values. It is clear that all the values of ξ may be expressed in the form $a + kh$, where $a = 0$ and $h = 1$. However, the span h is not maximal, since all the possible values of ξ may also be expressed in the form $b + kh_1$, with $b = 1$ and $h_1 = 2$.

The conditions for a distribution span to be maximal can be expressed in other terms.

(1) A distribution span h is maximal if and only if the greatest common divisor of the differences of every pair of possible values of the random variable ξ divided by h equals 1.

(2) A distribution span h is maximal if and only if the absolute value of the characteristic function less than 1 in the interval $0 < |t| < 2\pi/h$ or is equal to 1 for $t = 2\pi/h$.

The last assertion follows immediately from the lemma just proved. In fact, if for $0 < t_1 < 2\pi/h$

$$|f(t_1)| = 1,$$

then, as was shown above, the quantity $2\pi/t_1$ must be a distribution span, but since $h < 2\pi/t_1$, the span h cannot be maximal.

From this, one may draw the following conclusion: If h is a maximal distribution span, then for any $\varepsilon > 0$ there can be found a number $c_0 > 0$ such that the inequality

$$|f(t)| \leqslant e^{-c_0} \tag{1}$$

holds for all t in the interval $\varepsilon \leqslant |t| \leqslant 2\pi/h$.

Suppose now that $\xi_1, \xi_2, \ldots, \xi_n, \ldots$ are mutually independent, lattice random variables having the same distribution function $F(x)$. Let us consider the sum

$$\zeta_n = \xi_1 + \xi_2 + \cdots + \xi_n.$$

It is clear that the sum is also a lattice random variable and its possible values can be written in the form $na + kh$. Let $P_n(k)$ denote the probability of the equality

$$\zeta_n = na + kh;$$

in particular,

$$P_1(k) = \mathbf{P}\{\xi_1 = a + kh\} = p_k.$$

Next, denote

$$z_{nk} = \frac{an + kh - A_n}{B_n},$$

where $A_n = \mathbf{E}\zeta_n$, $B_n^2 = \mathbf{D}\zeta_n = n\mathbf{D}\xi_1$.

It is possible now to prove the following statement, which is evidently a generalization of the DeMoivre-Laplace Local Limit Theorem.

THEOREM. *Let $\xi_1, \xi_2, \ldots, \xi_n, \ldots$ be independent lattice random variables with identical distribution function $F(x)$ and their mathematical expectation and variance are finite. Then the relation*

$$\frac{B_n}{h} P_n(k) - \frac{1}{\sqrt{2\pi}} e^{-z_{nk}^2/2} \to 0,$$

holds uniformly in k $(-\infty < k < +\infty)$ as $n \to \infty$ if and only if the distribution span h is maximal[2].

PROOF: The necessity of the condition is almost obvious. Indeed, if the span h were not maximal, then the possible values of the sum

$$\zeta_n = \sum_{k=1}^{n} \xi_k$$

would be systematically omitted because the difference between the closest possible values of the sum could not be less than dh, where d is the largest common divisor of the differences of the possible values of ζ_n divided by h. If h were not the maximal span, then $d > 1$ for all values of n.

The proof of the sufficiency of the condition of the theorem requires a somewhat more sophisticated argument.

[2] This theorem has been proved by the author. -*I.U.*

The characteristic function of the random variable ξ_k ($k = 1, 2, 3,\ldots$) is

$$f(t) = \sum_{k=-\infty}^{\infty} p_k e^{iat+itkh} = e^{iat} \sum_{k=-\infty}^{\infty} p_k e^{itkh},$$

and the characteristic function of the sum ζ_n is

$$f^n(t) = e^{iant} \sum_{k=-\infty}^{\infty} P_n(k) e^{itkh}.$$

Multiplying both sides of this equation by $e^{-iant-itkh}$ and integrating from $-\pi/h$ to π/h, one obtains:

$$\frac{2\pi}{h} P_n(k) = \int_{-\pi/h}^{\pi/h} f^n(t) e^{-iant-itkh}\,dt.$$

Note that

$$hk = B_n z_{nk} + A_n - an$$

(let us use z below instead of z_{nk}) and then write:

$$\frac{2\pi}{h} P_n(k) = \int_{-\pi/h}^{\pi/h} f^{*n}(t) e^{-itzB_n}\,dt,$$

where

$$f^*(t) = e^{-itA_n/n} f(t).$$

Finally, by setting $x = tB_n$ one obtains:

$$\frac{2\pi B_n}{h} P_n(k) = \int_{-\pi B_n/h}^{\pi B_n/h} e^{-izx} f^{*n}\left(\frac{x}{B_n}\right) dx.$$

It is easy to find that

$$\frac{1}{\sqrt{2\pi}} e^{-z^2/2} = \frac{1}{2\pi} \int e^{-izx - x^2/2}\,dx.$$

Let us represent the difference

$$R_n = 2\pi \left[\frac{B_n}{h} P_n(k) - \frac{1}{\sqrt{2\pi}} e^{-z^2/2}\right]$$

as the sum of four integrals

$$R_n = J_1 + J_2 + J_3 + J_4,$$

where

$$J_1 = \int_{-A}^{A} e^{-itx}\left[f^{*n}\left(\frac{x}{B_n}\right) - e^{-x^2/2}\right]dx,$$

$$J_2 = -\int_{|x|>A} e^{-itx - x^2/2}\,dx,$$

$$J_3 = \int_{\varepsilon B_n \leqslant |x| \leqslant \pi B_n/h} e^{-itx} f^{*n}\left(\frac{x}{B_n}\right)dx,$$

$$J_4 = \int_{A \leqslant |x| < \varepsilon B_n} e^{-itx} f^{*n}\left(\frac{x}{B_n}\right)dx,$$

where $A > 0$ is sufficiently large and $\varepsilon > 0$ is sufficiently small (more precise values for them will be chosen later).

By virtue of the corollary of the theorem proved in the preceding section, the relation

$$f^{*n}\left(\frac{t}{B_n}\right) \to e^{-t^2/2}\,(n \to \infty).$$

is satisfied uniformly for t in any finite interval. But from this it follows that

$$J_1 \to 0 \quad (n \to \infty).$$

whatever may be the value of the constant A. The integral J_2 can be estimated by means of the inequality

$$|J_2| \leqslant \int_{|x|>A} e^{-x^2/2}\,dx \leqslant \frac{2}{A}\int_A^{\infty} xe^{-x^2/2}\,dx = \frac{2}{A}e^{-A^2/2}.$$

By choosing A sufficiently large, one can make J_2 arbitrarily small. According to the inequality (1), it follows:

$$J_3 \leqslant \int_{\varepsilon B_n \leqslant |x| \leqslant \pi B_n/h} \left|f^{*}\left(\frac{x}{B_n}\right)\right|^n dx \leqslant e^{-nc_0}\left(\frac{\pi}{h} - \varepsilon\right).$$

From this it is clear that as $n \to \infty$,

$$J_3 \to 0.$$

To estimate the integral J_4, note that the existence of the variance implies the existence of the second derivative of the function $f^*(t)$. Therefore by the use of (3) of Section 32, one can write the expansion

$$f^*(t) = 1 - \frac{\sigma^2 t^2}{2} + o(t^2),$$

which is valid in the neighborhood of the point $t = 0$. Then for $|t| \leqslant \varepsilon$ if ε is sufficiently small, one obtains

$$|f^*(t)| < 1 - \frac{\sigma^2 t^2}{4} < e^{-\sigma^2 t^2/4}.$$

If $|x| \leqslant \varepsilon B_n$ then

$$\left|f^*\left(\frac{x}{B_n}\right)\right|^n < e^{-n\sigma^2 t^2/4B_n^2} = e^{-t^2/4}.$$

Therefore

$$|J_4| \leqslant 2\int_A^{\varepsilon B_n} e^{-t^2/4}\,dt < 2\int_A^{\infty} e^{-t^2/4}\,dt.$$

By choosing A sufficiently large, one can make the integral J_4 arbitrarily small. This proves the theorem.

There exists another case in which it is natural to investigate the local behavior of distribution functions of sums. This is the case of continuous distributions.

The problem which can be posed here is: When do the density functions of normalized sums converge to the normal density function if the corresponding distribution functions converge to the normal distribution?

It was shown that for identically distributed independent random variables with finite variance the sufficient condition is the condition of integrability of density function of summands in some power $S > 1$.

If one rejects this condition then it is possible to demonstrate examples of random variables which have densities and exist only in a bounded interval but for which in any case the local theorem does not hold (see monograph by Gnedenko and Kolmogorov).

49 EXERCISES

1. Prove that as $n \to \infty$

$$\frac{1}{\Gamma\left(\frac{n}{2}\right)}\sqrt{\left(\frac{n}{2}\right)^n}\int_0^{1+t\sqrt{2/n}} z^{(n/2)-1} e^{-nz/2}\,dz \to \frac{1}{\sqrt{2\pi}}\int_{-\infty}^{t} e^{-1/2z^2}\,dz.$$

Hint. Use the Liapounov's Theorem applied to χ^2-distribution.

2. The random variables

$$\xi_n = \begin{cases} -n^\alpha, & \text{with probability } 0.5 \\ +n^\alpha, & \text{with probability } 0.5 \end{cases}$$

are independent. Prove that for $\alpha > -0.5$ the Liapounov's Theorem is applicable to them.

3. Prove that as $n \to \infty$

$$e^{-n}\sum_{k=0}^{n}\frac{n^k}{k!} \to \frac{1}{2}.$$

Hint. Apply the Liapounov's Theorem to a sum of random variables distributed according to the Poisson law with parameter $\lambda = 1$.

4. Let p_i denote the probability of occurrence of the event A in the i-th trial and μ be the number of times A occurs in n independent trials. Prove that

$$\mathbf{P}\left\{\frac{\mu - \sum_{k=1}^{n} p_k}{\sqrt{\sum_{k=1}^{n} p_k q_k}} < x\right\} \to \frac{1}{\sqrt{2\pi}}\int_{-\infty}^{x} e^{-z^2/2}\,dz$$

if and only if

$$\sum_{i=1} p_i q_i = \infty$$

5. Prove that under the conditions of the preceding example the requirement

$$\sum_{i=1}^{\infty} p_i q_i = \infty$$

is sufficient not only for the Integral Theorem but also for the Local Theorem.

Chapter 10

THE THEORY OF INFINITELY DIVISIBLE DISTRIBUTIONS

For a long time, the central problem of probability theory was to discover conditions under which distributions for sums of independent random variables would converge to the normal distribution. Very general sufficient conditions for this convergence were found by Liapounov (see Chapter 9). Attempts to broaden the Liapounov conditions were successful only in recent years, with the determination of conditions that were not only sufficient but, under quite natural restrictions, also necessary.

In parallel with the solutions of the classical problems, there was created and developed a new theoretical approach to limit theorems for sums of independent random variables. This direction was closely connected with the introduction and development of the theory of stochastic (random) processes. In the first place, the question arose: what distribution, except the normal one, could be the limit distributions for sums of independent random variables. It was found that the normal distribution did not exhaust the class of limit distributions. The problem thereupon arose of determining the conditions that had to be imposed on the summands in order that the distribution functions for sums might converge to this or some other limit distribution. In the present chapter, we make it our object to present some of the recent investigations concerning limit theorems for sums of independent random variables. In doing so, we confine ourselves to the case in which the summands have finite variances. To consider this problem without this restriction would require more difficult computation. The complete solution to this problem the reader can find in the monograph of Gnedenko and Kolmogorov already mentioned. As a simple consequence of the general theorems presented, the necessary and sufficient conditions for convergence of the distribution functions of sums to the normal distribution will be found.

The last section of this chapter deals with the new direction of research, summation of a random number of random variables.

50 INFINITELY DIVISIBLE DISTRIBUTIONS AND THEIR FUNDAMENTAL PROPERTIES

A distribution $\Phi(x)$ is called infinitely divisible if, for any natural number n, its characteristic function can be presented as the power n of some other characteristic function.

The investigations of recent years have shown that infinitely divisible distributions play an important role in varied problems of probability theory. In particular, the class of limit distributions for sums of independent random variables has been found to coincide with the class of infinitely divisible distributions.

Let us now consider some properties of infinitely divisible distributions that will be needed in the sequel. The discussion begins with a proof of the fact that the normal and the Poisson distributions are infinitely divisible. Indeed, the characteristic function for the normal distribution with mathematical expectation a and variance σ^2 is given as

$$\varphi(t) = e^{iat-(1/2)\sigma^2 t^2}.$$

For any n, the n-th root of $\varphi(x)$ is again the characteristic function for a normal distribution, but with mathematical expectation a/n and variance σ_2/n.

Let us slightly generalize the concept of the Poisson distribution which was encountered earlier. The random variable ξ is said to be distributed according to the Poisson distribution if it can assume only the values $ak+b$, where a and b are real constants and $k=0,1,\ldots$ and

$$\mathsf{P}\left\{\xi = ak+b\right\} = \frac{e^{-\lambda}\lambda^k}{k!}, \tag{1}$$

where λ is a positive constant. As can easily be shown, the characteristic function for the distribution (1) is given by

$$\varphi(t) = e^{\lambda(e^{iat}-1)+ibt}.$$

Thus, for any n the n-th root of $\varphi(t)$ is again the characteristic function for a Poisson distribution, but with different parameters: $\alpha, \lambda/n$, and b/n.

THEOREM 1. *The characteristic function for an infinitely divisible distribution never vanishes.*

PROOF: Let $\Phi(x)$ be an infinitely divisible distribution and $\varphi(t)$ its characteristic function. Then for any n, one has by definition

$$\varphi(t) = \{\varphi_n(t)\}^n, \tag{2}$$

where $\varphi_n(t)$ is some characteristic function. By virtue of the continuity of the function $\varphi(t)$ there exists an interval $|t| \leqslant a$ in which $\varphi(t) \neq 0$. It is clear that $\varphi_n(t) \neq 0$ also holds in this interval. For sufficiently large n, the quantity

$$|\varphi_n(t)| = \sqrt[n]{|\varphi(t)|}$$

can be made arbitrarily close to 1 uniformly in $t(|t| \leqslant a)$.

Let us now take two mutually independent random variables η_1 and η_2 each of which is distributed according to some distribution $F(x)$, and let us consider their difference $\eta = \eta_1 - \eta_2$. The characteristic function of the variable η is

$$f^*(t) = \mathbf{E}e^{it(\eta_1 - \eta_2)} = |\mathbf{E}e^{it\eta_1}|^2 = |f(t)|^2.$$

Thus the square of the absolute value of any characteristic function is a characteristic function.

Further, since a real characteristic function has the form

$$f(t) = \int \cos xt \, dF(x),$$

then one may consequently write the inequality

$$1 - f(2t) = \int (1 - \cos 2xt) \, dF(x) = 2 \int \sin^2 xt \, dF(x)$$

$$= 2 \int (1 - \cos xt)(1 + \cos xt) \, dF(x)$$

$$\leqslant 4 \int (1 - \cos xt) \, dF(x) = 4(1 - f(t)).$$

From this one sees that the function $|\varphi(t)|^2$ satisfies the inequality

$$1-|\varphi_n(2t)|^2 \leqslant 4(1-|\varphi_n(t)|^2).$$

As it follows from the last inequality, if n is sufficiently large, so that $1-|\varphi(t)|^2<\varepsilon$, then in the same interval

$$1-|\varphi_n(2t)| \leqslant 1-|\varphi_n(2t)|^2 \leqslant 4(1-|\varphi_n(t)|^2) \leqslant 8(1-|\varphi_n(t)|) < 8\,\varepsilon.$$

Thus, in the interval $|t| \leqslant 2a$

$$1-|\varphi(t)|^2 < 8\varepsilon.$$

Hence, for sufficiently large values of n the function $\varphi_n(t)$ does not vanish in the interval $|t| \leqslant 2a$, and therefore, neither does $\varphi(t)$.

Similarly, one can show that $\varphi(t) \neq 0$ in the interval $|t| < 4a$, etc.

This proves the theorem.

THEOREM 2. *The distribution function for a sum of independent random variables having infinitely divisible distribution functions is itself infinitely divisible.*

PROOF: It obviously suffices to prove the theorem for the case of two summands. If $\varphi(t)$ and $\psi(t)$ are the characteristic functions for the summands then by assumption one has for any n:

$$\varphi(t)=\{\varphi_n(t)\}^n, \quad \psi(t)=\{\psi_n(t)\}^n,$$

where $\varphi(t)$ and $\psi(t)$ are characteristic functions. Therefore, the characteristic function for the sum satisfies the equality

$$\chi(t)=\varphi(t)\psi(t)=\{\varphi_n(t)\cdot\psi_n(t)\}^n$$

for any n.

THEOREM 3. *The limit distribution function (in the sense of weak convergence) of a sequence of infinitely divisible distribution functions is itself infinitely divisible.*

PROOF: Let the sequence $\Phi^{(k)}(x)$ of infinitely divisible distribution functions converge weakly to the distribution function $\Phi(x)$. Then

$$\lim_{k\to\infty} \varphi^{(k)}(t)=\varphi(t) \tag{3}$$

uniformly in any finite interval t. By assumption of the theorem, the functions

$$\varphi_n^{(k)}(t)=\sqrt[n]{\varphi^{(k)}(t)} \tag{4}$$

are characteristic functions for any n[1]. From (3) one can conclude that for any n

$$\lim_{k\to\infty} \varphi_n^{(k)}(t)=\varphi_n(t). \tag{5}$$

The continuity of $\varphi_n^{(k)}(t)$ implies the continuity of $\varphi_n(t)$. By the limit theorem for characteristic functions, $\varphi_n(t)$ is a characteristic function. From (3), (4) and (5) it follows that for any n equality

$$\varphi(t)=\{\varphi_n(t)\}^n$$

holds. Q.E.D.

51 CANONICAL REPRESENTATION OF INFINITELY DIVISIBLE DISTRIBUTIONS

In the following, the discussion will be restricted by an investigation of infinitely divisible distributions with finite variance. The purpose of this section is to give the proof of the following theorem which completely characterizes the class of distribution functions under investigation. This theorem was formulated and proved in 1932 by Kolmogorov.

THEOREM. *In order that a distribution function $\Phi(x)$ with a finite variance be infinitely divisible it is necessary and sufficient that the logarithm of its characteristic function be given by*

$$\ln\varphi(t)=i\gamma t+\int\left\{e^{itx}-1-itx\right\}\frac{1}{x^2}dG(x), \tag{1}$$

where γ is a real constant and $G(x)$ is a non-decreasing function with bounded variation.

[1]Here the principal value of the root is considered.

PROOF: Let us suppose first of all that $\Phi(x)$ is an infinitely divisible distribution and that $\varphi(t)$ is its characteristic function. Then for any n

$$\varphi(t) = \{\varphi_n(t)\}^n,$$

where $\varphi_n(t)$ is some characteristic function. Since $\varphi(t) 0$, this equation is equivalent to the following[2]

$$\ln \varphi(t) = n \ln \varphi_n(t) = n \ln[1 + \varphi_n(t) - 1)].$$

For any T as $n \to \infty$

$$\varphi_n(t) \to 1,$$

uniformly in the interval $|t| < T$ and therefore, in any finite interval of the t-axis the quantity $|\varphi_n(t) - 1|$ can be made smaller than any given number if n is taken sufficiently large. Consequently, one can use the relation

$$\ln[1 + (\varphi_n(t) - 1] = (\varphi_n(t) - 1)(1 + 0(1)),$$

which yields:

$$\ln \varphi(t) = \lim_{n \to \infty} n(\varphi_n(t) - 1) = \lim_{n \to \infty} n \int (e^{itx} - 1) d\Phi_n(x), \tag{2}$$

Where $\Phi(x)$ is the distribution function having $\varphi_n(t)$ as its characteristic function. From the definition of mathematical expectation and the relation between the functions $\varphi_n(t)$ and $\varphi(t)$ it follows that

$$n \int x d\Phi_n(x) = \int x d\Phi(x).$$

Denote this quantity by γ. Then equation (2) may be written in the following form:

$$\ln \varphi(t) = i\gamma t + \lim_{n \to \infty} n \int \left\{ e^{itx} - 1 - itx \right\} d\Phi_n(x).$$

Let us now define

$$G_n(x) = n \int_{-\infty}^{x} u^2 d\Phi_n(u).$$

[2]It is understood that the principal value of the logarithm is to be taken here.

Evidently, the functions $G_n(z)$ do not decrease with increasing argument, and $G_n(-\infty)=0$. Moreover, the functions $G_n(x)$ are uniformly bounded. The last assertion follows from the properties of the variance and the relation between the functions $\Phi(x)$ and $\Phi_n(x)$. In fact,

$$G_n(+\infty)=n\int u^2\,d\Phi_n(u)$$
$$=n\left[\int u^2\,d\Phi_n(u)-\left(\int u\,d\Phi_n(u)\right)^2\right]$$
$$+n\left(\int u\,d\Phi_n(u)\right)^2=\sigma^2+\frac{1}{n}\gamma^2, \tag{3}$$

where σ^2 is the variance of the distribution $\Phi(x)$. In new terms one has (see property 6 of the Stieltjes integral in Section 22):

$$\ln\varphi(t)=i\gamma t+\lim_{n\to}\int(e^{itx}-1-itx)\frac{1}{x^2}\,dG_n(x).$$

According to the Helly's First Theorem, one may select from the sequence of functions $G_n(x)$ a subsequence that converges to some limit function $G(x)$. If $A<0$ and $B>0$ are points of continuity of the function $G(x)$ then by virtue of the Helly's Second Theorem,

$$\int_A^B(e^{itx}-1-itx)\frac{1}{x^2}\,dG_{n_k}(x)\to\int_A^B(e^{itx}-1-itx)\frac{1}{x^2}\,dG(x). \tag{4}$$

as $k\to\infty$. It is known that

$$|e^{itx}-1-itx|\leqslant|e^{itx}-1|+|tx|\leqslant|tx|+|tx|=2|t|\cdot|x|,$$

and therefore

$$\left|\int_{-}^{A}+\int_B^\infty(e^{itx}-1-itx)\frac{1}{x^2}\,dG_{n_k}(x)\right|$$
$$\leqslant\int_{-}^{A}+\int_B^\infty\frac{|e^{itx}-1-itx|}{x^2}\,dG_{n_k}(x)$$
$$\leqslant 2|t|\left(\int_{-\infty}^{A}+\int_B^\infty\frac{1}{|x|}\,dG_{n_k}(x)\right)\leqslant\frac{2|t|}{\Gamma}\left(\int_{-\infty}^{A}+\int_B^\infty dG_{n_k}(x)\right)$$
$$\leqslant\frac{2|t|}{\Gamma}\max_{1\leqslant k<\infty}\int dG_{n_k}(x),$$

where $\Gamma = \min(|A|, B)$. Since the variations of the functions $G_{n_k}(u)$ are uniformly bounded, for any $\varepsilon > 0$ one can obtain the inequality

$$\left|\int_{-\infty}^{A} + \int_{B}^{\infty} (e^{itx} - 1 - itx)\frac{1}{x^2}\,dG_{n_k}(x)\right| < \frac{\varepsilon}{2} \tag{5}$$

for all t in some finite interval and for all k, by choosing A and B sufficiently large.

It follows from (4) and (5) that for any $\varepsilon > 0$ and for all t in an arbitrary finite interval the inequality

$$\left|\int (e^{itx} - 1 - itx)\frac{1}{x^2}\,dG_{n_k}(x) - \int (e^{itx} - 1 - itx)\frac{1}{x^2}\,dG(x)\right| < \varepsilon$$

holds for sufficiently large values of n. In other words,

$$\lim_{k\to\infty} \int (e^{itx} - 1 - itx)\frac{1}{x^2}\,dG_{n_k}(x) = \int (e^{itx} - 1 - itx)\frac{1}{x^2}\,dG(x).$$

Thus it has been shown that the logarithm of the characteristic function of any infinitely divisible distribution is expressible in the form (1). One has now to prove the converse statement that any function whose logarithm is representable by formula (1) is the characteristic function of some infinitely divisible distribution.

For any ε $(0 < \varepsilon < 1)$, the integral

$$\int_{\varepsilon}^{1/\varepsilon} (e^{itx} - 1 - itx)\frac{1}{x^2}\,dG(x) \tag{6}$$

by the definition of the Stieltjes integral, is the limit of the sum

$$\sum_{s=1}^{n} (e^{it\bar{x}_s} - 1 - it\bar{x}_s)\frac{1}{\bar{x}_s^2}(G(x_{s+1}) - G(x_s)),$$

where $x_1 = \varepsilon$, $x_{n+1} = 1/\varepsilon$, $x_s \leqslant \bar{x}_s \leqslant x_s + 1$, and $\max(x_{s+1} - x_s) \to 0$. Each term of this sum is the logarithm of the characteristic function for some Poisson distribution. By Theorems 2 and 3, the integral (6) is the logarithm of the characteristic function of some infinitely divisible distribution. By taking the limit as $\varepsilon \to \infty$, we can see that the same is true for the integral

$$\int_{x>0} (e^{itx} - 1 - itx)\frac{1}{x^2}\,dG(x). \tag{7}$$

In an analogous way, one can show that the integral

$$\int_{x<0} (e^{itx} - 1 - itx)\frac{1}{x^2}\,dG(x) \tag{8}$$

is the logarithm of the characteristic function of some infinitely divisible distribution. The integral on the right-hand side of the expression (1) is the sum of the integrals (7) and (8) and the quantity

$$i\gamma t - \frac{1}{2}t^2(G(+0) - G(-0)).$$

This last term is the logarithm of the characteristic function of the normal distribution. It follows from Theorem 2 that the function $\varphi(t)$ given by the expression (1) is the characteristic function of some infinitely divisible distribution.[3] It now remains to show that the representation of log $\varphi(t)$ in expression (1) is unique, i.e., that the function $G(x)$ and the constant γ are uniquely determined by specifying $\varphi(t)$.

By differentiation of (1), one finds:

$$\frac{d^2}{dt^2}\ln\varphi(t) = -\int e^{itx}\,dG(x). \tag{9}$$

From the theory of characteristic functions, one knows that the function $G(x)$ in this integral is uniquely determined by (d^2/dt^2) $\log\varphi(t)$. In the course of proving the theorem, one saw that the constant γ was the mathematical expectation and hence it too is uniquely determined by the function $\varphi(t)$.

Finally, one could indicate the probabilistic significance of the total variation of the function $G(x)$. One knows that if a random variable ξ

[3]It has now been proven that every infinitely divisible distribution is either the composition of a finite number of Poisson and normal distributions or the limit of a uniformly convergent sequence of such distributions. Thus it was shown that the normal and the Poisson distributions are those fundamental elements that go to make up any given infinitely divisible distribution.

is distributed according to the distribution $\Phi(x)$, then (see (6) of Section 33)

$$\mathsf{D}\xi = -\left[\frac{d^2}{dt^2}\ln\varphi(t)\right]_{t=0};$$

and it therefore follows from (9) that

$$\mathsf{D}\xi = \int dG(x) = G(+\infty).$$

As examples, let us consider the canonical representations of the normal and the Poisson distributions.

For the normal distribution with variance σ^2 and mathematical expectation a,

$$\gamma = a \quad \text{and} \quad G(x) = \begin{cases} 0 & \text{for} \quad x<0 \\ \sigma^2 & \text{for} \quad x>0 \end{cases}$$

Indeed, this function and the constant γ lead to the distribution in question, since

$$\int\left\{e^{itx} - 1 - itx\right\}\frac{1}{x^2}\,dG(x)$$

$$= \lim_{u\to 0}\frac{e^{itu} - 1 - itu}{u^2}[G(+0) - G(-0)] = -\frac{t^2\sigma^2}{2},$$

and by virtue of the uniqueness of the canonical representation, no other function $G(x)$ can yield the normal distribution.

In an analogous way, it is easy to show that the $G(x)$ corresponding to the Poisson distribution with the characteristic function

$$\varphi(t) = e^{\lambda(e^{ita} - 1) + ibt}$$

is the function having a single jump at the point a:

$$G(x) = \begin{cases} 0 & \text{for} \quad x<0 \\ a^2\lambda & \text{for} \quad x>0 \end{cases}$$

and $\gamma = b + a\lambda$.

52 A LIMIT THEOREM FOR INFINITELY DIVISIBLE DISTRIBUTIONS

It has been already shown that if a sequence of infinitely divisible distributions converges to a limit distribution law, then this limit distribution is itself infinitely divisible. Let us now indicate conditions which are sufficient for the convergence of a given sequence of infinitely divisible distribution functions to the limit.

THEOREM. *In order for a sequence* $\{\Phi_n(x)\}$ of infinitely divisible distribution functions to converge to some distribution function $\Phi(x)$ as $n \to \infty$ *and for their respective variances to converge to the variance of the limit distribution, it is necessary and sufficient that there exist a constant* γ *and function* $G(x)$ such that as $n \to \infty$

(1) $G_n(x)$ *converges weakly to* $G(x)$,
(2) $G_n(\infty) \to G(\infty)$
(3) $\gamma_n \to \gamma$

where γ_n *and* $G_n(x)$ *are defined by formula* (1) *of Section 43 for the distribution* $\Phi_n(x)$. *The constant* γ *and the function* $G(x)$ *define the limit distribution* $\Phi(x)$ *by the same formula.*

PROOF: The sufficiency of the conditions of the theorem immediately follows from the Helly's Second Theorem. In fact, from these assumptions and formula (1) of Section 43, it follows that

$$\ln \varphi_n(t) \to \ln \varphi(t)$$

as $n \to \infty$ uniformly in every finite interval t.

In the preceding section it was shown that the integrals

$$\int dG_n(u) \quad \text{and} \quad \int dG(u)$$

were the variances of the distributions $\Phi_n(x)$ and $\Phi_n(x)$, respectively. Therefore the second condition of the theorem is nothing other than a requirement of the variances convergence.

Now suppose that it is known that

$$\Phi_n(x) \to \Phi(x) \tag{1}$$

as $n \to \infty$ and that the variances of the distributions $\Phi_n(x)$ converge to the variance of the limit distribution $\Phi_n(x)$. It will be proved that these requirements imply all of the conditions of the theorem. As has just been noted, no additional arguments are needed as regards condition 2. From this condition it follows that the total variations of the functions $G_n(u)$ are uniformly bounded. One can therefore apply the Helly's First Theorem and from the sequence of functions $G_n(u)$ select a subsequence $G_{n_k}(u)$ which converges to some limit $G_{\infty}(u)$ as $k \to \infty$. The goal is to prove that

$$G_{\infty}(u) = G(u).$$

To do this, let us first establish that

$$J_k = \int \left\{e^{itu} - 1 - itu\right\} \frac{1}{u^2}\, dG_{n_k}(u) \to J_\infty = \int \left\{e^{itu} - 1 - itu\right\} \frac{1}{u^2}\, dG_\infty(u) \quad (2)$$

as $k \to \infty$. Let $A < 0$ and $B > 0$ be points of continuity of the function $G_{\infty}(u)$. Then by the Helly's Second Theorem

$$\int_A^B \left\{e^{itu} - 1 - itu\right\} \frac{1}{u^2}\, dG_{n_k}(u) \to \int_A^B \left\{e^{itu} - 1 - itu\right\} \frac{1}{u^2}\, dG_{\infty}(u) \quad (3)$$

as $k \to \infty$. On the other hand, from the inequality

$$|e^{itx} - 1 - itx| \leqslant 2|tx|$$

follows that

$$\begin{aligned} L_k &= \left| \int_{-\infty}^{A} + \int_B^\infty \left\{e^{itu} - 1 - itu\right\} \frac{1}{u^2}\, dG_{n_k}(u) \right| \\ &\leqslant 2|t| \left| \int_{-\infty}^{A} + \int_B^\infty \frac{1}{|u|}\, dG_{n_k}(u) \right| \\ &\leqslant \frac{2|t|}{2} \left(\int_{-}^{A} + \int_B^\infty dG_{n_k}(u) \right) \leqslant \frac{2|t|}{\Gamma} \int dG_{n_k}(u), \end{aligned}$$

where $\Gamma = \min(-A, B)$. By virtue of the uniform boundedness of the total variations of the functions $G_n(u)$, for any $\varepsilon > 0$ it is possible to choose absolute values of A and B sufficiently large that

$$L_k < \varepsilon. \quad (4)$$

Similarly, the inequality

$$\left|\int_{-\infty}^{A} + \int_{B}^{\infty}\left\{e^{itu} - 1 - itu\right\}\frac{1}{u^2}dG_\infty(u)\right| < \varepsilon \tag{5}$$

holds for any $\varepsilon > 0$ if B and A are sufficiently large. From (3), (4) and (5), one concludes that for any $\varepsilon > 0$

$$|J_k - J_\infty| < 3\varepsilon$$

for sufficiently large values of k. This proves the relation (2). From (1), one sees that

$$\lim_{n\to\infty} \ln \varphi_n(t) = \lim_{n\to\infty}\left(i\gamma_n t + \int\left\{e^{itu} - 1 - itu\right\}\frac{1}{u^2}dG_n(u)\right)$$

$$= \ln \varphi(t) = i\gamma t + \int\left\{e^{itu} - 1 - itu\right\}\frac{1}{u^2}dG(u),$$

or

$$\lim_{n\to\infty}\left(i\gamma_{n_k} + \int\left\{e^{itu} - 1 - itu\right\}\frac{1}{tu^2}dG_{n_k}(u)\right)$$

$$= i\gamma + \int\left\{e^{itu} - 1 - itu\right\}\frac{1}{tu^2}dG(u). \tag{6}$$

From the inequality

$$|e^{itu} - 1 - itu| \leqslant t^2u^2/2$$

and the uniform boundedness of the total variations of the functions $G_{n_k}(u)$ one concludes that as $t \to 0$

$$\left|\int\left(e^{itu} - 1 - itu\right)\frac{1}{tu^2}dG_{n_k}(u)\right| \leqslant \left|t\int\left(e^{itu} - 1 - itu\right)\frac{1}{t^2u^2}dG(u)\right| \to 0$$

uniformly in n. Therefore, for $t \to 0$ expression (6) yields:

$$\lim_{k\to} \gamma_{n_k} = \gamma,$$

but on the other hand, by (2) and (7),

$$\ln \varphi(t) = i\gamma t + \int\left\{e^{iut} - 1 - iut\right\}\frac{1}{u^2}dG_\infty(u). \tag{7}$$

Since formula (1) of Section 43 gives a unique representation for infinitely divisible distributions, one concludes that $G_{\ }(u)=G(u)$. Thus, any convergent subsequence of functions $G_{n_k}(u)$ converges to the function $G(u)$, and at the same time the sequence of constants γ_{n_k} converges to γ. It is now easy to show that the sequence $G_n(u)$ itself also converges to $G(u)$, which implies that, at the same time, $\lim_{n\to\infty}\gamma_n=\gamma$. If this were not so then one could find a point of continuity of the function $G(u)$, say, $u=c$, and a subsequence of functions $G_{n_k}(u)$ which converges at the point $u=c$ to a value distinct from $G(c)$. By the Helly's First Theorem, one can select from this subsequence a convergent subsequence $G_{n_{k_r}}(u)$.

From the above, it follows that

$$\lim_{r\to\infty} G_{n_{k_r}}(u)=G(u)$$

at all points of continuity of the function $G(u)$. This contradicts the assumption above. Thus, at all points of continuity of the function $G(u)$,

$$\lim_{n\to\infty} G_n(u)=G(u);$$

and, as it has been shown, it immediately follows from this that

$$\lim_{n\to\infty} \gamma_n=\gamma.$$

This completes the proof.

53 LIMIT THEOREMS FOR SUMS: FORMULATION OF THE PROBLEM

Let there be given the sequence of series

$$\left.\begin{matrix} \xi_{11}, & \xi_{12}, & \dots & \xi_{1k_1}, \\ \xi_{21}, & \xi_{22}, & \dots & \xi_{2k_2}, \\ \dots & \dots & \dots & \dots \\ \xi_{n1}, & \xi_{n2}, & \dots & \xi_{nk_n}, \\ \dots & \dots & \dots & \dots \end{matrix}\right\} \qquad (1)$$

in which the random variables in each series (row) are independent. The question is: To what limit distribution function can the distribution functions of the sequence of sums

$$\zeta_n = \xi_{n1} + \xi_{n2} + \cdots + \xi_{nk_n}$$

converge as $n \to \infty$, and under what conditions ?

In what follows, let us confine ourselves to the study of *elementary systems*, i.e., to such sequences (1) which satisfy the following conditions:

(1) The variables ξ_{nk} have finite variances;

(2) The variances of the sums ζ_n are bounded from above by a constant C which is independent of n;

(3)

$$\beta_n - = \max_{1 \leqslant k \leqslant n} D\xi_{nk} \to 0$$

as $n \to \infty$.

The last requirement means that the effect of the individual terms in the sum becomes smaller and smaller with increasing n.

· The limit theorems for sums considered earlier are clearly contained in this general scheme. Thus, in the DeMoivre-Laplace and Liapounov Theorems one had the following sequence:

$$\xi_{n1}, \xi_{n2}, \ldots, \xi_{nn},$$

where

$$\xi_{nk} = \frac{\xi_k - \mathbf{E}\xi_k}{\sqrt{\sum_{k=1}^{n} \mathbf{D}\xi_k}} \qquad (1 \leqslant k \leqslant n;\ n = 1, 2, \ldots).$$

In the theorems concerning the law of large numbers in the form of Bernoulli, Chebyshev, and Markov, the quantities ξ_{nk} were represented by

$$\xi_{nk} = \frac{\xi_k - \mathbf{E}\xi_k}{n}.$$

54 LIMITS THEOREMS FOR SUMS

Let there be given an elementary system. Denote the distribution function of the random variable ξ_{nk} by $F_{nk}(x)$ and the distribution of the random variable $\bar{\xi}_{nk} = \xi_{nk} - E\xi_{nk}$ by $\bar{F}_{nk}(x)$.
It is clear that

$$\bar{F}_{nk}(x) = F_{nk}(x + \mathsf{E}\xi_{nk}).$$

THEOREM 6. *The distribution functions of the sequence of sums*

$$\zeta_n = \xi_{n1} + \xi_{n2} + \cdots + \xi_{nk_n} \tag{1}$$

converge to a limit distribution function as $n \to \infty$ if and only if the sequence of infinitely divisible distributions whose characteristic functions have logarithms given by the formula

$$\psi_n(t) = \sum_{k=1}^{k_n} \left\{ it\mathsf{E}\xi_{nk} + \int (e^{itx} - 1)d\bar{F}_{nk}(x) \right\}^{*)} \tag{2}$$

converge to a limit distribution.

The limit distributions of both sequences coincide.[4]

PROOF: The characteristic function of the sum (1) is

$$f_n(t) = \prod_{k=1}^{k_n} f_{nk}(t) = e^{it\sum_{k=1}^{k_n} \mathsf{E}\xi_{nk}} \prod_{k=1}^{k_n} \bar{f}_{nk}(t), \tag{3}$$

where $f_{nk}(t)$ is the characteristic function of the random variable ξ_{nk} and $\bar{f}_{nk}(t)$ is that of the random variable $\bar{\xi}_{nk}$.

---

[4] If we introduce the notation

$$\gamma_n = \sum_{k=1}^{k_n} E\xi_{nk}, \qquad G_n(x) = \sum_{k=1}^{k_n} \int_{-\infty}^{x} x^2 d\bar{f}_{nk}(x)$$

and to notice that $\int x d\bar{F}_{nk}(x) = 0$ then the function $\Psi(t)$ can be written in the form

$$\psi_n(t) = i\gamma_n t + \int \left\{ e^{itu} - 1 - itu \right\} \frac{1}{u^2} dG_n(u).$$

This means that $\chi_n(t)$ is thje logarithm of the characteristic function of some infinitely divisible distribution.

The variance of ζ_n coincides with that of the infinitely divisible distribution (2).

It is known that the distribution functions of the sequence of sums (1) converges to a limit $\Phi(x)$, if and only if $f_n(t) \to \varphi(t)$ as $n \to \infty$ where $\varphi(t)$ is a continuous function. Then $\varphi(t)$ turns out to be the characteristic function of the distribution $\Phi(x)$.

Let us set

$$\alpha_{nk} = \bar{f}_{nk}(t) - 1.$$

For the quantities ξ_{nk} one has

$$\alpha_n = \max_{1 \leqslant k \leqslant k_n} |\alpha_{nk}| \to 0. \tag{4}$$

uniformly in every finite interval t. Indeed,

$$\alpha_{nk} = \int (e^{itx} - 1) dF_{nk}(x) = \int (e^{itx} - 1 - itx) dF_{nk}(x),$$

since

$$\mathbf{E}\bar{\xi}_{nk} = \int x dF_{nk}(x) = 0.$$

One knows that for all real values of α

$$|e^{i\alpha} - 1 - i\alpha| \leqslant \alpha^2/2;$$

therefore

$$|\alpha_{nk}| \leqslant \frac{t^2}{2} \int x^2 dF_{nk}(x) = \frac{t^2}{2} \mathbf{D}\xi_{nk}. \tag{5}$$

From (5) and the third condition for an elementary system there follows (4).

From (4), one first of all concludes that for arbitrary T, large enough n and $|t| \leqslant T$ the inequality

$$|\alpha_{nk}| < 1/2 \tag{6}$$

holds. In virtue of this one can apply the series expansion of the logarithm

$$\ln \bar{f}_{nk}(t) = \ln(1 + \alpha_{nk}) = \alpha_{nk} - \frac{\alpha_{nk}^2}{2} + \frac{\alpha_{nk}^3}{3} - \cdots = \alpha_{nk} + r_{nk}.$$

It is evident that

$$R_n = \left| \ln f_n(t) - \sum_{n=1}^{k_n} (i + \mathsf{E}\,\xi_{nk} + \alpha_{nk}) \right| = \left| \sum_{n=1}^{k_n} (\ln \bar{f}_{nk}(t) - \alpha_{nk}) \right|$$

$$\leqslant \sum_{k=1}^{k_n} \sum_{n=1} \frac{|\alpha_{nk}|^s}{s} \leqslant \frac{1}{2} \sum_{k=1}^{k_n} \frac{|\alpha_{nk}|^2}{1 - |\alpha_{nk}|}. \tag{7}$$

Formulas (5) and (6) lead to the inequality

$$R_n \leqslant \max_{1 \leqslant k \leqslant k_n} |\alpha_{nk}| \sum_{k=1}^{k_n} |\alpha_{nk}| \leqslant \frac{Ct^2}{2} \max_{1 \leqslant k \leqslant k_n} |\alpha_{nk}|.$$

By virtue of (4), one can conclude that

$$|\ln f_n(t) - \psi_n(t)| \to 0 \tag{8}$$

uniformly in every finite interval t as $n \to \infty$.

Thus the distribution functions of the sums ζ_n in any elementary system and the infinitely divisible distributions defined by (2) asymptotically converge as $n \to \infty$. This proves Theorem 6.

This theorem allows one to replace the investigation of sums of random variables (1) having, generally speaking, arbitrary distribution functions, by the investigation of infinitely divisible distributions. The latter turns out, as will be shown, to be quite simple in many cases.

THEOREM 7. *Every limit distribution of the sequence of distribution functions of the sums in an elementary system is infinitely divisible and has a finite variance and, conversely, every infinitely divisible distribution with a finite variance is the limit of the distribution functions of the sums in some elementary system.*

PROOF: From the preceding theorem follows that the limit distribution for the distribution functions of the sums (1) is the limit of infinitely divisible distributions and, consequently, by Theorem 3, this means that the limit distribution is infinitely divisible. Its variance is finite, since the variances of the sums are uniformly bounded according to the second condition for an elementary system. The converse statement, that every infinitely divisible distribution with a finite variance is the limit of distribution functions for sums follows immediately from the definition of infinitely divisible distributions.

THEOREM 8. *The distribution function of the sequence of sums* (1) *converge to some limit distribution function as* $n \to \infty$ *and their variances converge to the variance of the limit distribution if and only if there exist a function* $G(u)$ *and constant* γ *such that as* $n \to \infty$

$$\sum_{k=1}^{k_n} \int_{-}^{u} x^2 \, dF_{nk}(x) \to G(u) \tag{1}$$

at the points of continuity of $G(u)$,

$$\sum_{k=1}^{k_n} \int x^2 \, dF_{nk}(x) \to G(+\infty), \tag{2}$$

$$\sum_{k=1}^{k_n} \int x \, dF_{nk}(x) \to \gamma. \tag{3}$$

The logarithm of the characteristic function of the limit distribution is determined by expression (1) *of Section 43 where* $G(u)$ *and* γ *are just defined.*

PROOF: By introducing the notation

$$G_n(u) = \sum_{k=1}^{k_n} \int_{-}^{u} x^2 \, dF_{nk}(x)$$

and

$$\gamma_n = \sum_{k=1}^{k_n} \int x \, dF_{nk}(x),$$

one returns to the conditions of Theorem 5. This proves the theorem.

By slightly modifying the formulation of the latter theorem, one can obtain not only conditions for the existence of the limit distribution but also conditions for convergence to any given limit distribution.

THEOREM 9. *The distribution functions of the sequence of sums* (1) *converge to a given distribution function* $\Phi(x)$ *as* $n \to \infty$ *and the variances of the sums converge to the variance of the limit distribution if and only if the following conditions are satisfied for* $n \to \infty$

$$\sum_{k=1}^{k_n} \int_{-}^{u} x^2 \, dF_{nk}(x) \to G(u) \tag{1}$$

at the points of continuity of the function $G(x)$;

$$\sum_{k=1}^{k_n} \int x^2 \, dF_{nk}(x) \to G(\infty), \tag{2}$$

$$\sum_{k=1}^{k_n} \int x \, dF_{nk}(x) \to \gamma, \tag{3}$$

where the function $G(u)$ *and the constant* γ *are determined by formula* (1) *of Section 43 for the function* $\Phi(x)$.

55   CONDITIONS FOR CONVERGENCE TO THE NORMAL AND POISSON DISTRIBUTIONS

Let us now apply the results of the preceding section in deriving conditions for the convergence of sequences of distribution functions of sums to the normal and Poisson distributions.

THEOREM 10. *Let there be given an elementary system of independent random variables. The sequence of distribution functions for the sums*

$$\zeta_n = \xi_{n1} + \xi_{n2} + \cdots + \xi_{nk_n} \tag{1}$$

converges to the distribution

$$\Phi(x) = \frac{1}{\sqrt{2\pi}} \int_{-}^{x} e^{-x^2/2} \, dx,$$

as $n \to \infty$ *if and only if the conditions*

$$\sum_{k=1}^{k_n} \int x \, dF_{nk}(x) \to 0. \tag{1}$$

$$\sum_{k=1}^{k_n} \int_{|x|>\tau} x^2\, dF_{nk}(x) \to 0. \tag{2}$$

$$\sum_{k=1}^{k_n} \int_{|x|<\tau} x^2\, dF_{nk}(x) \to 1. \tag{3}$$

are satisfied for $n \to \infty$ (here τ is an arbitrary positive constant).

PROOF: From Theorem 9 it follows that the required conditions consist in the relations

$$\sum_{k=1}^{k_n} \int x\, dF_{nk}(x) \to 0,$$

$$\sum_{k=1}^{k_n} \int_{-\infty}^{u} x^2\, dF_{nk}(x) \to \begin{cases} 0 & \text{for} \quad u<0, \\ 1 & \text{for} \quad u>0, \end{cases}$$

$$\sum_{k=1}^{k_n} \int x^2\, dF_{nk}(x) \to 1.$$

being satisfied for $n \to \infty$. The first of these conditions coincides with the first assumption of the theorem and the remaining two are obviously equivalent to the second and third assumptions of the theorem.

This theorem assumes a particularly simple form if the elementary system in question is normalized beforehand by the conditions

$$\sum_{k=1}^{k_n} \int x^2\, dF_{nk}(x) = 1,$$

$$\int x\, dF_{nk}(x) = 0 \qquad (1 \leqslant k \leqslant k_n;\ n = 1, 2, \ldots). \tag{2}$$

THEOREM 11. *If the elementary system is normalized by relations* (2), *then the sequence of distribution functions of the sums* (1) *converges to the normal distribution if and only if for all $\tau > 0$*

$$\sum_{k=1}^{k_n} \int_{|x|>\tau} x^2\, dF_{nk}(x) \to 0 \tag{3}$$

for $n \to \infty$.

The proof of this theorem is obvious.

Condition (3) is called the Lindeberg condition, since in 1923 Lindeberg proved that it is sufficient for the distribution functions of sums to converge to the normal distribution. In 1935, Feller showed the necessity of this condition.

As another illustration of how the general limit theorems of the preceding section can be used, let us consider the convergence of distribution functions for elementary systems to the Poisson distribution

$$P(x)=\begin{cases}0 & \text{for } x\leqslant 0,\\ \sum\limits_{0\leqslant k<x} e^{-\lambda}\dfrac{\lambda^k}{k!} & \text{for } x>0.\end{cases} \tag{4}$$

If ξ is a random variable with the distribution (4), then $E\xi=\mathbf{D}\xi=\lambda$.

Let us confine ourselves to elementary systems for which

$$\left.\begin{aligned}\sum_{k=1}^{k_n}\mathbf{E}\xi_{nk}\to\lambda,\\ \sum_{k=1}^{k_n}\mathbf{D}\xi_{nk}\to\lambda.\end{aligned}\right\} \tag{5}$$

THEOREM 12. *Let there be given an elementary system satisfying the conditions* (5). *The distribution functions of the sums*

$$\zeta_n=\xi_{n1}+\xi_{n2}+\cdots+\xi_{nk_n}$$

converge to the distribution (4) *if and only if for any* $\tau>0$

$$\sum_{k=1}^{k_n}\int_{|x-1|>\tau} x^2\,dF_{nk}(x+\mathbf{E}\xi_{nk})\to 0 \quad (n\to\infty).$$

The proof of this theorem is left to the reader.

In Section 13 the Poisson Theorem was proved. It is easy to see that for $np_n=\lambda$ this theorem is a special case of the statement just proved. Indeed, let $\xi_{nk}(1\leqslant k<n)$ be a random variable which assumes the values 0 or 1 depending on whether the event A under the observation will or will not occur in the trial k of the series n of the trials. Here

$$\mathbf{P}\left\{\xi_{nk}=1\right\}=\frac{\lambda}{n} \quad\text{and}\quad \mathbf{P}\left\{\xi_{nk}=0\right\}=1-\frac{\lambda}{n}.$$

It is clear that the sum

$$\mu_n = \xi_{n1} + \xi_{n2} + \cdots + \xi_{nn}$$

is the number of times the event A occurs in the series n of trials.

According to the Poisson Theorem, the distribution functions of the variables μ_n converge to the Poisson distribution (4) as $n \to \infty$. This result also follows from the theorem just formulated, since all of the conditions of the theorem are satisfied in the this case.

General theorems concerning the approach of distribution functions of the sums (1) to some infinitely divisible distribution function, proved under broader assumptions than considered, also enable one to obtain necessary and sufficient conditions for the law of large numbers (in the case of independent summands). For this, see the monograph by Gnedenko and Kolmogorov, referred to above.

56 SUM OF A RANDOM NUMBER OF RANDOM VARIABLES

In many practical cases one meets with the problem of summation of a random number of random variables. Consider examples.

Consumers come to buy varied goods at a shop. Each of them spends a random amount of money for the total purchase. The number of customers is also random. Thus the day's profit can be represented as a sum of random number of random variables.

Return to the problem concerning reliability of a duplicate system considered in Section 38. Consider the random variable ζ which is the time of failure-free operation of the duplicated system. Depending on chance this random variable might take different values:

$\zeta = \xi_1 + \xi_2$ is the sum of failure-free operation of main and redundant units if $\eta_1 > \xi_2$ (the failed main unit was still being repaired when the redundant unit also failed);

$\zeta = \xi_1 + \xi_2 + \xi_3$ is the sum of failure-free operation of main, redundant, and again main units if $\eta_1 < \xi_2$ but $\eta_2 > \xi_3$ (the repaired main unit failed before the redundant unit was repaired);

In general, for $n \geq 2$ one has $\zeta = \xi_1 + \xi_2 + \cdots + \xi_n$ with the probability $\alpha^{n-2}(1-\alpha)$ (first $n-2$ times a unit under current repair has been renewed before the failure of the unit at the operating place).

This is another example of summation of random number of random variables. The number of summands is distributed by a geometric law.

The theorem of convergence of the sum described can be formulated using the following limit theorem.

THEOREM. *Let random variables of the sequence $\{\xi_n\}$ be independent and identically distributed with the distribution function $F(x)$ and mathematical expectation $a>0$. Further, let $\{\nu_n\}$ be a sequence of integer random variables and ν_n distributed geometrically with parameter α_n, i.e., $P\{\nu_n=k\}=(1-\alpha_n)\alpha_n^{k-1}$. Then if $\alpha_n\to\infty$ the distribution function of the sum*

$$\zeta_n=\xi_1+\xi_2+\cdots+\xi_{\nu_n}$$

converges to the exponential distribution.

Consider the following example. A Geiger counter (see Example 4 of Section 8) is switched on at $t=0$. Find the time when a particle registration will be missed the first time. Let us generalize the formulation of the problem in comparison with that in Chapter 1: let the time intervals between particle arrivals ξ_k be independent random variables distributed in accordance to $F(x)$. Discharging time is a random variable η_k with the distribution function $G(x)$. All of the considered random variables are independent.

Let ζ_n denote the length of the time interval of interest. It is clear that $\zeta_n=\xi_1+\xi_2+\cdots+\xi_\nu$ where ν is an integer random variable independent of ξ_k and distributed geometrically.

Consider now the sequence $\{\xi_{nk}\}$ of independent random variables which are identically distributed for each n. Introduce the notation

$$F_n(x)=\mathrm{P}\{\xi_{nk}<x\},\quad f_n(t)=\int_{-\infty}^{\infty}e^{itx}\,dF_n(x).$$

Further consider an infinitely increasing sequence of integer positive numbers $\{k_n\}$ and sequence of integer positive random variables $\{\nu_n\}$ which possess the following property: for each n they are independent of ξ_{nk}, $1\leqslant k<\infty$.

The goal is to find conditions under which the convergence of the distribution functions of the sums (as $n\to\infty$)

$$s_n=\sum_{k=1}^{k_n}\xi_{nk}$$

implies the convergence of the distribution functions of the sums (as $n \to \infty$)

$$s_{\nu_n} = \sum_{k=1}^{\nu_n} \xi_{nk}.$$

THEOREM. *If as* $n \to \infty$

$$\mathbf{P}\left\{\sum_{k=1}^{k_n} \xi_{nk} < x\right\} \to \Phi(x) \tag{A}$$

and

$$\mathbf{P}\left\{\frac{\nu_n}{k_n} < x\right\} \to A(x) \tag{B}$$

(let us assume that $A(+0) = 0$*), then*

$$\mathbf{P}\left\{\sum_{k=1}^{\nu_n} \xi_{nk} < x\right\} \to \Psi(x). \tag{C}$$

Distribution function $\psi(x)$ *is defined via its characteristic function* $\psi(t)$ *by the formula*

$$\psi(t) = \int_0^\infty \varphi^z(t)\, A(z),$$

where $\varphi(t)$ *is the characteristic function for* $\Phi(x)$.

PROOF: The characteristic function of the sum $\sum_{k=1}^{\nu_n} \xi_{kn}$ is

$$\psi_n(t) = \sum_{j=0} p_{nj}(f_n(t))^j,$$

where $p_{nj} = P\{\nu_n = j\}$.

Introduce $A_n(x) = P\{\nu_n < x\}$, and then it is clear that

$$\psi_n(t) = \int_0 f_n^x(t)\, dA_n(x).$$

Now let

$$\bar{A}_n(z) = \mathbf{P}\left\{\frac{\nu_n}{k_n} < x\right\} = A_n(k_n z).$$

Then

$$\psi_n(t)=\int_0 [f_n^{k_n}(t)]^z \, d\bar{A}_n(z).$$

From calculus theory one knows the theorem which states that if the sequence of uniformly bounded continuous functions $f_n(x)$ converges to the function $f(x)$ for all real x and monotone bounded functions $A_n(x)$ converges to the function $A(x)$ for all real x, then as $n\to\infty$

$$\int_0 f_n(x)\,dA_n(x)\to\int_0 f(x)\,dA(x).$$

By virtue of this theorem and conditions (A) and (B) as $n\to\infty$ one has

$$\psi_n(t)\to\psi(t)=\int_0 \varphi^z(t)\,dA(z).$$

The theorem has been proven.

Let us now consider some corollaries of this theorem.

COROLLARY 1. *Let the function* $\Phi(x)$ be infinitely divisible and $\varphi(t)$. *Then the function*

$$\psi(t)=\frac{1}{1-\ln\varphi(t)}$$

is also a characteristic function (and later it will be shown that it is also infinitely divisible).

Indeed, one knows that for any infinitely divisible function $\Phi(x)$ it is possible to find such a sequence $\{\xi_{nk}\}$ of independent random variables that the condition (A) holds. Let us now choose v_n in such a way that $A(x)=1-e^{-x}$. This could be done several ways, for instance, by choosing v_n distributed geometrically with a proper given parameter. This satisfies at the same time the condition (B). But then by virtue of (B) the function

$$\psi(t)=\int_0 [\varphi(t)]^z\,dA(z)=\int_0 e^{-z(1-\ln\varphi(t))}\,dz=\frac{1}{1-\ln\varphi(t)}$$

is a characteristic function.

In particular, if the function $\Phi(x)$ is normal, i.e., $\varphi(t)=\exp(-t^2/2)$, then $\psi(t)=2/2+t^2$. The corresponding distribution function $\psi(x)$ is given by the formula

$$\psi(x)=\begin{cases}0,5\,e^{x\sqrt{2}} & \text{for} \quad x\leqslant 0,\\ 1-0,5\,e^{-x\sqrt{2}} & \text{for} \quad x\geqslant 0.\end{cases}$$

Usually the Laplace distribution is defined by its density function which in this case equals $p(x)=\exp(-|x|\sqrt{2})$.

Later it will be shown that for $A(x)=1-e^{-x}$ the summation of a random number of summands plays the same role as the normal distribution plays in classical limit theorems for the sum of independent and identically distributed random variables.

COROLLARY 2. *The conditions which determine the convergence of the distribution function of the sums*

$$s_n=\xi_{n1}+\xi_{n2}+\cdots+\xi_{nk_n}$$

of independent and identically distributed random variables to the limit distribution $\Phi(x)$ *are necessary and sufficient for the convergence of the distributions of the sum*

$$s_{\nu_n}=\xi_{n1}+\xi_{n2}+\cdots+\xi_{n\nu_n}$$

to the distribution $\Psi(x)$.

The corollary directly follows from the theorem.

COROLLARY 3. *Let* $\{\xi_n\}$ *be the sequence of independent and identically distributed random variables and* $\xi_{nk}=(\xi_k-a)/B_n$ *where real constants a and* $B_n>0$ *such that the functions of distribution of the sums*

$$\bar{s}_n=\frac{1}{B_n}\sum_{k=1}^{k_n}(\xi_k-a)$$

converge to $\Phi(x)$. *Further, let the condition (B) of the theorem hold. Then for a and* B_n *as chosen above the distribution functions of the sums*

$$\bar{s}_n=\frac{1}{B_n}\sum_{k=1}^{k_n}(\xi_k-a)$$

also converge to the limit distribution $\Psi(x)$.

This corollary follows directly from the theorem.

One should emphasize that for any possible limit distributions $A(x)$ normalizing multipliers b_n and centering coefficients a can be chosen once and forever.

Feller's Remark. In his outstanding book (vol. 2) W. Feller noted that if in formula (1) the function $A(x)$ is infinitely divisible then the function $\Box(x)$ is also infinitely divisible.

The proof of this statement follows from the fact that a random variable v with the distribution function $A(x)$ can be represented as the sum $v = v_1 + v_2$ of two independent random variables v_1 and v_2 then $\chi(t) = \chi_1(t)\chi_2(t)$. The latter follows from the definition of the infinite divisibility.

57 EXERCISES

1. Prove that the distributions of
   (a) Pascal (Exercise 1a of Chapter 5),
   (b) Polya (Exercise lb of Chapter 5),
   (c) Cauchy (Example 5 of Section 24)

are infinitely divisible.

2. Prove that the random variable having the density function

$$p(x) = \begin{cases} 0 & \text{for } x \leqslant 0, \\ \dfrac{\beta^{\alpha}}{\Gamma(\alpha)} x^{\alpha-1} e^{-\beta x} & \text{for } x > 0. \end{cases}$$

where $\alpha > 0$, $\beta > 0$ are constants, is infinitely divisible.

Note: In particular, it follows from this that the Maxwell distribution and the χ^2-distribution (for any value of n) are infinitely divisible.

3. Prove that for any constants $\alpha > 0$, $\beta > 0$

$$\varphi(t) = \left(1 + \frac{t^2}{\beta^2}\right)^{-\alpha}$$

is an infinitely divisible characteristic function.

Note: In particular, it follows from this that the Laplace distribution (Exercise 6 of Chapter 5) is infinitely divisible.

4. Determine the function $G(x)$ and the parameter γ in Kolmogorov's formula for the logarithm of an infinitely divisible characteristic

function for the following distributions:

(a) That of Exercise 2;
(b) The Laplace distribution.

5. Prove that if the sum of two independent infinitely divisible random variables is distributed according to

(a) the Poisson distribution,
(b) the normal distribution

then each term is distributed in case (a), according to the Poisson distribution, and in case (b) according to the normal distribution.

6. Determine conditions under which the distribution functions of sums of random variables comprising an elementary system converge to the distribution (a) of Exercise 2, (b) of Laplace.

Chapter 11

THE THEORY OF STOCHASTIC PROCESSES

58 INTRODUCTION

Demands of physical statistics and a number of engineering fields have posed a great number of new problems for probability theory outside of the framework of the classical theory. The physicists and engineers were interested in studying processes, i.e., phenomena that changed with time. Probability theory had neither general methods for these problems nor special schemes for solving the problems that arose in the investigation of such phenomena. There appeared a real necessity for the development of a general theory of random processes, i.e., a theory that could be used in the study of random variables depending on one or several continuously varying parameters.

Let us enumerate several problems that illustrate the need for constructing a theory of random processes.

Imagine that one intends to follow the motion of some molecule of gas or fluid. At random moments of time, the molecule collides with other molecules and, as a result, changes its velocity and position. Thus the state of the molecule is subject to random change at each moment of time. Many physical phenomena require for their analysis that one should be able to compute the probability that some specified number of molecules will be displaced a certain distance in a given interval of time. Thus, for example, if two fluids make contact, there begins a mutual penetration of the molecules of one fluid into another fluid, i.e., diffusion occurs. How fast does the process of diffusion take place and following what laws ? When does the mixture of them become practically homogeneous? The answers to these and many other questions are given by the statistical theory of diffusion, the basis for which is the theory of stochastic processes. An analogous problem obviously arises in chemistry when the process of a chemical reaction is investigated. What proportion of the molecules has already entered into the reaction? How does the reaction progress in time? When will the reaction practically be complete?

A very important class of phenomena takes place in accordance with the principle of radioactive decay. Atoms of a radioactive material decompose and are converted into atoms of some other element. The decay of each atom occurs instantaneously, similar to an explosion, with the release of a certain amount of energy. Numerous observations show that the decay of the various atoms occurs at moments of time which, to the observer, are random. Moreover, these moments of time are independent in a probabilistic sense. For studying the process of radioactive decay, it is necessary to determine the probability that a certain number of atoms will decompose in a specified time interval. From a formal point of view, if one is concerned exclusively with the mathematical model of a phenomenon, then there are other phenomena that also proceed in a similar way: the number of calls that are received at a telephone central office in a given interval of time (the traffic at the telephone central office), the breaking of threads on a ring spinning frame or, in Brownian motion, the fluctuation in the number of particles appearing in a given region of space at some moment of time. In this chapter, a simple solution to the mathematical problems of this kind is presented.

Note that the first examples of physical problems related to stochastic processes were those considered by a number of eminent physicists at the beginning of this century. Let us show how Planck and Fokker derived the differential equation of diffusion theory from a consideration of a highly schematic random walk problem. Suppose that a particle, located initially at point $x=0$, undergoes independent random impacts at the times $k\tau$ $(k=1,2,3,\ldots)$ as a result of which it moves each time a distance h to the right, with probability p, or a distance h to the left with probability $q = 1-p$. Let $f(x,t)$ denote the probability that at the time $t(t=n\tau)$ after n impacts the particle will occur at the position x. (It is clear that for an even number of impacts x, the distance, h, must be equal to an even number of steps, and for an odd number of impacts to an odd number of steps.) If m and $n-m$ denote the number of steps the particle has taken to the right and to the left, respectively, then according to Bernoulli's formula

$$f(x,t) = C_n^m p^m q^{n-m}.$$

It is clear that m, n, x, and h are related by the equation

$$m - (n - m) = x/h.$$

It is easy to find by a direct computation that function $f(x,t)$ satisfies the following difference equation

$$f(x,t+\tau) = pf(x-h,t) + q(x+h,t) \tag{1}$$

and initial conditions

$$f(0,0) = 1, \quad f(x,0) = 0 \text{ for } x \neq 0.$$

Let us see how this difference equation will transform if $h \to 0$ and $\tau \to 0$ simultaneously. The physical nature of the problem imposes certain restrictions on the quantities of h and τ. Similarly, the quantities p and q cannot be arbitrary. If these conditions do not hold, the particle could go off to infinity with probability 1 in a finite interval of time. In order to avoid such a possibility, one imposes the following requirements:

$$x = nh, \quad t = n\tau, \quad \frac{h^2}{\tau} \to 2D, \quad \frac{p-q}{h} \to \frac{c}{D} \tag{2}$$

where c and D are certain constants. The quantity c is called the *drift* and D *the diffusion coefficient.*

Subtracting $f(x, t)$ from both sides of equation (1), one finds that

$$f(x,t+\tau) - f(x,t) = p[f(x-h,t) - f(x,t)] + q[f(x+h,t) - f(x,t)]. \tag{3}$$

Suppose now that $f(x, t)$ is differentiable by t and double differentiable with respect to x. Then

$$f(x,t+\tau) - f(x,t) = \tau\frac{\partial f(x,t)}{\partial t} + o(\tau),$$

$$-f(x,t) + f(x-h,t) = -h\frac{\partial f(x,t)}{\partial x} + \frac{1}{2}h^2\frac{\partial^2 f(x,t)}{\partial x^2} + o(h^2),$$

$$f(x+h,t) - f(x,t) = h\frac{\partial f(x,t)}{\partial x} + \frac{1}{2}h^2\frac{\partial^2 f(x,t)}{\partial x^2} + o(h^2).$$

After substituting these equivalences into (3), one obtains

$$\tau\frac{\partial f(x,t)}{\partial t}+o(\tau)=-(p-q)h\frac{\partial f(x,t)}{\partial x}+\frac{h^2}{2}\frac{\partial^2 f(x,t)}{\partial x^2}+o(h^2).$$

From here, by virtue of relations (2), one finds in the limit that

$$\frac{\partial f(x,t)}{\partial t}=-2c\frac{\partial f(x,t)}{\partial x}+D\frac{\partial^2 f(x,t)}{\partial t^2}.$$

Thus one derived the equation which is called in diffusion theory the Fokker–Planck equation.

It is interesting to note that a physically meaningful and definitive result was obtained on the basis of some artificial formulation. Moreover, this result reflects the true picture of diffusion well. Later on, the general equations that are satisfied by the distributions of stochastic processes under very broad assumptions will be derived.

The general theory of stochastic processes had its origin in the fundamental papers of the Russian mathematicians Kolmogorov and Khinchine in the early nineteen thirties. In Kolmogorov's article "On analytical methods in the theory of probability" there is given a systematical and rigorous presentation of the foundations of the theory of stochastic processes without aftereffect or, as they are often called, Markov processes. The so-called theory of stationary processes originated in a series of papers by Khinchine.

Before some natural phenomenon or technical process can be studied mathematically, it must be formalized. The reason for this lies in the fact that mathematical analysis can be used for analyzing the process of a system state change only if it is supposed that every possible state of the system is fully defined by means of some definite mathematical methods. Of course, such a mathematical model is not the reality of the system itself but is merely a scheme suitable for its describing. For instance, one meets such a situation in mechanics where it is supposed that the state of the system, y, at any moment t is completely determined by its state x at any preceding moment t_0. The state of a mechanical system is understood as the set of positions and the velocities of the particles in the system.

In all of modern physics one has to deal with more complicated situations than described above wherein a knowledge of the state of a system at some moment of time t_0 does not uniquely determine the states of the system at succeeding moments of time, but only determines the probability that the system will be in one of the states of some subset of the system states. If x denotes the system's state at time t_0 and E is some class of the system's states, then for the processes just described the probability $\mathrm{P}\{t_0, x; t, E\}$ that the system being in state x at moment t_0 will go at moment t to one of the states belonging to set E.

If any additional information concerning the states of the system at the times $t<t_0$ does not affect this probability, then it is natural to refer to this class of stochastic processes as *processes without aftereffect*, or, because of their analogy to Markov chains, Markov processes.

The general concept of a stochastic process based on the axiomatic approach presented earlier can be introduced in the following way. Let Ω be the set of elementary events and t be a continuous parameter. The function of two arguments

$$\xi(t) = \varphi(\omega, t) \qquad (\omega \in \Omega)$$

is called a *stochastic process.*

For each value of the parameter t, the function $\varphi(\omega,t)$ is a function of ω alone and, consequently, is a random variable. For each fixed value of ω, i.e., for each elementary event, function $\varphi(\omega,t)$ depends only on t and is therefore an ordinary function of one real variable. Each such function is called a realization of the stochastic process $\xi(t)$. A random process may be understood either as a collection of random variables $\xi(t)$ depending on the parameter t or as a collection of realizations of the process $\xi(t)$. It is natural to assign a probability measure in the function space of its realizations to define a process.

Almost all this chapter will be dedicated to the stochastic processes without aftereffect. Only in the last section will stationary processes be considered.

59 THE POISSON PROCESS

Let us study in detail an example of a stochastic process without aftereffect that plays an important role in physics, theory of communication and reliability theory before presenting some of the general results. It seems that this kind of process was first investigated in detail by physicians A. Einstein and M. Smolukhovsky in context of the Brownian movement.

Let us suppose that a certain event occurs at random instants of time. One is interested in the number of occurrences of this event in the time interval 0 to t. Let us denote this number by $\xi(t)$. The process of the occurrence of the event is assumed to be characterized by the parameters: it is (1) stationary, (2) without aftereffect, and (3) ordinary. These three terms have the following meaning.

1. A process is called stationary if for any finite set of non-intersected intervals of time, the probability of occurrence of a specified number of events during each of these intervals depends only on these numbers and on the length of the intervals. This probability is unchanged by a shifting of all of the intervals by the same amount. In particular, the probability of occurrence of k events during the interval of time from τ to $\tau + t$ does not depend on τ and is a function only of t and k.

2. No aftereffect means that the probability of occurrence of k events during the time interval $(\tau, \tau + t)$ does not depend on knowledge of how many events have occurred earlier or how they occurred. This statement means that the conditional probability of occurrence of k events in the interval $(\tau, \tau + t)$ under any assumption about the occurrences of the events up to the time τ coincides with the unconditional probability. In particular, no aftereffect means that the occurrence of events in any non-intersecting intervals of time are mutually independent.

3. Ordinariness expresses the requirement that the occurrence of two or more events in an infinitesimally small time interval Δt is practically impossible. Let $P_{>1}(\Delta t)$ denote the probability of occurrence of more than one event in the interval of time Δt. Then the strong formulation of the ordinariness is as follows:

$$P_{>1}(\Delta t) = o(\Delta t).$$

Let us now determine the probability $P_k(t)$ that k events will occur in the interval of time t. By assumption of stationarity, this probability does not depend on where the interval of time is located. To solve the problem, let us first show that for small Δt the following equality holds:

$$P_1(\Delta t) = \lambda \Delta t + o(\Delta t),$$

where λ is a constant.

Indeed, consider a time interval of duration 1, and let p denote the probability that no event occurs during this period. Now let the time interval be split up into n equal non-intersecting subintervals. By the stationarity and absence of aftereffect, one has the equality $p = P_0(1/n)^n$, or $P_0(1/n) = p^{1/n}$.
From this, it follows that for any integer k

$$P_0(k/n) = p^{k/n}.$$

Now let t be a non-negative number. For any n, one can find such a k that $(k-1)/n \leqslant t < k/n$. Since the probability $P_0(t)$ is a decreasing function of time, then

$$P_0\left(\frac{k-1}{n}\right) \geqslant P_0(t) \geqslant P_0\left(\frac{k}{n}\right).$$

Thus, $P_0(t)$ satisfies the inequality

$$p^{(k-1)/n} \geqslant P_0(t) \geqslant p^{k/n}.$$

Let now k and n tend to infinity in such a way that

$$\lim_{n \to \infty} \frac{k}{n} = t.$$

From the above it is clear that for any t

$$P_0(t) = p^t.$$

Since $P_0(t)$ as a probability satisfies the inequality $0 \leqslant P_0 \leqslant 1$, any of the following three cases can be considered: (1) $p = 0$, (2) $p = 1$, (3) $0 < p < 1$. The first two cases are of little interest. In the first one $P_0(t) = 0$ for any t. It means that the probability that at least one event will occur during an arbitrary time interval is equal to 1. In other words, an infinite number of events will occur in an interval of time of arbitrary length with probability 1. In the second case, $P_0(t) = 1$, and therefore no events occur. The only interesting case is the third one, in which let us put $p = e^{\lambda}$ where λ is a positive quantity ($\lambda = -\log p$).

Thus, from the assumptions of stationarity and no aftereffect, it is found that for any $t > 0$

$$P_0(t) = e^{-\lambda t}. \tag{1}$$

It is clear that for any value of t, the following relation holds:

$$P_0(t)+P_1(t)+P_{>1}(t)=1.$$

From equation (1), it follows that for small values of t

$$P_0(t)=1-\lambda t+o(t).$$

Hence, by the ordinariness property one has

$$P_1(t)=\lambda t+o(t). \tag{2}$$

It is now possible to derive the formulas for $P_k(t)$ for $k\geqslant 1$. For this purpose one defines the probability that the event will occur exactly k times in the interval $t+\Delta t$. This may happen in $k+1$ different ways, namely:

(1) All of the k events occur in the interval t, and none occurs in the interval Δt;

(2) $k-1$ events occur in the interval t and a single event occurs in the interval Δt;

....

$(k+1)$ The event does not occur in the interval t, but k events have occurred the interval Δt.

By the formula on total probability, one has

$$P_k(t+\Delta t)=\sum_{j=0}^{k}P_j(t)P_{k-j}(\Delta t)$$

(here the conditions of stationarity and no aftereffect are taken into account). Let us set

$$R_k=\sum_{j=0}^{k-2}P_j(t)P_{k-j}(\Delta t).$$

It is apparent that

$$R_k\leqslant\sum_{j=0}^{k-2}P_{k-j}(\Delta t)=\sum_{s=2}^{k}P_s(\Delta t)\leqslant\sum_{s=2}P_s(\Delta t)=P_{>1}(\Delta t)=o(\Delta t),$$

by the condition of ordinariness.

Thus,

$$P_k(t+\Delta t) = P_k(t)\,P_0(\Delta t) + P_{k-1}(t)\,P_1(\Delta t) + o(\Delta t).$$

Furthermore, by (2)

$$P_0(\Delta t) = e^{-\lambda\Delta t} = 1 - \lambda\Delta t + o(\Delta t), \quad P_1(\Delta t) = \lambda\Delta t + o(\Delta t),$$

and therefore

$$P_k(t+\Delta t) = (1-\lambda\Delta t)P_k(t) + \lambda\Delta t P_{k-1}(t) + o(\Delta t).$$

From this, one has

$$\frac{P_0(t+\Delta t) - P_k(t)}{\Delta t} = -\lambda P_k(t) + \lambda P_{k-1}(t).$$

Since the right-hand side of this equation has a limit for $\Delta t \to 0$, so does the left-hand side. As a result, one obtains the equation

$$\frac{dP_k(t)}{dt} = -\lambda P_k(t) + \lambda P_{k-1}(t) \tag{3}$$

for the determination of $P_k(t)$. The initial conditions are chosen as follows:

$$P_0(0) = 1, \quad P_k(0) = 0 \quad \text{for} \quad k \geqslant 1. \tag{4}$$

The solution of equation (3) can be obtained by substitution

$$P_k(t) = e^{-\lambda t} v_k(t) \tag{5}$$

where $v_k(t)$ is the new desired function. Note that by (1) $v_0(t) = 1$. The relations (4) imply the following initial conditions:

$$v_0(0) = 1 \quad \text{and} \quad v_k(0) = 1 \quad \text{for} \quad k \geqslant 1. \tag{6}$$

Substituting (5) in (3) leads to the equation

$$\frac{dv_k(t)}{dt} = \lambda v_{k-1}(t). \tag{7}$$

In particular,

$$\frac{dv_1(t)}{dt} = \lambda. \tag{7'}$$

Sequentially solving (7) and (7′) and taking into account the initial conditions, one arrives at the relations

$$v_1(t) = \lambda t, \quad v_2(t) = \frac{(\lambda t)^2}{2}, \quad v_3(t) = \frac{(\lambda t)^3}{3!}$$

and, in general,

$$v_k(t) = \frac{(\lambda t)^k}{k!}.$$

Thus,

$$P_k(t) = \frac{(\lambda t)^k}{k!} e^{-\lambda t} \tag{8}$$

for any $k \geqslant 0$ and the problem is solved.

The conditions imposed on the process of the event occurrence considered above are satisfied with astonishing accuracy in many scientific and technological phenomena. As an example, one could mention the number of spontaneously disintegrating atoms of a radioactive substance in a given interval of time or the number of cosmic particles falling in a specified area in a fixed interval of time. Another example can be represented by observation of unit failures of a complex electronics system consisting of a large number of independent components. The number of components that fail in some interval of time $(0, t)$ is a stochastic process which is usually successfully described by the Poisson process just considered. The number of such examples can be increased easily.

The interval of time between occurrences of two neighboring events under investigation is a random variable which will be denoted by τ. Let us find the distribution of this random variable. Since the event $\tau > t$ is obviously equivalent to the non-occurrence of the event in the interval t, one can say that

$$\mathsf{P}\{\tau > t\} = e^{-\lambda t}.$$

Thus, the required distribution function is given by the formula

$$\mathsf{P}\{\tau < t\} = 1 - e^{-\lambda t}. \tag{9}$$

The result obtained can be interpreted physically in a variety of ways. For example, one can look upon it as the time distribution of the free motion of a molecule or as the time distribution between two failures of a complex electronic system.

Note that the theory developed in this section is applicable even when the parameter t does not represent time. Consider the following example.

Example. Points are distributed in space in accordance with the following rules:

1) The probability of k points being in a region G depends only on the volume v of the region G and not on its shape nor its position in space. This probability is denoted by $p_k(t)$;

2) The numbers of points falling in non-overlapping regions are independent random variables;

3)

$$\sum_{k=2} p_k(\Delta v) = o(\Delta v). \tag{3}$$

The conditions formulated are equivalent to those of stationarity, no aftereffect, and ordinariness. Hence, one has

$$p_k(v) = \frac{(av)^k}{k!} e^{-av}.$$

If small particles of a substance are suspended in a fluid, these particles are found to be in a continuous chaotic motion (Brownian motion) as a consequence of collisions with the surrounding molecules. As a result, at every instant of time a random position of the particles in some given region of space will obey the Poisson law.

In Table 14, the empirical results obtained with particles of gold suspended in water, which we have adopted from an article by Smolukhovsky, are compared with the results computed according to the Poisson law.

The constant $\lambda = av$ defining Poisson's law has been chosen as the arithmetic average of the number of particles observed, i.e.,

$$\lambda \approx \frac{0.112 + 1.168 + 2.130 + 3.69 + 4.32 + 5.5 + 6.1 + 7.1}{518} \approx 1.54$$

Table 12

number of particles	number of cases	frequency m/518	$\lambda^n(n!)^{-1}e^{-\lambda}$	calculated number of cases
0	112	0.216	0,213	110
1	168	0.325	0,328	173
2	130	0.251	0,253	131
3	69	0.133	0,130	67
4	32	0.062	0,050	26
5	5	0.010	0,016	8
6	1	0.002	0,004	2
7	1	0.002	0,001	1
	$\Sigma = 518$			$\Sigma = 518$

60 DEATH AND BIRTH PROCESSES

In the beginning of the twentieth century the practical needs, first of all biology and telecommunication, generated a new interesting and useful mathematical model, the so-called Birth and Death Processes. The Poisson process considered in the preceding section represents a very particular case of this process. In spite of an evident narrowness of the initial suggestions the Birth and Death Processes find a wide application to varied application problems because they give the possibility of obtaining not only a schematic understanding but they also deliver some useful computational formulas.

Consider a system which can be in one of states E_0, E_1, E_2, ... in finite or infinite space. With time, the system's states change such that in time h the system changes state E_n for state E_{n+1} with the probability $\lambda_n h + o(h)$ and for state E_{n-1} with the probability $v_n h + o(h)$. The probabilities that the system will change its state E_n for state $E_{\pm k}$ for $k > 1$ are infinitesimally small in comparison with the probabilities mentioned above if h is small. Hence, the probability that the system preserves its state during interval h equals $1 - \lambda_n h - v_n h + o(h)$. Parameters λ_n and v_n are assumed to be dependent of n and independent of t and the way which the system used to reach the last state. The last assumption means that the process under investigation is Markovian. The theory laid out below can be applied also to the case where λ_n and v_n depend on t.

The stochastic process described above is called the Birth and Death Processes. If one assumes that E_n is the event that the population total equals n then the transition $E_n \to E_{n+1}$ means that it has increased by 1. Analogously, the transition $E_n \to E_{n-1}$ means that the population has lost 1 individual.

If $v_n = 0$ for any $n \geqslant 1$, i.e., only transitions of a type $E_n \to E_{n+1}$ or a type $E_n \to E_n$ then it is said about the Process of Birth (Pure Birth). Note that the Poisson process is just such a process. If $\lambda_n = 0$ for any $n \geqslant 1$ then it is refers to the Process of Death.

Let $p_k(t)$ denote the probability that the system of interest is in state E_k at moment t. By means of the same arguments which were used in the preceding section, one can easily write the system of equations governing the Process of Birth and Death

$$p_0'(t) = -\lambda_0 p_0(t) + v p_1(t) \tag{1}$$

and for $k \geqslant 1$

$$p_k'(t) = -(\lambda_k + v_k) p_k(t) + \lambda_{k-1} p_{k-1}(t) + v_{k+1} p_{k+1}(t). \tag{2}$$

One needs to specify the initial conditions, i.e., state E_i from which the system begins its process. Of course, complete information were in the probabilities $p_{ij}(t)$ which mean that the system at moment t will be in state E_j if it was at $t = 0$ in state E_i. For the Poisson process one assumes that at $t = 0$ the process is in state E_0.

Equations (1) and (2) are very simple in the case of the Process of Pure Birth or Pure Death. In the second case by the sequential integration one has (for different λ_n)

$$p_0(t) = e^{-\lambda_0 t},$$

$$p_1(t) = \frac{\lambda_0}{\lambda_1 - \lambda_0}[e^{-\lambda_0 t} - e^{-\lambda_1 t}],$$

$$p_2(t) = \frac{\lambda_0 \lambda_1}{\lambda_1 - \lambda_0}\left[\frac{1}{\lambda_2 - \lambda_0}(e^{-\lambda_0 t} - e^{-\lambda_2 t}) + \frac{1}{\lambda_2 - \lambda_1}(e^{-\lambda_1 t} - e^{-\lambda_2 t})\right].$$

The results are written for the condition that at $t = 0$ the system is in state E_0. A general solution can be obtained without difficulties. One finds that functions $p_k(t)$ are non-negative for all of k and t. However if values of λ_k grow too fast with k it might happen that

$$\sum_{k=0} p_k(t) < 1.$$

FELLER'S THEOREM. *The relation*

$$\sum_{k=0} p_k(t) = 1, \tag{3}$$

holds for $p_k(t)$ for all t if and only if the series

$$\sum_{k=0}' \lambda_k^{-1} \tag{4}$$

diverges.

PROOF: Consider a partial sum of the series (3)

$$s_n(t) = p_0(t) + p_1(t) + \cdots + p_n(t). \tag{5}$$

From the equations describing the process of Birth it follows that

$$s_n'(t) = -\lambda_n p_n(t).$$

This leads to

$$1 - s_n(t) = \lambda_n \int_0^t p_n(t)\,dt \tag{6}$$

Note that if instead of the initial condition $p_0(0) = 1$ one takes, for instance, $p_i(0) = 1$ then equality (1) holds for $n \geqslant i$.

Since all summands of the sum (5) are non-negative, for any fixed t the sum $s_n(t)$ does not decrease in n. Consequently, the limit

$$\lim_{n\to\infty} [1 - s_n(t)] = \mu(t) \tag{7}$$

exists.

By virtue of (6) one concludes that

$$\lambda_n \int_0^t p_n(t)\,dt > \mu(t).$$

It follows that

$$\int_0^t s_n(z)\,dz \geqslant \mu(t)\left(\frac{1}{\lambda_0} + \frac{1}{\lambda_1} + \cdots + \frac{1}{\lambda_n}\right).$$

Since for any t and n the inequality $s_n(t) \leqslant 1$ then

$$t \geqslant \mu(t)\left(\frac{1}{\lambda_0} + \frac{1}{\lambda_1} + \cdots + \frac{1}{\lambda_n}\right).$$

If the series (4) diverges then from the latter inequality it follows that for all of t it must be true that $\mu(t) = 0$. From (7) follows that the divergence of (4) that leads to (3).

It follows from (6) that

$$\lambda_n \int_0^t p_n(t)\, dt \leqslant 1$$

and, consequently,

$$\int_0^t s_n(z)\, dz \leqslant \frac{1}{\lambda_0} + \frac{1}{\lambda_1} + \cdots + \frac{1}{\lambda_n}.$$

In limit as $n \to \infty$, one obtains

$$\int_0^t [1 - \mu(z)]\, dz \leqslant \sum_{n=0}^{\infty} \frac{1}{\lambda_n}.$$

If $\mu(t) = 0$ for all of t then the left-hand side of the inequality equals t and because t is arbitrary, the series in the right-hand side diverges. Thus the theorem is proven.

In the preceding section there was $\lambda_n = \lambda$. Consequently, the series (4) diverges and for all of t the equality

$$\sum_{n=0}^{n} p_n(t) = 1.$$

The sum

$$\sum_{n=0}^{\infty} p_n(t)$$

can be considered as the probability that during time t there will occur only a finite number of changes of the system's state. Thus, the difference

$$1 - \sum_{n=0} p_n(t)$$

can be interpreted as the probability of an infinite number of changes of the system states during the same period of time. In radioactive decay such a possibility means an avalanche type of the decay.

Example 1. Redundancy without renewal. Consider a system consisting of one main and n redundant units. The main unit during interval $(t, t+h)$ has failed with the probability $\lambda h + o(h)$. Each redundant unit has failed with the probability $\lambda h + o(h)$ during this interval. After failure, the main unit is replaced instantly by one of the redundant ones. The system as a whole has failed when all of its units have failed. The problem is to find the probability that at the moment t there are is k failed units (event E_k).

In this case, one deals with the model of a "pure birth." For this case

$$\lambda_k = \lambda + (n-k)\lambda' \quad \text{for} \quad 0 \leqslant k \leqslant n,$$

$$\lambda_{n+k} = 0, \quad k \geqslant 1.$$

Routine transformations lead one to the equalities

$$p_k(t) = \frac{\lambda_0 \lambda_1 \cdots \lambda_{k-1}}{k! \lambda'^k} e^{-\lambda t}(1 - e^{-\lambda' t})^k, \quad 0 \leqslant k \leqslant n,$$

and

$$P_{n+1}(t) = \frac{\lambda_0 \lambda_1 \cdots \lambda_{n-1} \lambda}{n! \lambda'^n} \int_0^t e^{-\lambda z}(1 - e^{-\lambda' z})^n dz.$$

In particular, if $\lambda' = 0$ (the case of the so-called unloaded, or "cold" redundancy where redundant units cannot fail) one has

$$p_k(t) = \frac{\lambda^k t^k}{k!} e^{-\lambda t} \quad (0 \leqslant k \leqslant n), \quad P_{n+1}(t) = 1 - \sum_{k=0}^{n} \frac{(\lambda t)^k}{k!} e^{-\lambda t}.$$

For $\lambda = \lambda'$ (the case of the so-called loaded, or "hot" redundancy where redundant units are in the same regime as the main unit) one has

$$p_k(t) = C_{n+1}^k e^{-(n+1-k)\lambda t}(1 - e^{-\lambda t})^k.$$

Let ξ_k denote time to failure of the kth unit at the operational place. For unloaded redundancy, the time to the system failure is $\xi = \xi_0 + \xi_1 + \cdots + \xi_n$. Since the mean time to failure of a single unit equals

$$\int_0^\infty e^{-\lambda t}dt = \frac{1}{\lambda}$$

the mean time to failure of the system equals $(n+1)/\lambda$, i.e., is proportional to the total number of the system's units.

For loaded regime of redundant units, one can use the following arguments. Let $t_1, t_2, \ldots, t_{n+1}$ be the ordered moments of failures of the system's units. Introduce the notation: $\tau_1 = t_1$, $\tau_2 = t_2 - t_1$, $\tau_3 = t_3 - t_2, \ldots, t_{n+1} = t_{n+1} - t_n$. Since on the first interval all of the system's units are operating, the probability of the system's failure-free operation during time t equals $e^{-(n+1)\lambda t}$. The analogous probability for the second interval where one of the system's units has failed equals $e^{-n\lambda t}$. At last, this probability for the last interval where a single unit is still operating equals $e^{-n\lambda t}$. Now it is simple to calculate the mean time to the system's failure:

$$\sum_{k=1}^{n+1} \mathbf{E}\tau_k = \frac{1}{\lambda}\left(1 + \frac{1}{2} + \cdots + \frac{1}{n}\right).$$

If n is large then

$$1 + \frac{1}{2} + \cdots + \frac{1}{n} \sim \ln\, n + c,$$

where c is the Euler Constant, $c = 0.577215\ldots$.

Example 2. Queuing system with losses. Let us now consider a problem related to the queuing theory which was initiated by Dutch scientist Erlang who for many years worked for Copenhagen Telephone Company.

Assume that users' calls arrive at the Central Office. If a destination is free, the connection performs instantaneously and the call proceeds. If the destination is busy then the current client originating the call is rejected. It is said that the call has been lost.

For further consideration one must take into account the two following assumptions. First, calls arrive at the Central Office in random and independently. Second, the duration of each call is a random variable.

Assume that there are n identical lines at each destination and if at least one of them is free then the connection is performed instantaneously. Each line is accessible for each call and each call occupies only one line. The probability that the call arrives during the interval $(t, t+h)$ equals $\lambda h + o(h)$. If at moment t all of k lines are busy then the probability that during interval $(t, t+h)$ one of them will be free equals $k\nu h + o(h)$.

This model is embedded into the scheme of the Birth and Death Processes. In this case $\lambda_k \lambda$, $\theta_k = k\nu$ for $1 \leqslant k \leqslant n$ and $\upsilon_k = 0$ for $k > n$. This queuing system can be in one of the following states: $E_0, E_1, E_2, \ldots, E_n$.

Equations (1) and (2) in this case can be written in the form

$$p_0'(t) = -\lambda p_0(t) + \nu p_1(t), \tag{8}$$

for $1 \leqslant k \leqslant n$

$$p_k'(t) = -(\lambda + k\nu) p_k(t) + \lambda p_{k-1}(t) + (k+1)\nu p_{k+1}(t) \tag{9}$$

and for $k = n$

$$p_n'(t)\, \lambda p_{n-1}(t) - n\nu p_n(t). \tag{10}$$

One must add one more equation

$$\sum_{k=0}^{n} p_k(t) = 1,$$

which has a very simple meaning: at any moment of time the system must be in one of its states.

One is usually interested in analysis of the stationary regime of the stochastic process, i.e., its behavior as $t \to \infty$. As it will be shown below, there exist limits

$$p_k = \lim_{k \to \infty} p_k(t)$$

and these limit probabilities satisfy the following system of algebraic equations obtained from (8)–(10) by substituting functions $p_k(t)$ by constant p_k and derivatives $p_k'(t)$ by zeros:

$$-\lambda p_0 + \nu p_1 = 0,$$

$$\lambda p_{k-1} - (\lambda + k\nu) p_k + (k+1)\nu p_{k+1} = 0, \quad 1 \leqslant k < n, \tag{11}$$

$$\lambda p_{n-1} - n\nu p_n = 0,$$

$$\sum_{k=0}^{n} p_k = 1.$$

Introducing $z_k = \lambda p_{k-1} - k\nu p_k$, one reduces the initial algebraic system to the following:

$$z_1 = 0, \ z_k - z_{k-1} = 0 \quad \text{for} \quad 1 \leqslant k < n, \ z_n = 0.$$

It follows that

$$k\nu p_{k-1} = \lambda p_{k-1} \qquad k = 1, 2, \ldots, n.$$

After simple transformations one has

$$p_k = \frac{\rho^k}{k!} p_0 \qquad (k \geqslant 1, \ \rho = \lambda/\nu).$$

Equation (11) allows one to find the norm multiplier p_0:

$$p_0 = \left[\sum_{k=0}^{n} \frac{\rho^k}{k!} \right]^{-1}.$$

Finally,

$$p_k = \frac{\rho^k}{k!} \left[\sum_{k=0}^{n} \frac{\rho^k}{k!} \right]^{-1}, \qquad 0 \leqslant k \leqslant n.$$

These expressions are called Erlang's Formulas. They have a wide use in the theory of teletraffic. The condition $k = n$ means that all of the lines are busy, and the probability p_k corresponds to the probability of loss which can be written as

$$p_n = \frac{\rho^n}{n!} \left[\sum_{k=0}^{n} \frac{\rho^k}{k!} \right]^{-1}.$$

Let us now pay attention to some general results of the theory of the Birth and Death Processes which will be presented without proofs. For the pure Birth Process the system (1)–(2) can be solved by simple sequential integration since these equations were represented by simple recurrent relations. General equations have a different structure and sequential determination of functions $p_k(t)$ becomes impossible. The conditions of existence and uniqueness of the solutions in general were investigated in detail by Feller, Reuter, Carlin, and

McGregor. It was found that the equation

$$\sum_{k=0} p_k(t) = 1$$

holds for all t in the series

$$\sum_{k=1} \prod_{i=1}^{k} \frac{v_i}{\lambda_i} \tag{12}$$

diverges. If in addition the series

$$\sum_{k=1} \prod_{i=1}^{k} \frac{\lambda_{i-1}}{v_i}, \tag{13}$$

then for t the limits

$$p_k = \lim_{t \to \infty} p_k(t) \tag{14}$$

exist.

This condition, incidentally, holds in all cases where beginning with some k_0 the inequality

$$\lambda_k / v_{k+1} \leqslant \alpha < 1$$

holds. Intuition hints that the speed of arrival of demands must not be higher than the speed of their being served.

For calculation of the limits in (13) one can use the following simple rule. One compiles from (1)–(2) a system of algebraic equations by substituting $p_k(t)$ by p_k and $p_k'(t)$ by 0. This system has the form:

$$-\lambda_0 p_0 + v_1 p_1 = 0,$$

$$-(\lambda_k + v_k) p_k + \lambda_{k-1} p_{k-1} + v_{k+1} p_{k+1} = 0 \qquad (k \geqslant 1).$$

Using the notation

$$z_k = -\lambda_k p_k + v_{k+1} p_{k+1}, \qquad k = 0, 1, 2, \ldots$$

one reduces the initial system into the following

$$z_0 = 0, \; z_{k-1} - z_k = 0 \qquad (\text{for } k \geqslant 1).$$

From this it follows that $z_k = 0$ for all k.

Hence

$$p_k = \frac{\lambda_{k-1}}{\nu_k} p_{k-1} = \prod_{i=1}^{k} \frac{\lambda_{i-1}}{\nu_i} p_0. \tag{15}$$

The constant p_0 can be found from the norm condition

$$\sum_{k=0} p_k = 1$$

and equals

$$p_0 = \left[1 + \sum_{k=1} \prod_{i=1}^{k} \frac{\lambda_{i-1}}{\nu_i}\right]. \tag{16}$$

It is obvious that the Erlang Formulas are contained in this result.

Example 3. Queuing system with line. A flow of demands described by the Poisson process with parameter λ arrives at the n-channel queuing system. Each demand needs a random time for its serving. The distribution of this random time is $H(x) = 1 - e^{\nu x}$. The service begins immediately if there is at least one free channel. If all channels are busy the demand waits for service in a line. In this case after any channel is free, the serving of a demand from the line begins. Find the probability that a specified number of demands are in the system.

For this problem $\lambda_k = \lambda$ for all of k, $\nu_k = k\nu$ for $k \leqslant n$, and $\nu_k = n\nu$ for $k \geqslant n$.

By formulas (15) and (16) the stationary solution for the system under consideration is for $k \leqslant n$

$$p_k = \frac{\rho^k}{k!} p_0$$

and for $k \geqslant n$

$$p_k = \frac{\rho^k}{n!\, n^{k-n}} p_0,$$

where $\rho = \lambda/\nu$. The constant p_0 is defined by the equality

$$p_0 = \left[\sum_{k=0}^{n} \frac{\rho^k}{k!} + \frac{\rho^n}{n!} \sum_{k=n+1} \left(\frac{\rho}{n}\right)^{k-n}\right]^{-1}.$$

If $\rho < n$ then

$$p_0 = \left[1 + \sum_{k=1}^{n} \frac{\rho^k}{k!} + \frac{\rho^{n+1}}{n!(n-\rho)!}\right]^{-1}.$$

If $\rho \geqslant n$ then the series in braces diverges and $p_0 = 0$. From the just-written formulas one obtains that $p_k = 0$ for all of k. This result is very important. In a verbal form it can be formulated as follows: If $\rho \geqslant n$ then the line for service increases infinitely.

Example 4. Maintenance of machines by a group of servicemen. A group of r servicemen maintains n identical machines. Failure of a machine attracts the attention of a free serviceman. A machine's failures are random and independent. The probability that failure of a single machine occurs in the time interval $(t, t+h)$ equals $\lambda h + o(h)$. The repair of a failed machine in the interval $(t, t+h)$ performs with the probability $vh + o(h)$. System of repair: one machine–one serviceman. Find the probability that in the stationary regime at a given moment of time there is a specified number of idle machines.

Let E_k be the event that at the moment there are k idle machines. It is clear that the system's states are $E_0, E_1, E_2, \ldots, E_n$. For the analyzed Process of Birth and Death $\lambda_k = (n-k)\lambda$ for $0 \leqslant k < n$ and $\lambda_0 = 0$ for $k \geqslant n$; $v_k = kv$ for $1 \leqslant k \leqslant r$ and $v_k = rv$ for $k \geqslant r$. Formulas (15) and (16) lead to the equations:
for $1 \leqslant k \leqslant r$ $(\rho = \lambda/v)$

$$p_k = \frac{n!}{k!(n-k)!}\rho^k p_0,$$

for $r \leqslant k \leqslant n$

$$p_k = \frac{n!}{r^{n-k} r!(n-k)!}\rho^k p_0$$

and

$$p_0 = \left[\sum_{k=0}^{r} \frac{n!}{k!(n-k)!}\rho^k + \sum_{k=r+1}^{n} \frac{n!}{r^{n-k} r!(n-k)!}\rho^k\right]^{-1}.$$

In particular, for $r = 1$

$$p_k = \frac{n!}{(n-k)!}\rho^k p_0,$$

$$p_0 = \left[\sum_{k=0}^{n} \frac{n!}{(n-k)!}\rho^k\right]^{-1}.$$

These examples show that the queuing theory can be used for preparatory solutions of practical problems and allows one to obtain reasonable recommendations.

61 CONDITIONAL DISTRIBUTION FUNCTIONS AND BAYES' FORMULA

For further consideration, we must generalize the concept of conditional probability (introduced in the first chapter) to the case of an infinite number of conditions. In particular, we must introduce the concept of a conditional distribution function with respect to some random variable.

Consider some event B and a random variable ξ with a distribution function $F(x)$. Let $A_{\alpha\beta}$ denote the event that

$$x - \alpha \leqslant \xi < x + \beta.$$

Using the definitions of Chapter 1, one can write

$$\mathsf{P}\{BA_{\alpha\beta}\} = \mathsf{P}\{A_{\alpha\beta}\} \cdot \mathsf{P}\{B|A_{\alpha\beta}\} = [F(x+\beta) - F(x-\alpha)]\mathsf{P}\{B|A_{\alpha\beta}\},$$

from which

$$\mathsf{P}\{B|A_{\alpha\beta}\} = \frac{\mathsf{P}\{BA_{\alpha\beta}\}}{F(x+\beta) - F(x-\alpha)}.$$

If the limit

$$\lim_{\alpha,\beta \to 0} \frac{\mathsf{P}\{BA_{\alpha\beta}\}}{F(x+\beta) - F(x-\alpha)},$$

exists it is called the conditional probability of the event B under the condition that $\xi = x$. It is denoted by the symbol $P\{B/x\}$. (This limit exists for almost all values of x in the sense of the measure defined by the function $F(x)$.) It is clear that for fixed x, $P\{B/x\}$ will be a finitely additive function of the event B defined over some field of events.

Under certain conditions, which practically always are fulfilled, $P\{B/x\}$ has all the properties of an ordinary probability corresponding to Axioms 1–3 of Section 8.

If η is a random variable and B denotes the event that $\eta < y$, then the function $\Phi(y|x) = \mathbf{P}\{\eta < y/x\}$ is a distribution function and is called the conditional distribution function of the variable η under the condition that $\xi = x$.

It is obvious that if $F(x, y)$ is the distribution function of the pair of random variables ξ and η then

$$\Phi(y|x) = \lim_{\alpha,\beta \to 0} \frac{F(x+\beta, y) - F(x-\alpha, y)}{F(x+\beta, \infty) - F(x-\alpha, \infty)},$$

if only this limit exists.

If the function $\mathbf{P}\{B/x\}$ is integrable with respect to $F(x)$, then the formula for total probability holds

$$\mathbf{P}\{B\} = \int \mathbf{P}\{B|x\} dF(x).$$

For deriving this formula, one divides the interval of definition of the variable ξ into the subintervals $x_i \leqslant \xi < x_{i+1}$ by means of the points x_i $(i = 0, \pm 1, \pm 2, \ldots)$. Let A_i denote the event $x_i \leqslant \xi < x_{i+1}$. By virtue of the extended axiom of addition , one has:

$$\mathbf{P}\{B\} = \sum_{i=-\infty}^{\infty} \mathbf{P}\{BA_i\} = \sum_{i=-\infty}^{\infty} \mathbf{P}\{B|A_i\}[F(x_{i+1}) - F(x_i)].$$

Let us now subdivide the intervals (x_i, x_{i+1}) into subintervals, in such a way that the length of the largest one approaches 0. By the definition of conditional probability and using the Stieltjes integral, one obtains:

$$\mathbf{P}\{B\} = \int \mathbf{P}\{B|x\} dF(x).$$

In particular,

$$\Phi(y) = \mathbf{P}\{\eta < y\} = \int \Phi(y \mid x) dF(x). \tag{1}$$

If a probability density function exists for the variable η, then

$$\varphi(y) = \int \varphi(y \mid x) dF(x), \tag{1'}$$

where $\varphi(y/x)$ is the conditional density function of the variable η.

Example. For illustration of formula (1), let us consider the following problem in the theory of ballistics. In firing at a certain target, two kinds of errors are possible: (1) an error in determining the position of the target; and (2) errors in shooting which occur for a large number of diverse reasons (fluctuations in the amount of charge in the shells, irregularities in the shell production, errors due to sighting, small variations in atmospheric conditions, and so on). Errors of the second kind are referred to as the technical dispersion. (Here the determination of the target position and the technical dispersion are assumed independent.)

After a single fix of the target position, let n independent shots be fired at it. Find the probability of at least one target hit.

For the sake of simplicity, let us restrict our discussion to a one-dimensional target of length $2a$, and the shell will be assumed to be a point. Let $f(x)$ denote the density function of the position of the target and $\varphi_i(x)$ the density function of the point of impact of the i-th shell.

If the center of the target is at the point z, then the probability of hitting the target in the i-th round equals the probability of a hit in the interval $(z-\alpha, z+\alpha)$, i.e., equals

$$\int_{z-\alpha}^{z+\alpha} \varphi_i(x)\, dx.$$

The conditional probability of a miss in the i-th round given that the target center is located at the point z is

$$1 - \int_{z-\alpha}^{z+\alpha} \varphi_i(x) dx.$$

The conditional probability of a miss in all n rounds (given the same condition) is

$$\prod_{i=1}^{n}\left(1-\int_{z-a}^{z+a}\varphi_i(x)dx\right).$$

From this one concludes that the probability of at least one hit under the condition that the target center is at the point z is

$$1-\prod_{i=1}^{n}\left(1-\int_{z-a}^{z+a}\varphi_i(x)dx\right).$$

The unconditional probability of at least one target hit (according to formula (1)) is thus

$$P=\int f(z)\left[1-\prod_{i=1}^{n}\left(1-\int_{z-a}^{z+a}\varphi_i(x)dx\right)\right]dz.$$

If the firing conditions do not change from round to round, then $\varphi_i(x)=\varphi(x)$ ($i=1, 2, \ldots, n$) and therefore

$$P=\int f(z)\left[1-\left(1-\int_{z-a}^{z+a}\varphi(x)dx\right)^n\right]dz.$$

As before, let A_i denote the event that $x_i \leqslant \xi < x_{i+1}$. According to the classical Bayes' Theorem,

$$\mathbf{P}\{A_i \mid B\}=\frac{\mathbf{P}\{A_i\}\mathbf{P}\{B \mid A_i\}}{\mathbf{P}\{B\}}$$

If $F(x)=\mathbf{P}\{\xi<x\}$ and $\mathbf{P}\{\xi<x/B)$ have continuous derivatives with respect to x, then using Lagrange Theorem, one obtains

$$\mathbf{P}\{A_i \mid B\}=p_\xi(\bar{x}_i \mid B)(x_{i+1}-x_i)=\frac{F'(\bar{x}_i')\mathbf{P}\{B \mid A_i\}}{\mathbf{P}\{B\}}(x_{i+1}-x_i),$$

where

$$x_i<\bar{x}_i<x_{i+1} \quad \text{and} \quad x_i<\bar{x}_i'<x_{i+1}.$$

In the limit, as $x_i \to x$ and $x_{i+1}x$, one obtains:

$$p_\xi(x \mid B)=\frac{p(x)\mathbf{P}\{B \mid x\}}{\mathbf{P}\{B\}},$$

or

$$p_\xi(x \mid B) = \frac{p(x)\mathbf{P}\{B \mid x\}}{\int \mathbf{P}\{B \mid x\} p(x) dx}. \tag{2}$$

It is natural to call this relation Bayes' formula.

Suppose now that B is the event that some random variable η assumes a value lying between $y - \alpha$ and $y + \beta$ and that the conditional distribution function $\Phi(y/x)$ of the variable η has a continuous density $p_\eta(y/x)$ for each value of x. Then, as follows from equation (2), if

$$\frac{1}{\alpha + \beta} \mathbf{P}\{B \mid x\}$$

approaches $p_\eta(y/x)$ uniformly in x as α and β tend to 0, then the relation

$$p_\xi(x \mid y) = \frac{p(x) p_\eta(y \mid x)}{\int p_\eta(y \mid x) p(x) dx}$$

holds. This formula will be widely used in the next chapter.

62 THE GENERALIZED MARKOV EQUATION

Let us now pass to a study of stochastic processes without after-effect restricting ourselves to the simplest problems only. In particular, assume that the class of possible states of a system is the set of real numbers. Thus, a stochastic process is represented by a collection of random variables $\xi(t)$ depending on a single real parameter t. One usually considers the parameter t as time, so the state of a system at a given moment of time is under consideration.

A complete probabilistic characterization of a process without after-effect can be obtained by giving a function $F(t, x; \tau, y)$ which is the probability that the random variable $\xi(\tau)$ will assume a value less than y at time τ if it is known that at time $t (t < \tau)$ the equation $\xi(t) = x$ holds. Additional knowledge concerning the states of a system at earlier instants of time before t for a process without aftereffect does not change the function $F(t, x; \tau, y)$.

Let us now note some conditions that have to be satisfied by the function $F(t, x; \tau, y)$. First of all, as is true of any distribution function, it has to satisfy the relations

(1)

$$\lim_{y\to -} F(t, x; \tau, y) = 0, \quad \lim_{y\to +} F(t, x; \tau, y) = 1;$$

for arbitrary values of x, t and τ;

(2) the function $F(t, x; \tau, y)$ is continuous on the left in the argument y.

Assume that the function $F(t, x; \tau, y)$ is continuous with respect to t, τ, and cx. Let us consider the instants of time t, s, and τ, with $t < s < t$. Since the system passes from the state x at time t to one of the states in the interval $(z, z + dz)$ with probability $d_z F(t, x; s, z)$ and from the state z at time s to a state less than y at time τ with a probability $F(s, z; \tau, y)$, then according to formula (1) of the previous section one finds that

$$F(t, x; \tau, y) = \int F(s, z; \tau, y) d_z F(t, x; s, z).$$

It is natural to call this equation the generalized Markov equation, since it extends identity (1) of Section 17 (the theory of Markov chains) to the theory of stochastic processes. The role of this identity in the theory of stochastic processes is as important as the identity mentioned is in the theory of Markov chains.

The probability $F(t, x; \tau, y)$ has only been defined for $\tau > t$.

Let us complete the definition by taking

$$\lim_{\tau\to t+0} F(t, x; \tau, y) = \lim_{t\to \tau-0} F(t, x; \tau, y) = E(x, y) = \begin{cases} 0 & \text{for} \quad y \leqslant x, \\ 1 & \text{for} \quad y > x. \end{cases}$$

If a density

$$f(t, x; \tau, y) = \frac{\partial}{\partial y} F(t, x; \tau, y)$$

exists, then it satisfies the following obvious equalities:

$$\int_{-}^{y} f(t, x; \tau, z) dz = F(t, x; \tau, y),$$

$$\int f(t,x;\tau,z)dz = 1.$$

In this case, the generalized Markov equation is expressible in the following form:

$$f(t,x;\tau,y) = \int f(t',z;\tau,y)f(t,x;t',z)dz.$$

63 CONTINUOUS STOCHASTIC PROCESSES. KOLMOGOROV'S EQUATIONS

A stochastic process $\xi(t)$ is continuous if it can obtain an appreciable increment in a short time interval with only a small probability. A more stronger statement is that a stochastic process $\xi(t)$ is continuous if for any $\delta(\delta > 0)$ the relation

$$\lim_{\Delta t \to 0} \frac{1}{\Delta t} \int_{|y-x|\geq\delta} dF(t-\Delta t, x; t, y) = 0 \tag{1}$$

holds.

The next problem is to derive the differential equations which, under certain conditions, are satisfied by the function $F(t,x;\tau,y)$ governing a continuous stochastic process without aftereffect. These equations were first rigorously derived by Kolmogorov (although the second equation was to be found before this in physics). They are reffered to as Kolmogorov's Equations.

Assume the following:

(1) The partial derivatives

$$\frac{\partial F(t,x;\tau,y)}{\partial x} \quad \text{and} \quad \frac{\partial^2 F(t,x;\tau,y)}{\partial x^2}$$

exist and are continuous for arbitrary values of t, x, and $\tau > t$;

(2) For arbitrary $\delta > 0$ the limits

$$\lim_{\Delta t \to 0} \frac{1}{\Delta t} \int_{|y-x|<\delta} (y-x)d_y F(t-\Delta t, x; t, y) = a(t,x) \tag{2}$$

and

$$\lim_{\Delta t \to 0} \frac{1}{\Delta t} \int_{|y-x|<\delta} (y-x)^2 d_y F(t-\Delta t, x; t, y) = b(t,x), \tag{3}$$

exist and the convergence is uniform in x.[1]

The left-hand sides of equations (2) and (3) depend on δ. However, in view of the definition of continuity of a process, i.e., by (1), this dependence is merely apparent.

THE FIRST KOLMOGOROV EQUATION: If conditions (1) and (2) just formulated are fulfilled then the function $F(t, x; \tau, y)$ satisfies the equation

$$\frac{\partial F(t,x;\tau,y)}{\partial t} = -a(t,x)\frac{\partial F(t,x;\tau,y)}{\partial x} - \frac{b(t,x)}{2}\frac{\partial^2 F(t,x;\tau,y)}{\partial x^2}. \tag{4}$$

PROOF: According to the generalized Markov equation

$$F(t-\Delta t, x; \tau, y) = \int F(t, z; \tau, y) d_z F(t-\Delta t, x; t, z).$$

Moreover, by the properties of a distribution function,

$$F(t, x; \tau, y) = \int F(t, x; \tau, y) d_z F(t-\Delta t, x; t, z).$$

From these equations, one concludes that

$$\frac{F(t-\Delta t, x; \tau, y) - F(t, x; \tau, y)}{\Delta t}$$
$$= \frac{1}{\Delta t}\int [F(t, z; \tau, y) - F(t, x; \tau, y)] d_z F(t-\Delta t, x; t, z).$$

By Taylor's formula, the expression

$$F(t, z; \tau, y) = F(t, x; \tau, y) + (z-x)\frac{\partial F(t,x;\tau,y)}{\partial x}$$
$$+ \frac{1}{2}(z-x)^2 \frac{\partial^2 F(t,x;\tau,y)}{\partial x^2} + o((z-x)^2).$$

---

[1]Kolmogorov has shown that the limits $a(t, x)$ and $b(t, x)$ exist under the sufficiently general assumptions. – *B. V.*

holds under the assumptions made above. The following analytical transformations require no special explanations:

$$\frac{F(t-\Delta t,x;\tau,y)-F(t,x;\tau,y)}{\Delta t}$$

$$=\frac{1}{\Delta t}\int_{|z-x|\geq\delta}[F(t,z;\tau,y)-F(t,x;\tau,y)]d_zF(t-\Delta t,x;t,z)$$

$$+\frac{1}{\Delta t}\int_{|z-x|<\delta}[F(t,z;\tau,y)-F(t,x;\tau,y)]d_zF(t-\Delta t,x;t,z)$$

$$=\frac{1}{\Delta t}\int_{|z-x|\geq\delta}[F(t,z;\tau,y)-F(t,z;\tau,y)]d_zF(t-\Delta t,x;t,z)$$

$$+\frac{1}{2}\frac{\partial F(t,x;\tau,y)}{\partial x}\cdot\frac{1}{\Delta t}\int_{|z-x|<\delta}(z-x)d_zF(t-\Delta t,x;t,z)$$

$$+\frac{\partial^2 F(t,x;\tau,y)}{\partial^2 x}\cdot\frac{1}{\Delta t}\int_{|z-x|<\delta}[(z-x)^2+o(z-x)^2]$$

$$\times d_zF(t-\Delta t,x;t,z). \tag{5}$$

Consider now the limit, letting $\Delta t\to 0$. The first term on the right-hand side has the limit 0, by virtue of (1). The second term by (2) equals

$$a(t,x)\frac{\partial F}{\partial x}$$

in the limit. Finally, the third term can only differ from

$$\frac{1}{2}b(t,x)\frac{\partial^2 F}{\partial x^2}$$

only by a term which approaches 0 as $\delta\to 0$. But since the left-hand side of the last equation is independent of δ, and the limiting values just indicated also do not depend on δ, it follows that the limit of the right-hand side exists and equals

$$a(t,x)\frac{\partial F(t,x;\tau,y)}{\partial x}+\frac{1}{2}b(t,x)\frac{\partial^2 F(t,x;\tau,y)}{\partial x^2}.$$

Hence, one concludes that the limit

$$\lim_{\Delta t\to 0}\frac{F(t-\Delta t,x;\tau,y)-F(t,x;\tau,y)}{\Delta t}=\frac{\partial F(t,x;\tau,y)}{\partial t}$$

exists.

Equation (5) thus leads to equation (4).

Assuming the existence of a density function

$$f(t,x;\tau,y)=\frac{\partial}{\partial y}F(t,x;\tau,y),$$

straightforward differentiation of (4) shows that the density $f(t, x; \tau, y)$ satisfies the equation

$$\frac{\partial f(t,x;\tau,y)}{\partial t}+a(t,x)\frac{\partial f(t,x;\tau,y)}{\partial x}+\frac{1}{2}b(t,x)\frac{\partial^2 f(t,x;\tau,y)}{\partial x^2}=0. \quad (4')$$

Let us now proceed to derive the second Kolmogorov equation. Here some assumptions which are not essential for the considered subject are made. Besides the assumptions already made, one imposes the following further restrictions on the function $F(t, x; \tau, y)$

(3) A probability density function

$$f(t,x;\tau,y)=-\frac{\partial F(t,x;\tau,y)}{\partial y}$$

exists;

(4) The derivatives

$$\frac{\partial f(t,x;\tau,y)}{\partial \tau},\quad \frac{\partial}{\partial y}[a(\tau,y)f(t,x;\tau,y)],\quad \frac{\partial^2}{\partial y^2}[b(\tau,y)f(t,x;\tau,y)]$$

exist and are continuous.

THE SECOND KOLMOGOROV EQUATION:[2] If conditions (1)–(4) are satisfied then for any continuous stochastic process without aftereffect the density $f(t, x; \tau, y)$ is a solution of the equation

$$\frac{\partial f(t,x;\tau,y)}{\partial \tau} = -\frac{\partial}{\partial y}[a(\tau,y)f(t,x;\tau,y)] + \frac{1}{2}\frac{\partial^2}{\partial y^2}[b(\tau,y)f(t,x;\tau,y)]. \quad (6)$$

PROOF: Let a and b $(a<b)$ be certain numbers and $R(y)$ a non-negative continuous function having continuous derivatives up to and including those of the second order. Additionally, it is required that

$$R(y) = 0 \quad \text{for} \quad y<a \quad \text{and} \quad y>\text{b}.$$

From the condition that the function $R(y)$ and its derivatives are continuous, one concludes that

$$R(a) = R(b) = R'(a) = R'(b) = R''(a)R''(b) = 0 \quad (7)$$

First of all, note that

$$\int_a^b \frac{\partial f(t,x;\tau,y)}{\partial \tau} R(y)dy = \frac{\partial}{\partial \tau}\int_a^b f(t,x;\tau,y)R(y)dy$$
$$= \lim_{\Delta\tau\to 0}\int \frac{f(t,x;\tau+\Delta\tau,y)-f(t,x;\tau,y)}{\Delta\tau}R(y)dy.$$

According to the generalized Markov equation,

$$f(t,x;\tau+\Delta\tau,y) = \int f(t,x;\tau,z)f(\tau,z;\tau+\Delta\tau,y)dz,$$

and therefore

$$\int_a^b \frac{\partial f(t,x;\tau,y)}{\partial \tau}R(y)dy = \lim_{\Delta\tau\to 0}\frac{1}{\Delta\tau}\Bigg[\iint f(t,x;\tau,z)f(\tau,z;\tau+\Delta\tau,y)$$
$$R(y)dz\,dy - \int f(t,x;\tau,y)\,R(y)dy\Bigg]$$

[2]The second Kolmogorov equation was obtained by the physicists Fokker and Planck in connection with diffusion theory.

$$= \lim_{\Delta\tau\to 0} \frac{1}{\Delta\tau}\left[\int f(t,x;\tau,z)\int f(\tau,z;\tau+\Delta\tau,y)\,R(y)dy\,dz - \int f(t,x;\tau,y)\,R(y)dy\right]$$

$$= \lim_{\Delta t\to 0} \frac{1}{\Delta\tau}\int f(t,x;\tau,y)\left[\int f(\tau,y;\tau+\Delta\tau,z)\,R(z)dz - R(y)\right]dy.$$

The transformations performed are obvious: the first time, one interchanged the order of integration, and the second time, one changed the notation for the variables of integration (y was replaced by z and z by y).

By Taylor's formula

$$R(z) = R(y) + (z-y)R'(y) + (z-y)^2 R''(y) + o[(z-y)^2].$$

Since, by the boundedness of the function $R(z)$ and the condition (1),

$$\int_{|y-z|\geq\delta} f(\tau,y;\tau+\Delta\tau,z)R(z)dz = o(\Delta\tau)$$

and

$$\int_{|y-z|<\delta} f(\tau,y;\tau+\Delta\tau,z)dz = 1 + o(\Delta\tau),$$

then

$$\int f(\tau,y;\tau+\Delta\tau,z)R(z)dz - R(y) = R'(y)\int_{|y-z|<\delta}(z-y)f(\tau,y;\tau+\Delta\tau,z)dz$$
$$+\frac{1}{2}R''(y)\int_{|y-z|<\delta}[(z-y)^2+o(z-y)^2]f(\tau,y;\tau+\Delta\tau,z)dz + o(\Delta\tau).$$

Thus,

$$\int_a^b \frac{\partial f(t,x;\tau,y)}{\partial\tau}R(y)dy =$$

$$= \lim_{\Delta\tau\to 0}\int f(t,x,\tau,y)\left\{R'(y)\int_{|y-z|<\delta}(z-y)f(\tau,y;\tau+\Delta\tau,z)dz\right.$$
$$\left.+\frac{1}{2}R''(y)\int_{|y-z|<\delta}[(z-y)^2+o(z-y)^2]\times f(\tau,y\tau+\Delta\tau,z)dz + o(\Delta\tau)\right\}dy.$$

Let us now pass to the limit, letting $\Delta\tau \to 0$. By the assumption that the limits in (2) and (3) are uniform in x, one concludes that the previous equation can be expressed in the form

$$\int_a^b \frac{\partial f(t,x;\tau,y)}{\partial \tau} R(y)\,dy = \int f(t,x;\tau,y)\left[a(\tau,y)R'(y) + \frac{1}{2}b(\tau,y)R''(y)\right]dy. \quad (8)$$

Since $R'(y) = R''(y) = 0$ for $y \leqslant a$ and $y \geqslant b$ then it follows that

$$\int_a^b \frac{\partial f(t,x;\tau,y)}{\partial \tau} R(y)\,dy = \int_a^b f(t,x;\tau,y)\left[a(\tau,y)R'(y) + \frac{1}{2}b(\tau,y)R''(y)\right]dy.$$

Using integration by parts and equations (7), one finds that

$$\int_a^b f(t,x;\tau,y)a(\tau,y)R'(y)\,dy = -\int_a^b R(y)\frac{\partial}{\partial y}[a(\tau,y)f(t,x;\tau,y)]\,dy,$$

$$\int_a^b f(t,x;\tau,y)b(\tau,y)R''(y)\,dy = \int_a^b R(y)\frac{\partial^2}{\partial y^2}[b(\tau,y)f(t,x;\tau,y)]\,dy.$$

After substitution of these expressions into (8), one obtains:

$$\int_a^b \frac{\partial f(t,x;\tau,y)}{\partial \tau} R(y)\,dy = \int_a^b \left\{-\frac{\partial}{\partial y}[a(\tau,y)f(t,x;\tau,y)] + \frac{1}{2}\frac{\partial^2}{\partial y^2}[b(\tau,y)f(t,x;\tau,y)]\right\} R(y)\,dy.$$

This equation, obviously, is expressible in the following form:

$$\int_a^b \left\{\frac{\partial f(t,x;\tau,y)}{\partial \tau} + \frac{\partial}{\partial y}[a(\tau,y)f(t,x;\tau,y)] - \frac{1}{2}\frac{\partial^2}{\partial y^2}[b(\tau,y)f(t,x;\tau,y)]\right\} R(y)\,dy = 0. \quad (9)$$

Since the function $R(y)$ is arbitrary, equation (6) follows from this last identity. However, suppose that this were not so. Then there exists a quadruplet of numbers $(t, x; \tau, y)$ such that the expression within the braces in (9) is different from 0. Under assumptions made-above, this expression represents a continuous function. Therefore, an interval

$\alpha < y < \beta$ can be found in which its algebraic sign is preserved. If $a \leqslant \alpha$ and $b \geqslant \beta$ then one sets $R(y) = 0$ for $y \leqslant \alpha$ and $y \geqslant \beta$ and $R(y) > 0$ for $\alpha < y < \beta$. With $R(y)$ chosen in such a way the integral on the left-hand side of equation (9) must differ from 0. Thus, it leads to a contradiction, i.e., the assumption made above is false. Hence equation (6) follows from (9).

Naturally, the main task is not to show that some given function $f(t, x; \tau, y)$ satisfies the Kolmogorov equations but rather to find an unknown function $f(t, x; \tau, y)$ from these equations if coefficients $a(t, x)$ and $b(t, x)$ are assumed known.

In addition, one does not look for just any solution of the Kolmogorov equations but only for those which satisfy the following conditions:

(1) $f(t, x; t, y) = 0$ for all t, x, τ, y.

(2)

$$\int f(t, x; \tau, y)\, dy = 1$$

for any $\delta > 0$.

(3)

$$\lim_{\tau \to t} \int_{|y-x| \geqslant \delta} f(t, x: \tau, y)\, dy = 0. \tag{10}$$

We shall not clarify the conditions that have to be imposed on the functions $a(t, x)$ and $b(t, x)$ for which a solution of the Kolmogorov equations exists and is unique.

Let us slightly strengthen the conditions of continuity in order to give a physical interpretation of the coefficients $a(t, x)$ and $b(t, x)$. For this purpose assume that instead of (1) for any $\delta > 0$, the relation

$$\lim_{\Delta t \to 0} \frac{1}{\Delta t} \int_{|y-x| > \delta} (y - x)^2\, d_y F(t - \Delta t, x; t, y) = 0 \tag{1'}$$

holds. One sees that (1) follows from (1'). Conditions (2) and (3) can be now expressed in different way, namely:

$$\lim_{\Delta t \to 0} \frac{1}{\Delta t} \int (y - x)\, d_y F(t - \Delta t, x; t, y) = a(t, x) \tag{2'}$$

and

$$\lim_{\Delta t\to 0}\frac{1}{\Delta t}\int (y-x)^2 d_y F(t-\Delta t, x; t, y) = b(t,x). \tag{3'}$$

The remaining requirements, as well as the final conclusions, are not changed by the replacement of (1) by (1′). Since

$$\int (y-x) d_y F(t-\Delta t, x; t, y) = \mathbf{E}[\xi(t) - \xi(t-\Delta t)]$$

is the mathematical expectation of the increment in $\xi(t)$ in time Δt and

$$\int (y-x)^2 d_y F(t-\Delta t, x; t, y) = \mathbf{E}[\xi(t) - \xi(t-\Delta t)]^2$$

is the mathematical expectation of the square of this increment, and is therefore proportional to the kinetic energy[3] it is clear from (2′) and (3′) that $a(t,x)$ is the average speed of change of $\xi(t)$ and that $b(t,x)$ is proportional to the average kinetic energy of the system under investigation.

This section is concluded with a discussion of a special case of the Kolmogorov equations where the function $f(t,x;\tau,y)$ depends on t,τ, and $y-x$, but not on x and y themselves. Physically this means that the process is homogeneous in space, i.e., the probability of the increment $\Delta = y - x$ does not depend on the position x the system was in at moment t. Obviously, in this case, the functions $a(t,x)$ and $b(t,x)$ are independent of x and are functions of only t:

$$a(t) = a(t,x); \quad b(t) = b(t,x).$$

The Kolmogorov's equations can be then rewritten in the following form:

$$\left.\begin{aligned} \frac{\partial f}{\partial t} &= -a(t)\frac{\partial f}{\partial x} - \frac{1}{2}b(t)\frac{\partial^2 f}{\partial x^2}, \\ \frac{\partial f}{\partial \tau} &= -a(\tau)\frac{\partial f}{\partial y} + \frac{1}{2}b(\tau)\frac{\partial^2 f}{\partial y^2}. \end{aligned}\right\} \tag{11}$$

[3]Here it is assumed that $\xi(t)$ is the coordinate of the position of a point moving about under influence of random factors. – *B. V.*

First consider the special case $a(t)=0$ and $b(t)=1$. Equations (11) reduce to the equation of heat conduction

$$\frac{\partial f}{\partial \tau}=\frac{1}{2}\frac{\partial^2 f}{\partial y^2}$$

and its adjoins

$$\frac{\partial f}{\partial t}=-\frac{1}{2}\frac{\partial^2 f}{\partial x^2}. \tag{12}$$

From the general theory of heat conduction, it is known that the unique solution of these equations satisfying conditions (10) is the function

$$f(t,x;\tau,y)=\frac{1}{\sqrt{2\pi(\tau-t)}}e^{-(y-x)^2/2(\tau-t)}.$$

By changing the variables

$$x'=x-\int_a^t a(z)dz, \quad y'=y-\int_a^\tau b(z)dz,$$

$$t'=\int_a^t b(z)dz, \quad \tau'=\int_a^\tau b(z)dz$$

equations (11) can be reduced to equations (12). This makes it possible to express the desired solution of equations (11) in the form

$$f(t,x;\tau,y)=\frac{1}{\sigma\sqrt{2\pi}}e^{-(y-x-A)^2/2\sigma^2},$$

where one uses the notation

$$A=\int_t^\tau a(z)dz, \quad \sigma^2=\int_t^\tau b(z)dz.$$

65 PURELY DISCONTINUOUS STOCHASTIC PROCESS. THE KOLMOGOROV—FELLER EQUATIONS

In modern physics, an important part is played by processes in which the changes of system states happen not continuously, but by jumps.

Examples of problems of this kind were represented in the beginning of this chapter.

One says that a stochastic process $\xi(t)$ is ***purely discontinuous*** if during any time interval $(t, t+\Delta t)$ the value of the variable $\xi(t)$ remains equal to x with probability $1-p(t,x)\Delta t + o(\Delta t)$ and may undergo a change only with probability $p(t,x)\Delta t + o(\Delta t)$. It is assumed here that the probability of more than one change of the process $\xi(t)$ in the interval Δt is $o(\Delta t)$.

Naturally, inasmuch as the consideration is restricted to processes without aftereffect, the distribution function for the changes of $\xi(t)$ that occur after a jump does not depend on what values $\xi(t)$ had at moments prior to the jump.

Let $P(t,x,y)$ denote the conditional distribution function of $\xi(t)$ under the condition that a jump has occurred at time t and that instantly before the jump $\xi(t)$ was equal to x, i.e., $\xi(t-0)=x$.

The distribution function $F(t,x;\tau,y)$ can be easily expressed in terms of the functions $p(t,x)$ and $P(t,x,y)$ as:

$$F(t,x;\tau,y) = [1-p(t,x)(\tau-t)]\,E(x,y) + (\tau-t)\,p(t,x)\,P(t,x,y) + o(\tau-t). \quad (1)$$

According to definition, the functions $p(t,x)$ and $P(t,x,y)$ are nonnegative and $P(t,x,y)$ being a distribution function satisfies the conditions

$$P(t,x,-\infty)=0, \quad P(t,x,+\infty)=1.$$

In addition, let us assume that $p(t,x)$ is bounded and that both functions $p(t,x)$ and $P(t,x,y)$ are continuous in t and x.
(Actually, it suffices to assume their Borel measurability in x).

No special assumptions concerning the function $F(t,x;\tau,y)$ are made except retaining its definition at $t=\tau$:

$$\lim_{\tau\to t+0} F(t,x;\tau,y) = \lim_{t\to\tau-0} F(t,x;\tau,y) = E(x,y) = \begin{cases} 0 & \text{for} \quad y \leqslant x, \\ 1 & \text{for} \quad y > x. \end{cases}$$

One of the goals of the current section is to prove the following theorem.

THEOREM. *The distribution function $F(t,x;\tau,y)$ for a purely discontinuous process without aftereffect satisfies the following two* *integro-differential equations:*

$$\frac{\partial F(t,x;\tau,y)}{\partial t} = p(t,x)\left[F(t,x;\tau,y) - \int F(t,z;\tau,y)\,d_zP(t,x,z)\right], \qquad (2)$$

$$\frac{\partial F(t,x;\tau,y)}{\partial \tau} = -\int_{-}^{y} p(t,z)\,d_zF(t,x;\tau,y)$$

$$+ \int p(\tau,z)\,P(\tau,z,y)\,d_zF(t,x;\tau,y). \qquad (3)$$

Equation (2) *was obtained by Kolmogorov in* 1931. *Both equations* (2) *and* (3) *were derived by Feller in* 1937 *under assumptions made above. These two facts make it natural to call equations* (2) *and* (3) *the Kolmogorov-Feller equations.*

PROOF: From the generalized Markov equation it follows that

$$F(t,x;\tau,y) = \int F(t+\Delta t,z;\tau,y)\,d_zF(t,x;t+\Delta t,z).$$

Substituting here $F(t,x;t+\Delta t,z)$ given by formula (1) one finds:

$$F(t,x;\tau,y) = \int F(t+\Delta t,z;\tau,y)\,d_z[1-p(t,x)\Delta(t)$$

$$+ o(\Delta t)]\,E(x,z) + \int F(t+\Delta t,z;\tau,y)\,d_z[p(t,x)\,\Delta t + o(\Delta t)]\,P(t,x,z).$$

Since

$$\int F(t+\Delta t,z;\tau,y)\,d_zE(x,z) = F(t+\Delta t,x;\tau,y),$$

it follows that

$$F(t,x;\tau,y) = [1-p(t,x)\,\Delta t]\,F(t+\Delta t,x;\tau,y)$$

$$+ \Delta t\,p(t,x)\int F(t+\Delta t,z;\tau,y)\,d_zP(t,x,z) + o(\Delta t).$$

Hence

$$\frac{F(t+\Delta t,x;\tau y)-F(t,x;\tau,y)}{\Delta t}=p(t,x)F(t+\Delta t,x;\tau,y)$$

$$+p(t,x)\int F(t+\Delta t,z;\tau,y)d_zP(t,x,z)+o \quad (1).$$

Passing to the limit leads to (2).

The Markov equation, equation (1), and also the definition of the function $E(x,z)$ enable one to write the following chain of equalities:

$$F(t,x;\tau+\Delta\tau,y)=\int F(\tau,z;\tau+\Delta\tau,y)\,d_zF(t,x;\tau,z)$$

$$=\int\{[1-p(\tau,z)\Delta\tau]\,E(z,y)+\Delta\tau p(\tau,z)\,P(\tau,z,y)$$

$$+o(\Delta\tau)\}\,d_zF(t,x;\tau,z)$$

$$=\int_{-}^{y}d_zF(t,x;\tau,z)-\Delta\tau\int_{-}^{y}p(\tau,z)\,d_zF(t,x;\tau,z)$$

$$+\Delta t\int p(\tau,z)\,P(\tau,z,y)\,d_zF(t,x;\tau,z)+o(\Delta\tau).$$

By standard reasoning, it follows from this that the partial derivative $\partial F/\partial\tau$ exists and that equation (3) holds.

The further problem is important for applications: What is the probability that a system changes its state n times ($n=0,1,2,\ldots$) in a time interval from t to τ ($\tau>t$)?

Let $p_n(t,x,t)$ denote the probability that starting from the state x at time t, the system changes its state n times by the moment τ. Consider first the case $n=0$.

Let us write down the following equation:

$$p_0(t,x,\tau)=p_0(t,x,\tau+\Delta\tau)+p_0(t,x,\tau)[1-p_0(\tau,x,\tau+\Delta\tau)], \quad (4)$$

which expresses the fact that the absence of changes in state of the system in a time interval (t,τ) can occur in the following two mutually exclusive ways: (1) The system has not changed its state in a greater interval of time $(\tau, \tau+\Delta\tau)$, and (2) The system has not changed its state up to the time τ but in the time interval $(\tau, \tau+\Delta\tau)$ it has done so. Since, by the definition of a purely discontinuous process,

$$p_0(\tau, x, \tau+\Delta\tau) = 1 - p(\tau, x)\,\Delta\tau + o(\Delta\tau),$$

equation (4) can also be written as:

$$\frac{p_0(t, x, \tau+\Delta\tau) - p_0(t, x, \tau)}{\Delta\tau} = -p_0(t, x, \tau)\,p(\tau, x) + o(1).$$

From here, by letting $\Delta\tau \to 0$, one finds that the derivative

$$\frac{\partial p_0(t, x, \tau)}{\partial\tau}$$

exists and that

$$\frac{\partial p_0(t, x, \tau)}{\partial\tau} = -p_0(t, x, \tau)\,p(\tau, x).$$

Solving this differential equation, one obtains:

$$p_0(t, x, \tau) = Ce^{-\int_t^\tau p(u,x)\,du}.$$

Since

$$p_0(\tau, x, \tau) = 1,$$

it follows that $C = 1$ and

$$p_0(t, x, \tau) = e^{-\int_t^\tau p(u,x)\,du}. \tag{5}$$

Let us now show that a knowledge of functions $p_0(t, x, \tau)$ and $P(t, x, y)$ allows one to compute the probability $p_n(t, x, \tau)$ for any n. In fact, an n-fold change of state occurs in the following way: (1) Up to the moment $s(t < s < \tau)$ the system does not undergo a change of state and the probability of this event is $p_0(t, x, s)$; (2) In the interval $(s, s+\Delta s)$ the system changes state and the probability of this is $p_1(s, x, s+\Delta s) = p(s, x)\Delta s + o(\Delta s)$; (3) The probability that the new state of the system will be between y and $y+\Delta y$ is $P(s, x, y+\Delta y) - P(s, x, y) = \Delta_y P(s, x, y)$; (4) Finally, in the time interval $(s+\Delta s, \tau)$ the system changes its state $n-1$ times and the probability of this event is $p_{n-1}(s+\Delta s, y, \tau)$.

By the Theorem of Multiplication, the probability that all four of these events will occur equals

$$p_0(t, x, s)\,[p(s, x) + o(1)]\,\Delta s \cdot \Delta_y P(s, x, y) \cdot p_{n-1}(s + \Delta s, y, \tau).$$

Since s and y are arbitrary ($t < s < \tau$ and $-\infty < y < +\infty$), from the formula on total probability it follows that

$$p_n(t, x, \tau) = \int_t^\tau \int p_0(t, x, s)\, p(s, x)\, p_{n-1}(s, y, \tau)\, d_y P(s, x, y)\, ds$$

$$= \int_t^\tau p_0(t, x, s)\, p(s, x) \int p_{n-1}(s, y, \tau)\, d_y P(s, x, y)\, ds. \tag{6}$$

Hence, in particular,

$$p_1(t, x, \tau) = \int_t^\tau p_0(t, x, s)\, p(s, x) \int p_0(s, y, \tau)\, d_y P(s, x, y)\, ds. \tag{7}$$

The method of determination of the distribution $p_n(t, x, \tau)$ is obvious. By formula (5) one finds $p_0(t, x, t)$, by formula (7) one computes $p_1(t, x, t)$, and then sequentially $p_2(t, x, \tau)$, $p_3(t, x, \tau)$, and finally, $p_n(t, x, \tau)$.

Example 1. Let the variable $\xi(t)$ be the number of state changes from moment 0 up to t. Under the assumption that $p(t, x) = a$ where $a > 0$ is a constant, find $p_n(t, x, \tau)$.

In this case, the possible states of the system will be all of the non-negative integers ($x = 0, 1, 2, \ldots$). Since with each change of state, the variable $\xi(t)$ is increased by exactly 1, one has

$$P(t, x, y) = \begin{cases} 0 & \text{for} \quad y \leqslant x, \\ 1 & \text{for} \quad y > x. \end{cases}$$

By formula (5), one has

$$p_0(t, x, \tau) = e^{-a(\tau - t)}.$$

According to (7),

$$p_1(t, x, \tau) = \int_t^\tau p_0(t, x, s)\, p(s, x)\, p_0(s, x + 1, \tau)\, ds$$

$$= a \int_t^\tau e^{-(s-t)a} e^{-(\tau - s)a}\, ds = a(\tau - t)\, e^{-a(\tau - t)}.$$

By formula (6),

$$p_2(t,x,\tau)=\int_t^\tau p_0(t,x,s)\,p(s,x)p_1(s,x+1,\tau)ds)=\frac{[a(\tau-t)]^2}{2!}e^{-a(\tau-t)}.$$

Let us now assume that

$$p_{n-1}(t,x,\tau)=\frac{[a(\tau-t)]^{n-1}}{(n-1)!}e^{-a(\tau-t)}.$$

Finally, by equation (7),

$$p_n(t,x,\tau)=\int_t^\tau p_0(t,x,s)\,p(s,x)\,p_{n-1}(s,x+1,\tau)ds$$

$$=\int_t^\tau \frac{a[a(\tau-s)]^{n-1}}{(n-1)!}e^{-a(\tau-t)}ds=\frac{[a(\tau-t)]^n}{n!}e^{-a(\tau-t)}.$$

This proves that for any integer $n\geqslant 0$

$$p_n(t,x,\tau)=\frac{[a(\tau-t)]^n}{n!}e^{-a(\tau-t)}.$$

The solution of the problem under investigation is the Poisson law. In particular,

$$p_n(0,0,\tau)=\frac{(a\tau)^n}{n!}e^{-a\tau}.$$

It is easy to guess that the function

$$F(t,x;\tau,y)=\begin{cases}0 & \text{for } y\leqslant 0,\\ \sum\limits_{n<y}\frac{[a(\tau-t)]^n}{n!}e^{-a(\tau-t)} & \text{for } y>0\end{cases}$$

is the solution of the integro-differential equations (2) and (3).

Example 2. At the moment $t=0$ there are N radioactive atoms. The probability of decay of an atom in the interval of time $(t,t+\Delta t)$ equals $aN(t)\Delta t+o(\Delta t)$ where $a>0$ is a constant and $N(t)$ is the number of atoms which have not decayed up to the time t. Find the probability that n atoms will decay in the interval $[t, \tau]$.

This is a typical purely discontinuous stochastic process.

The random variable $\xi(t)$ can assume the values $0, 1, 2, \ldots, N(t)$.

By the condition of the problem,

$$p(t,x) = \begin{cases} 0 & \text{for } x \leqslant 0 \text{ and } x \geqslant N, \\ a(N-x) & \text{for } 0 < x \leqslant N, \end{cases}$$

and

$$P(t,x,y) = \begin{cases} 0 & \text{for } y \leqslant x, \\ 1 & \text{for } y > x. \end{cases}$$

To begin, let us evaluate the probability of n atom disintegrations during the time interval from 0 to t.

By equation (5)

$$p_0(0,0,\tau) = e^{-\int_0^\tau p(t,0)\,dt} = e^{-aN\tau}.$$

Similarly,

$$p_0(t,k\,\tau) = e^{-a(N-k)(\tau-t)}.$$

Further, by equation (7),

$$p_1(0,0,\tau) = \int_0^\tau p_0(0,0,s)\,p(s,0)\,p_0(s,1,\tau)\,ds$$

$$= \int_0^\tau e^{-aNs}aNe^{-a(N-1)(\tau-s)}ds$$

$$= Ne^{-aN\tau}\int_0^\tau ae^{a(\tau-s)}ds = Ne^{-aN\tau}[e^{a\tau}-1]. \tag{8}$$

By formula (6), it is easy to find in succession $p_2(0,0,t), p_3(0,0,\tau)$, etc., and to prove that

$$p_n(0,0,\tau) = C_N^n e^{-aN\tau}[e^{a\tau}-1]^n. \tag{9}$$

The proof of the fact is left to the reader.

Obviously that for $0 \leqslant n \leqslant N-k$ the equality

$$p_n(t,k,\tau) = C^n_{N-k} e^{-a(N-k)(\tau-t)}[e^{a(\tau-t)}-1]^n \tag{9'}$$

holds.

It is possible now to find the probability denoted by $p_n(t,\tau)$. By the formula on total probability and using (9) and (9′), one finds that

$$\begin{aligned} p_n(t,\tau) &= \sum_{k=0}^{N-n} = p_k(0,0,t)\cdot p_n(t,k,\tau) \\ &= \sum_{k=0}^{N-n} C^k_N e^{-aNt}[e^{at}-1]^k C^n_{N-k} e^{-a(N-k)(\tau-t)}[e^{a(\tau-t)}-1]^n \\ &= e^{-aN\tau}[e^{a(\tau-t)}-1]^n \sum_{k=0}^{N-n} C^k_N C^n_{N-k} e^{ak(\tau-t)}[e^{at}-1]^k. \end{aligned}$$

Since

$$C^k_N C^n_{N-k} = C^n_N C^k_{N-n}$$

and

$$\sum_{k=0}^{N-n} C^k_{N-n}[e^{a(\tau-t)}(e^{at}-1)]^k = [1+e^{a(2\tau-t)}-e^{a(\tau-t)}]^{N-n},$$

one finally has

$$p_n(t,\tau) = C^n_N[e^{-at}-e^{-a\tau}]^n[e^{-a\tau}+e^{a(\tau-t)}-e^{-a\tau}]^{N-n}.$$

It is easy to show that the function

$$F(t,x;\tau,y) = \begin{cases} 0 & \text{for} \quad y \leqslant x, \\ \sum\limits_{n<y} p_n(t,x,\tau) & \text{for} \quad y < N-x, \\ 1 & \text{for} \quad y > N-x \end{cases}$$

is the solution of the integro-differential equations (2) and (3).

65 HOMOGENEOUS STOCHASTIC PROCESSES WITH INDEPENDENT INCREMENTS

Let us now consider an important class of stochastic processes a complete characterization of which can be given in terms of characteristic functions.

By a ***homogeneous stochastic process with independent increments*** is understood the assembly of random variables $\xi(t)$ depending on a single real parameter t and satisfying the following two conditions:

(1) The distribution function of the quantity $\xi(t+t_0)-\xi(t)$ does not depend on t_0 (the process is homogeneous in time);

(2) For any finite number of non-overlapping intervals (a, b) of the parameter t the increments of the variable $\xi(t)$, i.e., the differences $\xi(b)-\xi(a)$, are mutually independent.

Before proceeding to obtain specific results, let us discuss several examples for which the above conditions may be taken as a working hypothesis. Of course, their admissibility can only be justified by the concordance between the theoretical deductions and experiment.

Example 1. Diffusion of a gas. Consider a molecule of a certain gas moving in space among other molecules of the same gas under conditions of constant temperature and density. Let us introduce into this space a Cartesian coordinate system, and let us observe how one of the molecule's coordinates (say, the x-coordinate) changes with time.

In consequence of the random collisions of the chosen molecule with the other molecules, this coordinate will change with time by obtaining random increments. The requirement of constant conditions of gas means that the process under study is homogeneous in time. In view of the large number of molecules in motion and of the weak dependence of their motion, the process is one with independent increments.

Example 2. Velocity of a molecule. Again consider a molecule of a certain gas which is moving in a volume filled with the other molecules under conditions of constant density and temperature. Let us again introduce a Cartesian coordinate system into the space, and let us follow how one of the components of velocity with respect to these axes changes in time. Again the molecule is moving and colliding with the other molecules. In consequence of these collisions, the velocity component will obtain random increments. Again a homogeneous stochastic process with independent increments is observed.

Example 3. Radioactive decay. It is well known that radioactivity is the phenomenon in which the atoms of a substance convert into atoms of some other substance with the liberation of a considerable amount of energy. Observations over comparatively large masses of a radioactive substance show that the rates of decay of various atoms are entirely independent of one another. It means that the number of atoms disintegrated in non-overlapping intervals of time are mutually independent. In addition, the probability that a certain number of disintegrations will occur in an interval of time of given length depends on the length of this interval but not on current time. (In reality, of course, the radioactivity of a substance gradually decreases in time as its mass decreases. However, for comparatively small intervals of time and/or not too vast amount of substance, this change can be neglected as insignificant).

One can readily present many other examples in which a natural phenomenon or technical process can be modelled by a homogeneous process with independent increments. One could refer to other examples: the number of cosmic particles falling on a given area in a given interval of time, the breaking of yarn on a ring spinning frame, the work load of a telephone operator (the number of calls in a given interval of time), and so on.

Now examine the characteristic property of homogeneous stochastic processes with independent increments.

Let the distribution function of the increment in the variable $\xi(t)$ for an interval of time of length x be denoted by $F(x, \tau)$. Then, if the intervals of time of length τ_1 and τ_2 do not overlap, it follows that

$$F(x; \tau_1 + \tau_2) = \int F(x - y; \tau_1) d_y F(y, \tau_2). \tag{1}$$

If $f(z, t)$ is the characteristic function, i.e., if

$$f(z, \tau) = \int e^{izx} d_x F(x; \tau),$$

then, in terms of characteristic functions, equation (1) takes on the following form:

$$f(z; \tau_1 + \tau_2) = f(z, \tau_1) \cdot f(z, \tau_2) \tag{1'}$$

In general, if the intervals $\tau_1, \tau_2, \ldots \tau_n$ are non-overlapping then

$$f\left(z; \sum_{k=1}^{n} \tau_k\right) = \prod_{k=1}^{n} f(z; \tau_k).$$

In particular, if $\tau_1 = \tau_2 = \cdots = \tau_n$ and

$$\sum_{k=1}^{n} \tau_k = \tau$$

then

$$f(z, \tau) = [f(z; \tau/n)]^n.$$

Thus, the distribution function of any homogeneous stochastic process with independent increments is infinitely divisible.

It should be noted that the study of homogeneous processes with independent increments led to the introduction of infinitely divisible distribution into probability theory. It was shown that the theory of infinitely divisible distribution laws has had a decisive effect on the development of the classical branch of probability theory concerning sums of random variables. As has been indicated, the early efforts of mathematicians were concentrated on finding the broadest conditions under which the law of large numbers was true and under which normalized sums would converge to the normal law. But after Kolmogorov had completely described the laws governing homogeneous stochastic processes without aftereffect, the more general questions which were considered in the preceding chapter arose. It was found that the fundamental distribution laws which were obtained earlier as asymptotic laws represent exact solutions of the corresponding functional equations in the theory of stochastic processes. Moreover, this new viewpoint enabled one to explain the reason why only two limit distributions were considered in the classical theory of probability, namely, the normal law and the Poisson law.

Since for arbitrary $\tau > 0$ for homogeneous processes with independent increments

$$f(z, \tau) = [f(z, 1)]^{\tau},$$

it follows that such processes are completely determined by the assignment of the characteristic function of the random variable $\xi(1) - \xi(0)$.

In Section 43 it was shown that the equation

$$\ln \varphi(z, 1) = i\gamma z + \int \{e^{izu} - l - izu\} \frac{1}{u^2} dG(u), \tag{2}$$

holds for any infinitely divisible law with a finite variance, where y is a real constant and $G(u)$ is a non-decreasing function of bounded variation. Let us confine ourselves to the consideration of this special case of homogeneous processes.

Introduce in (2) the following notation:

$$M(u) = \int_{-\infty}^{u} \frac{1}{x^2} dG(x) \quad \text{for } u < 0,$$

$$N(u) = \int_{u}^{\infty} \frac{1}{x^2} dG(x) \quad \text{for } u > 0,$$

$$\sigma^2 = G(+0) - G(-0);$$

Then formula (2) takes the form

$$\ln \varphi(z, l) = i\gamma z - \frac{\sigma^2 z^2}{2} + \int_{-\infty}^{0} \{e^{izu} - 1 - izu\}\, dM(u)$$

$$+ \int_{0}^{\infty} \{e^{izu} - 1 - izu\}\, dN(u).$$

The probability-theoretic meaning of the functions $M(u)$ and $N(u)$ is a subject of interest.

In deriving the canonical representation of infinitely divisible laws in Section 43 we introduced the function

$$G_n(u) = n \int_{-\infty}^{u} x^2 d\Phi_n(x).$$

Let us set

$$M_n(u) = \int_{-\infty}^{u} \frac{1}{x^2} dG_n(x) = n\,\Phi_n(u) \quad \text{for} \quad u < 0$$

and

$$N_n(u) = \int_{u}^{\infty} \frac{1}{x^2} dG_n(x) = n\,[1 - \Phi_n(u)] \quad \text{for} \quad u > 0.$$

Since

$$G_n(u) \to G(u)$$

at the points of continuity of the function $G(u)$ as $n \to \infty$, one concludes from the Second Helly Theorem that

$$M_n(u) = n\Phi_n(u) \to M(u)$$

at the points of continuity of the function $M(u)$.

From the viewpoint of stochastic processes, $\Phi(x)$ for $x < 0$ is the probability that the random variable $\xi(\tau)$ will acquire a decrement which in absolute value is larger than x in the interval of variation

$$\left(\frac{k}{n}, \frac{k+1}{n}\right)$$

of the parameter τ. Thus, $M_n(x)$ is the sum over all k from 0 to $n-1$ of the probabilities that the variable $\xi(t)$ will acquire the decrement under conditions above-described. Since $M(u)$ and $N(u)$ are the limits of the respective functions $M_n(u)$ and $N_n(u)$ as $n \to \infty$, they are referred to as jump functions.

If $M(u) \equiv 0$ for $u < 0$ and $N(u) \equiv 0$ for $u > 0$, i.e., if there are no jump functions, then from equation (2′) follows that, in this case, the stochastic process is governed by the normal law. One sees that a stochastic process governed by the normal law is continuous in the sense of probability theory. Let us now prove a stronger statement.

THEOREM. *A homogeneous stochastic process with independent increments and a finite variance*[4] *is governed by the normal law*[5] *if and only if for arbitrary $\varepsilon > 0$ the probability of the maximum of the absolute value of the increments in $\xi(\tau)$ for the intervals $((k-1)/n, k/n)$ $(k = 1, 2, 3, \ldots, n)$ exceeding ε tends to 0 with $1/n$*[6].

PROOF: As was just shown, a homogeneous stochastic process with independent increments is governed by the normal law if and only if

[4]The theorem is true even without the assumption that the variance exists. – *B. G.*

[5]In particular, by the normal law with a variance equal to 0, i.e., by a distribution of the form $F(x) = 0$ for $x \leqslant a$ and $F(x) = 1$ for $x > a$. – *B. G.*

[6]Thus, processes which are governed by the normal law, and such processes only are "uniformly continuous" in a probabilistic sense. – *B. G.*

for $x > 0$

$$M(-x) \equiv N(x) \equiv 0. \tag{3}$$

Since

$$M(u) = \lim_{n \to \infty} M_n(u) \quad \text{and} \quad N(u) = \lim_{n \to \infty} N_n(u),$$

condition (3) is equivalent to the following:

$$\lim_{n \to \infty} n\Phi_n(-u) = \lim_{n \to \infty} n[1 - \Phi_n(u)] = 0. \tag{4}$$

Let ξ_{nk} denote the increment in $\xi(\tau)$ in the interval $((k-1)/n, k/n)$. Then

$$p_{nk} = \Phi_n(-x) + 1 - \Phi_n(x+0) = \mathbf{P}\{|\xi_{nk}| > x\}.$$

Obviously, relations (4) are equivalent to the following:

$$\lim_{n \to \infty} \sum_{k=1}^{n} p_{nk} = 0.$$

From the inequalities

$$1 - \sum_{k=1}^{n} p_{nk} \leqslant \prod_{k=1}^{n} (1 - p_{nk}) \leqslant e^{-\sum_{1}^{n} p_{nk}} \leqslant 1,$$

one sees that relations (4) (4)are equivalent to the statement

$$\lim_{n \to \infty} \prod_{k=1}^{n} (1 - p_{nk}) = 1,$$

which signifies that the probability that the inequalities $|\xi_{nk}| < \varepsilon$ hold for all $k (1 \leqslant k \leqslant n)$ approaches 1 as $n \to \infty$. In other words, it has been shown that relations (3) hold if and only if

$$\mathbf{P}\left\{\max_{1 \leqslant k \leqslant n} |\xi_{nk}| \geqslant \varepsilon\right\} \to 0,$$

as $n \to \infty$, Q.E.D.

66 THE CONCEPT OF A STATIONARY STOCHASTIC PROCESS. KHINCHINE'S THEOREM ON THE CORRELATION COEFFICIENT

The Markov processes, or processes without aftereffect, which were examined in the preceding sections do not exhaust all of the requests submitted to probability theory by science and varied applications. In fact in many situations the past states of a system have a very significant effect on the probability of its future states, and the influences of the past cannot be neglected even in an approximate investigation of the problem. In principle, such a situation could be corrected by changing our conception of the state of the system by introducing new parameters. Thus, for example, if one considers the change in position of a particle in diffusion phenomenon or Brownian motion as a process without aftereffect, this would mean that the inertia of the particle is not taken into account. But of course, it plays an essential role in these phenomena. By including in our conception of the state of the system the velocity of the particle in addition to the coordinates of its position, one would correct the situation in this example. However, there are cases where such a correction does not facilitate the solution of the problems that have been posed. In the first place, one could point to statistical mechanics, in which specifying that a particle is in a certain point of phase space merely yields a probabilistic judgment concerning its future state. In this concrete case, a knowledge of the previous positions of the particle can essentially change the judgments concerning its future state. In this connection, Khinchine has separated an important class of stochastic processes with aftereffect, the so-called stationary processes, which behave homogeneously with time.

A stochastic process $\xi(t)$ is called *stationary* if the n-dimensional distribution functions for the two finite groups of variables $\xi(t_1), \xi(t_2), \ldots, \xi(t_n)$ and $\xi(t_1+u), \xi(t_2+u), \ldots, \xi(t_n+u)$ coincide, which means that they do not depend on u. Here the numbers n and u as well as the moments of time $t_1, t_2, \ldots, t_n$ may be chosen quite arbitrarily.

The theory of stationary processes has found considerable application in physics and engineering. For example, the study of a number of acoustical phenomena, random noise problems in radio engineering, as well as the finding of hidden periodicities of interest to astronomers, geophysicists, and meteorologists lead to the consideration of stationary processes.

It is clear that any numerical characteristic of a stationary process $\xi(t)$ is independent of t and if, for example, $\xi(t)$ has a finite variance, then the following equations are valid

$$\mathbf{E}\xi(t+u)=\mathbf{E}\xi(t)=\mathbf{E}\xi(0)=a,$$

$$\mathbf{D}\xi(t+u)=\mathbf{D}\xi(t)=\mathbf{D}\xi(0)=\sigma^2,$$

$$\mathbf{E}\{\xi(t+u)\,\xi(t)\}=\mathbf{E}\{\xi(u)\,\xi(0)\}.$$

This fact allows one to take $a=0$ and $\sigma=1$ without limiting the generality of the subsequent results (for this purpose, it obviously suffices to consider the ratio $(\xi(t)-a)/\sigma$ instead of $\xi(t)$).

Let us confine ourselves to an investigation of the correlation function, i.e., the correlation coefficient between random variables $\xi(t)$ and $\xi(t+u)$:

$$R(u)=\frac{\mathbf{E}[\xi(t+u)-\mathbf{E}\xi(t+u)]\,[\xi(t)-\mathbf{E}\xi(t)]}{\sqrt{\mathbf{D}\xi(t)\cdot\mathbf{D}\xi(t+u)}}.$$

By virtue of the assumption that $a=0$ and $\sigma=1$, the expression for $R(u)$ takes on the simpler form:

$$R(u)=\mathbf{E}\{\xi(u)\,\xi(0)\}.$$

The stationary stochastic process is called continuous if

$$\lim_{u\to 0}R(u)=1.$$

In the case of the stationary stochastic process the function $R(u)$ is a continuous function of u. Indeed,

$$|R(u+\Delta u)-R(u)|=|\mathbf{E}\{\xi(u+\Delta u)\,\xi(0)\}-\mathbf{E}\{\xi(u)\,\xi(0)\}|$$
$$=|\mathbf{E}\{\xi(0)\,[\xi(u+\Delta u)-\xi(u)]\}|.$$

By the Cauchy-Bouniakowsky Inequality

$$|\mathbf{E}\{\xi(0)\,[\xi(u+\Delta u)-\xi(u)]\}|\leqslant\sqrt{\mathbf{E}\xi^2(0)\cdot\mathbf{E}[\xi(u+\Delta u)-\xi(u)]^2}.$$

Since

$$\mathbf{E}\,\xi^2(0)=1$$

and

$$\mathsf{E}[\xi(u+\Delta u)-\xi(u)]^2 = 2[1-R(\Delta u)],$$

then finally it follows that

$$|R(u+\Delta u)-R(u)| \leqslant \sqrt{2(1-R(\Delta u))}.$$

This inequality proves the statement above.

In the theorem which will now be proved, the stationary property of the process $\xi(t)$ may be understood in the wide sense: the process $\xi(t)$ is *stationary in the wide sense* if the mathematical expectation and variance of $\xi(t)$ do not depend on t and the correlation coefficient between $\xi(t_1)$ and $\xi(t_2)$ is the function depending only on $|t_2-t_1|$.

KHINCHINE'S THEOREM. *The function $R(u)$ is the correlation function for a continuous stationary process if and only if it can be represented in the form*

$$R(u) = \int \cos ux \, dF(x),$$

where $F(x)$ is some distribution function.

PROOF: The condition of the theorem is necessary. Indeed, if $R(u)$ is the correlation function for a continuous stationary process, then it is continuous and bounded. Let us prove additionally that it is positive semidefinite. In fact, for any real numbers $u_1, u_2, \ldots, u_n$, complex numbers $\eta_1, \eta_2, \ldots, \eta_n$, and integer n, the following relation is valid:

$$0 \leqslant \mathsf{E}\left|\sum_{k=1}^{n} \eta_k \xi(u_k)\right|^2 = \mathsf{E}\left\{\sum_{i=1}^{n}\sum_{j=1}^{n} \eta_i \overline{\eta_j}\, \xi(u_i)\, \xi(u_j)\right\}$$

$$= \sum_{j=1}^{n}\sum_{i=1}^{n} R(u_i - u_j)\, \eta_i \overline{\eta_j}.$$

From here, by virtue of the Bochner-Khinchine Theorem (Section 36) it follows that $R(u)$ is representable in the form

$$R(u) = \int e^{iux} \, dF(x) \tag{2}$$

where $F(x)$ is a non-decreasing function of bounded variation. Since the function $R(u)$ is real one obtains that

$$R(u) = \int \cos ux \, dF(x). \tag{3}$$

Finally, taking into account the assumption that the process is continuous and hence that $R(+0) = 1$, one finds that $F(\infty) - F(-\infty) = 1$, i.e., $F(x)$ is a distribution function.

The condition is sufficient. It is given that $R(u)$ is a function of the form (3). One must prove that a stationary process $\xi(t)$ exists which has as its correlation function the function $R(u)$. For this purpose, for any integer n and any set of real numbers $t_1, t_2, \ldots, t_n$ a normally distributed n-dimensional vector $\xi(t_1), \xi(t_2), \ldots, \xi(t_n)$ is considered. This vector must possess the following properties:

$$\mathbf{E}\,\xi(t_1) = \mathbf{E}\,\xi(t_2) = \cdots = \mathbf{E}\,\xi(t_n) = 0,$$

$$\mathbf{D}\,\xi(t_1) = \mathbf{D}\,\xi(t_2) = \cdots = \mathbf{D}\,\xi(t_n) = 1,$$

for arbitrary i and j, the correlation coefficient between $\xi(t_i)$ and $\xi(t_j)$ equals $R(t_i - t_j)$, i.e.,

$$\mathbf{E}\,\xi(t_i)\,\xi(t_j) = R(t_i - t_j).$$

The form of the function $R(u)$ ensures the positive-semidefiniteness of the quadratic form appearing in the exponent of the n-dimensional normal distribution. The normal stochastic process thus defined is stationary both in the wide and in the strict senses.

This theorem plays a basic role in the theory of stationary processes and its applications in physics.

Example 1. Let

$$\xi(t) = \xi \cos \lambda t + \eta \sin \lambda t,$$

where ξ and η are uncorrelated[7] random variables for which $\mathbf{E}\xi = \mathbf{E}\eta = 0$, $\mathbf{D}\xi = \mathbf{D}\eta = 1$, and λ is a constant. Since

[7]The random variables ξ and η are said to be uncorrelated if $\mathbf{E}\xi\eta = \mathbf{E}\xi\mathbf{E}0\eta$. – *B. G.*

$$\begin{aligned}R(u) &= \mathbf{E}\,\xi(t+u)\,\xi(t)\\ &= \mathbf{E}\,[\xi\cos\lambda(t+u)+\eta\sin\lambda(t+u)]\cdot[\xi\cos\lambda t+\eta\sin\lambda t]\\ &= \mathbf{E}\,[\xi^2\cos\lambda t\cos\lambda(t+u)+\xi\eta\,(\sin\lambda(t+u)\cos\lambda t\\ &\qquad +\cos\lambda(t+u)\sin\lambda t)+\eta^2\sin\lambda t\sin\lambda(t+u)\\ &= \cos\lambda t\cos\lambda(t+u)+\sin\lambda t\sin\lambda(t+u)=\cos\lambda u,\end{aligned}$$

the process $\xi(t)$ is stationary in the wide sense. For this case one must set

$$F(x)=\begin{cases}0 & \text{for} \quad x\leqslant-\lambda,\\ 0,5 & \text{for} \quad -\lambda<x\leqslant\lambda,\\ 1 & \text{for} \quad x>\lambda.\end{cases}$$

Example 2. Let

$$\xi(t)=\sum_{k=1}^{n} b_k\xi_k(t),$$

where $\xi_k(t)=\xi_k\cos\lambda_k t+\eta_k\sin\lambda_k t$, λ_k are constants,

$$\sum_{k=1}^{n} b_k^2=1,$$

and the random variables ξ_k and η_k satisfy the following conditions:

$$\begin{aligned}&\mathbf{E}\,\xi_k=\mathbf{E}\,\eta_k=0, \quad \mathbf{D}\,\xi_k=\mathbf{D}\,\eta_k=1, \quad (1\leqslant k\leqslant n),\\ &\mathbf{E}\,\xi_i\xi_j=\mathbf{E}\,\eta_i\eta_j=0 \quad \text{for} \quad i\neq j,\\ &\mathbf{E}\,\xi_i\eta_j=0 \quad \text{for} \quad i,j=1,2,\ldots,n.\end{aligned}$$

It is easy to show that the correlation function of the process $\xi(t)$ is

$$R(u)=\sum_{k=1}^{n} b_k^2\cos\lambda_k u$$

and therefore that the process $\xi(t)$ is stationary in the wide sense. The function $F(x)$ in formula (3) increases only at some points and at each of them has jumps of magnitude $0.5b_k^2$.

Stochastic processes for which the function $F(x)$ increases only by jumps are called processes with a *discrete spectrum.*

It is easy to see that any process of the form

$$\xi(t) = \sum_{k=1}^{\infty} b_k \xi_k(t) \tag{4}$$

where $\Sigma_{k=1}^{\infty} b_k^2 < \infty$ and $\xi_k(t)$ has the same meaning as in Example 2, is stationary in the wide sense and has a discrete spectrum.

It is worth noting that E. Slutsky proved a deeper converse statement: Any stationary process with a discrete spectrum is representable in the form (4). The generalization of this Slutsky theorem to the case of an arbitrary spectrum will be formulated in the next section.

In parallel with the theory of stationary processes there has been developed a theory of stationary sequences. A sequence of random variables

$$\ldots, \xi_{-2}, \xi_{-1}, \xi_0, \xi_1, \xi_2, \ldots$$

is called stationary in the wide sense if for any of its terms the mathematical expectations and variances are constants independent of their location within the sequence

$$\cdots = \mathsf{E}\,\xi_{-2} = \mathsf{E}\,\xi_{-1} = \mathsf{E}\,\xi_0 = \mathsf{E}\,\xi_1 = \mathsf{E}\,\xi_2 = \cdots = a,$$

$$\cdots = \mathsf{D}\,\xi_{-2} = \mathsf{D}\,\xi_{-1} = \mathsf{D}\,\xi_0 = \mathsf{D}\,\xi_1 = \mathsf{D}\,\xi_2 = \cdots = \sigma^2,$$

and the correlation coefficient between ξ_i and ξ_j is the function of only $|i-j|$.

The reader can consider as an exercise the following problems:

(1) Using the results of Section 36 to find a general form of the correlation function for a stationary sequence;

(2) to prove the theorem: If for a stationary sequence

$$\lim_{s\to\infty} R(s) = 0,$$

where $R(s)$ is the correlation coefficient between ξ_i and ξ_j then for this sequence the law of large numbers holds, i.e., for $n \to l$

$$\mathsf{P}\left\{\left|\frac{1}{n}\sum_{k=1}^{n} \xi_k - a\right| < \varepsilon\right\} \to 1,$$

for any constant $\varepsilon > 0$.

67 THE NOTION OF A STOCHASTIC INTEGRAL. SPECTRAL DECOMPOSITION OF STATIONARY PROCESSES

The next discussion needs the notion of a stochastic integral. Let the stochastic process $\xi(t)$ and the function $f(t)$ be given in the interval $a \leqslant t \leqslant b$. Subdivide the interval $[a, b]$ by points $a = t_0 < t_1 < \cdots < t_n = b$ and consider the sum

$$J_n = \sum_{i=1}^{n} f(t_i)\xi(t_i)(t_i - t_{i-1}).$$

If for

$$\max_{1 \leqslant i \leqslant n} (t_i - t_{i-1}) \to 0$$

this sum approaches a limit (which is, in general, a random variable), then this limit is called the *integral of the stochastic process* $\xi(t)$ and is denoted by the symbol

$$J = \int_a^b f(t)\xi(t)\,dt.$$

The improper integral (for $a = -\infty$, $b = +\infty$) is defined in the usual way as the limit of the proper integral as $a \to -\infty$, $b \to +\infty$.

The convergence of the integral sums J_n is understood in the following sense: There exists a random variable J such that

$$\mathsf{E}(J_n - J)^2 \to 0 \tag{1}$$

as $n \to \infty$.

Based on known theorems of the theory of functions of a real variable, one can easily prove that the sequence of random variables J_n converges to a limit J in the sense of (1) if and only if

$$\mathsf{E}(J_m - J)^2 \to 0 \tag{2}$$

as $\min(m, n) \to \infty$. The proof of this theorem is omitted.

THEOREM 1. *In order for the integral*

$$J = \int_a^b f(t)\xi(t)\,dt$$

to exist it is sufficient that the integral

$$A = \int_a^b \int_a^b R(t-s) f(t) f(s)\, ds\, dt$$

exists and at the same time

$$A = \mathbf{E}\left[\int_a^b f(t)\xi(t)\, dt\right]^2.$$

PROOF: Indeed, for the proof of the first half of the theorem, it suffices to show that if the integral A exists, then relation (2) holds. One has:

$$\begin{aligned}
\mathbf{E}(J_n - J_m)^2 &= \mathbf{E}\left[\sum_{i=1}^{n} f(t_i)\xi(t_i)\Delta t_i\right]^2 \\
&\quad - 2\mathbf{E}\sum_{i=1}^{n}\sum_{j=1}^{m} f(t_i) f(s_j)\xi(t_i)\xi(s_j)\Delta t_i \Delta s_j \\
&\quad + \mathbf{E}\left[\sum_{j=1}^{m} f(s_j)\xi(s_j)\Delta s_j\right]^2 \\
&= \sum_{i=1}^{n}\sum_{k=1}^{n} f(t_i) f(\tau_k) R(t_i - \tau_k)\Delta t_i \Delta \tau_k \\
&\quad - 2\sum_{i=1}^{n}\sum_{j=1}^{m} f(t_i) f(s_j) R(t_i - s_j)\Delta t_i \Delta s_j \\
&\quad + \sum_{j=1}^{m}\sum_{k=1}^{m} f(s_j) f(\sigma_k) R(s_j - \sigma_k)\, ds_j \Delta \sigma_k .
\end{aligned}$$

Here the numerical values of t_i and τ coincide, as do the values of s_j and σ_j.

By the assumption that the integral A exists, one has

$$\begin{aligned}
A &= \lim \sum_{k=1}^{n}\sum_{i=1}^{n} f(t_k) f(\tau_i) R(t_k - \tau_i)\Delta t_i \Delta \tau_k \\
&= \lim \sum_{i=1}^{n}\sum_{k=1}^{m} f(t_i) f(s_j) R(t_i - s_j)\Delta t_i \Delta s_j \\
&= \lim \sum_{j=1}^{m}\sum_{k=1}^{m} f(s_j) f(\sigma_k) R(s_j - \sigma_k)\Delta s_j \Delta \sigma_k ,
\end{aligned}$$

if $\max(\Delta t_i, \Delta s_j) \to 0$. Thus,

$$\mathsf{E}(J_m - J_n)^2 \to 0$$

as $\min(m, n) \to \infty$.

For the proof of the second part of the theorem, it is necessary to note that

$$\mathsf{E}\left[\sum_{i=1}^{n} f(t_i)\xi(t_i)\Delta t_i\right]^2 = \mathsf{E}\sum_{j=1}^{n}\sum_{i=1}^{n} f(t_i)f(\tau_j)\xi(t_i)\xi(\tau_j)\Delta t_i\Delta\tau_j$$
$$= \sum_{j=1}^{n}\sum_{i=1}^{n} f(t_i)f(\tau_j)R(t_i-\tau_j)\Delta t_i\Delta\tau_j;$$

and that as $\max\Delta t_i \to 0$ the last sum approaches A.

Together with the notion of stochastic integral just introduced, it is also possible to consider a *stochastic Stieltjes integral*, which is defined as the limit of the sum

$$\sum_{k=1}^{n} f(t_k)[\xi(t_k) - \xi(t_{k-1})] \tag{3}$$

as $\max(t_i - t_{i-1}) \to 0$. Here as before $a = t_0 < t_1 < \cdots < t_n = b$ and the limit is taken in the sense of (1). If the limit of the sums (3)exists, then we denote it by the symbol

$$\int_a^b f(t)\,d\xi(t).$$

At the end of the preceding section the Slutsky theorem was formulated. This theorem characterizes the relation existing between stationary processes with a discrete spectrum and Fourier series with uncorrelated random coefficients. It can be shown that the following property holds for any process stationary in the wide sense: For any $\varepsilon \geqslant 0$ and for T as large as one wishes, there exist pairwise uncorrelated random variables ξ_k, η_k $(1 \leqslant k \leqslant n)$ and such real numbers λ_k $(1 \leqslant k \leqslant n)$[8] that for any t in the interval $-T \leqslant t \leqslant T$ the inequality

$$\mathsf{E}\left[\xi(t) - \sum_{k=1}^{n}(\xi_k\cos\lambda_k t + \eta_k\sin\lambda_k t)\right]^2 < \varepsilon$$

holds.

---

[8]Numbers n and λ_k as well as random variables ξ_k and η_k do not depend on ε and T. – *B. G.*

In particular, it follows from this that under the above conditions

$$\mathbf{P}\left\{\left|\xi(t)-\sum_{k=1}^{n}(\xi_k\cos\lambda_k t+\eta_k\sin\lambda_k t)>\eta\right\}\leqslant\varepsilon/\eta^2\right.$$

where η is some positive number specified in advance.

Let us present the following important theorem without a proof.

THEOREM 2. *Any process which is stationary in the wide sense is representable in the form*

$$\xi(t)=\int_0^\infty\cos\lambda t\,dZ_1(\lambda)+\int_0^\infty\sin\lambda t\,dZ_2(\lambda),\tag{4}$$

where the stochastic processes $Z_1(\lambda)$ and $Z_2(\lambda)$ have the following properties:

(a)

$$\mathbf{E}[Z_i(\lambda_1+\Delta\lambda_1)-Z_i(\lambda_1)]\cdot[Z_j(\lambda_j+\Delta\lambda_2)-Z_j(\lambda_2)]=0,$$

(where $i,\ j=1,2$) if $i\neq j$, but if the intervals $(\lambda_1,\ \lambda_1+\Delta\lambda_1)$ and $(\lambda_2,\ \lambda_2+\Delta\lambda_2)$ do not overlap then i and j can be equal;

(b)

$$\mathbf{E}[Z_1(\lambda+\Delta\lambda)-Z_1(\lambda)]^2=\mathbf{E}[Z_2(\lambda+\Delta\lambda)-Z_2(\lambda)]^2.$$

It is natural to refer to formula (4) as the spectral decomposition of the process $\xi(t)$.

The stochastic processes $Z_1(\lambda)$ and $Z_2(\lambda)$ figuring in formula (4) are determined by the equations

$$Z_1(\lambda)=\lim_{T\to\infty}\frac{1}{2\pi}\int_{-T}^{T}\frac{-\sin\lambda t}{t}\,\xi(t)\,dt$$

and

$$Z_2(\lambda)=\lim_{T\to\infty}\frac{1}{2\pi}\int_{-T}^{T}\frac{1-\cos\lambda t}{t}\,\xi(t)\,dt.$$

It is easy to show that both of the above integrals exist. It can also be proved that

$$F(\lambda+\Delta\lambda)-F(\lambda)=\mathbf{E}[Z_1(\lambda+\Delta\lambda)-Z_1(\lambda)]^2,$$

where $F(\lambda)$ is defined by the Khinchine Theorem.

The possibility of decomposing an arbitrary stochastic process which is stationary in the wide sense into the form (4) was indicated by Kolmogorov in 1940. He formulated this result in the terminology of the Geometry of Hilbert space and proved it by using the Spectral Theory of Operators. Afterwards many authors (Cramer, Karhunen, Loew, Blanc-Lapierre, etc.) devoted their works to the probabilistic interpretation of this decomposition.

68 THE BIRKHOFF-KHINCHINE ERGODIC THEOREM

In 1931, the American mathematician George Birkhoff proved a general theorem in statistical mechanics which, as was shown by Khinchine three years later, admits a broad probabilistic generalization. This theorem is as follows: If a continuous stationary process $\xi(t)$ has a finite mathematical expectation, then with probability 1 the limit

$$\lim_{T\to\infty} \frac{1}{T}\int_0^T \xi(t)\,dt$$

exists. The stationarity of the process is understood here in the strict sense.

Since this theorem is a special variant of the strong law of large numbers, it will be proved for stationary sequences rather than stationary processes for the purpose of continuing directly the results given in Chapter 6.

THEOREM. *If*

$$\ldots, \xi_{-1}, \xi_0, \xi_1, \ldots$$

is a stationary sequence of random variables for which the mathematical expectations exist, then with probability 1 the sequence of arithmetic means

$$\frac{1}{n}\sum_{i=1}^{n} \xi_i$$

converges to a limit.

PROOF: Let us introduce the notation

$$h_{ab}=\frac{\xi_a+\xi_{a+1}+\cdots+\xi_b}{b-a}.$$

One has to prove that with probability 1 the quantities h_{0b} approach a limit as $b\to\infty$. Let $-K$ denote the random event consisting in the existence of this limit. One must show that $\mathbf{P}(-K)=1$, or what is the same, $\mathbf{P}(K)=0$.

Let us suppose the converse is true and that the event K has a positive probability (i.e., that the quantity h_{0b} does not approach a limit as $b\to\infty$). Such an assumption leads to a contradiction.

Consider all of intervals (α_n, β_n) with rational endpoints such that $\alpha_n<\beta_n$. The set of all of such intervals is denumerable. If the

$$\lim_{b\to\infty} h_{0b}$$

does not exist, then there exists such an interval (α_n, β_n) for which

$$\limsup_{b\to\infty} h_{0b}>\beta_n$$

and

$$\limsup_{b\to\infty} h_{0b}<\alpha_n$$

(the event K_n). Thus, the event K decomposes into a denumerable number of mutually exclusive events K_n. Since by assumption $\mathbf{P}(K)>0$. One can find such n for which $\mathbf{P}(K_n)>0$.

Thus, it has been shown that if $\mathbf{P}(K)>0$ then two numbers α and β $(\alpha<\beta)$ exist such that the following inequalities simultaneously hold:

$$\left.\begin{aligned}\limsup h_{0b}&>\beta,\\ \liminf h_{0b}&<\alpha.\end{aligned}\right\}\qquad(1)$$

Let us now suppose that all of ξ_j have taken on some definite values. If (a, b) is such an interval that $h_{ab}>\beta$ but at the same time $h_{ab'}\leqslant\beta$ for all b' for which $a<b'<b$, then this interval will be called special (relative to β).

It is easy to see that two special intervals cannot overlap. Indeed, if (a, b) and (a_1, b_2) are two special intervals such that $a<a_1<b<b_1$

then from the equality

$$h_{ab} = \frac{(a_1 - a)\, h_{aa_1} + (b - a_1)\, h_{a_1 b}}{b - a}$$

and the inequality $h_{ab} > \beta$ it follows that either

$$h_{aa_1} > \beta$$

or

$$h_{a_1 b} > \beta .$$

However, the first of these two inequalities is impossible, because the interval (a, b) is special; and the second inequality is impossible, because the interval (a_1, b_1) is special.

Let the difference $b - a$ be called a rang of the interval (a, b). If the interval (a, b) is special, possesses a rang not exceeding s, and is not contained in any other special interval of rang not exceeding s, then such an interval will be called s-special.

Among all of the special intervals of rang not exceeding s and containing an arbitrary interval (a, b), there must be found one of maximum length. If two such intervals would exist, they have to overlap which is impossible as it was shown above. Thus each special interval of rang not exceeding s can lie inside only one s-special interval (or coincides with it). It follows from the definition that two s-special intervals must be completely disjoint.

Let K_s denote the event that the inequalities (1) are satisfied and that, moreover, there exists at least one $t \leqslant s$ such that $h_{0t} > \beta$. Then since K is a limit for K_s one can write

$$\mathsf{P}(K) = \lim_{s \to \infty} \mathsf{P}(K_s).$$

It follows from this that $\mathsf{P}(K_s) > 0$ for s sufficiently large. Hereafter, only such values of s will be considered.

Suppose that the event K_s takes place. Then among those values of $t \leqslant s$ for which $h_{0t} > \beta$ there exists a smallest, t'. The interval $(0, t')$ is special. Consequently, it lies in some s-special interval (a, b) (or is itself such an interval) for which $a \leqslant 0 < b$. The converse is also true: If there exists an s-special interval (a, b) such that $a \leqslant 0 < b$, then there exists such $t \leqslant s$ for which $h_{0t} > \beta$. For $a = 0$ it obviously suffices to put $t = b$.

However, if $a<0$, then from the identity

$$h_{ab}=\frac{-ah_{a0}+bh_{0b}}{b-a}$$

and the inequalities $h_{ab}>\beta$ and $h_{a0}\leqslant\beta$ it follows that $h_{0b}>\beta$. Thus in this case, we can also set $t=b$.

Introduce $p=-a$ and $q=b-a$. Since there can be only one s-special interval $(-p, -p+q)$, the event K_s can be decomposed into the mutually exclusive events K_{pq} corresponding to the s-special intervals $(-p, -p+q)$:

$$K_s=\sum_{p,q}K_{pq} \qquad (q=1,2,\dots,s,\ p=0,1,\dots,q-1).$$

Changing of index $i'=i+p$ reduces the event K_{0q} into the event K_{pq}. Therefore, since the sequence is stationary,[9] $\mathbf{P}(K_{0q})=\mathbf{P}(K_{pq})$ and $\mathbf{E}(\xi_0/K_{0q})=\mathbf{E}(\xi_0/K_{pq})$. Since

$$\mathbf{P}(K_s)\mathbf{E}(\xi_0|K_s)=\sum_{p,q}\mathbf{P}(K_{pq})\mathbf{E}(\xi_0|K_{pq})$$

$$=\sum_q\mathbf{P}(K_{0q})\sum_p\mathbf{E}(\xi_p|K_{0q})=\sum_q\mathbf{P}(K_{0q})\mathbf{E}(h_{0q}|K_{0q}),$$

and by taking into account that the inequality $h_{0q}>\beta$ holds for the event K_{0q} one finds:

$$\mathbf{P}(K_s)\mathbf{E}(\xi_0|K_s)>\sum_q\mathbf{P}(K_{0q})\beta=\beta\sum\mathbf{P}(K_{pq})=\beta\mathbf{P}(K_s).$$

By assumption $\mathbf{P}(K_s)\neq 0$ so that

$$\mathbf{E}(\xi_0|K_s)>\beta.$$

Since $K_s\to K$ one can write

$$\mathbf{E}(\xi_0|K)\geqslant\beta.$$

[9]Notice that the stationarity assumption was used only at this point. – *B. G.*

In the same way (if special intervals relative to α are considered), it can be shown that

$$\mathbf{E}(\xi_0 | K) \leqslant \alpha.$$

Thus one has arrived at a contradiction. It follows that $\mathbf{P}(K) = 0$. Q.E.D.

The investigation of the limit of h_{0n} approached as $n \to \infty$ requires supplementary considerations. Let us confine ourselves here to a proof of the following statement.

THEOREM. *If the random variables* ξ_k *are stationary, and have finite variances, and the correlation function* $R(k) \to 0$ as $k \to \infty$, then

$$\lim_{n \to \infty} \mathbf{P}\{h_{0n} \to a\} = 1 \qquad (a = \mathbf{E}\,\xi_k).$$

PROOF: Consider the variance of the quantity h_{0n}. Since the sequence is stationary, one has

$$\mathbf{D}\, h_{0n} = \mathbf{E}\left[\frac{1}{n}\sum_{k=1}^{n} (\xi_k - a)\right]^2 = \frac{\mathbf{D}\,\xi_n}{n^2}[n + 2 \sum_{1 \leqslant i < j \leqslant n} R(j-i)].$$

It is obvious that

$$\sum_{1 \leqslant i < j \leqslant n} R(j-i) = \sum_{k=1}^{n-1} (n-k)\, R(k).$$

Consider a value of m sufficiently large that the inequality

$$|R(k)| \leqslant \varepsilon \qquad (\varepsilon > 0)$$

holds for $k > m$. From this it follows that

$$\mathbf{D}\, h_{0n} \leqslant \frac{\mathbf{D}\,\xi_n}{n^2}\left[n + 2\sum_{k=1}^{m} (n-k)\, R(k) + 2\varepsilon \sum_{m+1}^{n-1} (n-k)\right].$$

This inequality can obviously be strengthened as follows:

$$\mathbf{D}\, h_{0n} \leqslant \frac{\mathbf{D}\,\xi_n}{n^2}[n + 2m(n-1) + \varepsilon\,(n-m-1)(n-m)].$$

Hence, it is clear that if n is sufficiently large, the right-hand side of this inequality can be made smaller than 3ε. Thus, as $n \to \infty$, the sequence of h_{0n} converges to a in probability. It leads to the theorem statement.

The above theorem is not only of significant theoretical interest but also finds wide application in statistical physics and engineering practice. The reason for this is that when a process is stationary such important characteristics as $\mathbf{E}\xi(t)$, $\mathbf{D}\xi(t)$, and $R(u)$ can be determined without knowledge of the distribution function of the possible values, which of course is needed to compute these characteristics from their respective formulas. The determination of these so-called spatial averages, as they are known in physics, requires the researcher to have information which is frequently unavailable. In any case, the experimental estimation of these quantities necessitates the repeated realization of trials for the process ξ. The Birkhoff-Khinchine Ergodic Theorem shows that with probability 1 we can confine ourselves (under certain conditions) to a single realization of the process $\xi(t)$.

Chapter 12

ELEMENTS OF STATISTICS

69 SOME PROBLEMS OF MATHEMATICAL STATISTICS

Probability theory gives the rules for operating with probabilities. But the question arises: how to find these initial probabilities, their distributions and various numerical parameters? This subject is the core of another science about mass random events which is known as mathematical statistics. As a separate branch of mathematics with its own problems and methods of investigation, mathematical statistics was formed, actually, only at the beginning of the twentieth century. However, some separate statistical problems were studied much earlier – in the nineteenth, eighteenth and even in seventeenth centuries.

The term statistics comes from the Latin word status which means "state." When statistics originally began to take shape as a scientific discipline in the eighteenth century, the term statistics was associated with the system of collecting facts describing the state of a country. It was not assumed that only phenomena of a mass nature were covered by statistics. At present, statistics has a larger and, at the same time, a more definite content.

Specifically, one can say that statistics incorporates the following three directions:

1) The collection of statistical data, i.e., data characterizing the individual items of mass populations;

2) The statistical analysis of the data obtained, consisting in discovering what laws can be established on the basis of the data of observation;

3) The development of methods of statistical control and analysis of statistical data.

The collecting of statistical data concerning populations's state and growth was begun long ago. It is known that the emperor Yao took a census of China's population in 2238 B.C. A census of the population had also been performed in ancient Egypt, in Persia, and in the Roman empire. It must be noted that sometimes that these censuses were extremely primitive. In China, for example, the population was counted by copying the lists of the previous censuses during 200 years. However, even such imperfect and incomplete censuses permitted the planning of important state measures.

A practical role of statistics in our day can hardly be overestimated. This role is not restricted by statistical data processing but expanded on varied technological processes. It also helps to confirm different scientific approaches by statistical analysis of experimental data.

A number of problems arise during a statistical analysis. Let us enumerate the main difficulties.

1) Estimation of unknown probability of a random event.

2) Estimation of an unknown distribution function. This problem may be posed as follows: As the result of n independent trials over a random value, ξ, the following realizations are obtained: $x_1, x_2, \ldots, x_n$. It is required to determine (at least approximately) the unknown distribution function, $F(x)$, for the variable ξ.

3) Estimation of the unknown parameters of a distribution. General theoretical considerations often allow one to make sufficiently definite inferences concerning the type of distribution function of the random variable under consideration. Thus, for example, Liapounov's Theorem enables us to say that a distribution is normal in certain cases. As a result, the determination of the unknown distribution function is reduced to estimating the unknown parameters a and σ on the basis of experiment.

The general problem can be formulated in the following way. The random variable ξ has a distribution function of a certain form, depending on k parameters whose values are unknown. On the basis of a series of observations over ξ, it is required to find the values of these parameters.

Obviously, the estimation of the unknown probability p of an event A is a special case of the problem formulated above, since we could consider a random variable ξ which takes on the value 1 if the event A occurs and the value 0 if A does not occur. The distribution function for ξ would depend on the single parameter p.

In Section 50 observations of golden particles suspended in water were analyzed. It was assumed that the number of these particles must follow Poisson distribution. One needs to estimate the value of parameter λ. This problem is included in the scope of considered problems.

The solution to the problem mentioned above will only be presented for the normal distribution

$$f(x|a,\sigma)=\frac{1}{\sigma\sqrt{2\pi}}e^{-(x-a)^2/2\sigma^2}.$$

In this case, the second problem can obviously be subdivided into three particular tasks:

1) The quantity σ is assumed to be known; it is required to estimate the unknown value of a.

2) The quantity a is assumed to be known; it is required to estimate the unknown value of σ.

3) Both of the parameters a and σ are unknown; it is required to estimate their values.

These questions can be formulated more precisely as follows. In n independent trials, the variable ξ has taken on the values

$$x_1, x_2, \ldots, x_n.$$

It is required to find functions $a- = a(x_1, x_2, \ldots, x_n)$ and $\sigma- = \sigma(x_1, x_2, \ldots, x_n)$ that would be reasonable approximations to the values of the quantities a and σ being estimated (in the first problem, a may also be a function of σ; in the second, σ may be a function of a). In addition, it is required to estimate the average accuracy of these approximations.

Sometimes rather than seek approximations for the unknown parameters a and σ in the form of the functions $a-$ and $\sigma-$, it is preferable to find functions of the observed values and of the known quantities, a' and a'' (σ' and σ''), such that it may be affirmed with sufficient practical assurance that

$$a' < a < a''$$

and, correspondingly,

$$\sigma' < \sigma < \sigma''.$$

The functions a' and a'' (σ' and σ'') are called confidence limits for a (σ). Two approaches to the solution of these problems will be offered below.

4) The testing of statistical hypotheses. The problem which is considered here can be formulated as follows. There are certain reasons for assuming the distribution function of the random variable ξ to be $F(x)$. The question is: Are the observed values consistent with the hypothesis that ξ has the distribution $F(x)$?

In particular, if the form of the distribution function is known *a priori* and only the values of certain parameters of the distribution require verification, then one may ask the question: Do the results of the observations confirm the hypothesis that the distribution parameters have the presumed values? This is the problem of *testing a simple hypothesis*. If the hypothesis in question is that the parameters assume values from certain specified sets rather than certain exact values (for example, the hypothesis $p < p_0$, in the case of the binomial distribution), then the hypothesis is called *composite*.

As the second example of testing statistical hypotheses one can consider a checking of the material homogeneity. One particular tasks here is the following. There are two sets of observations over random variables ξ and η with distribution functions $F_1(x)$ and $F_2(x)$, respectively. These sets of data are: $x_1, x_2, \ldots, x_n$ and $y_1, y_2, \ldots, y_m$. Functions $F_1(x)$ and $F_2(x)$ are unknown. One must estimate whether the hypothesis that $F_1(x) = F_2(x)$ is a likelihood.

5) Estimation of dependence. Two random variables } and η are tested simultaneously. The results of observations are represented by pairs: (x_1, y_1), (x_2, y_2), $\ldots$, (x_n, y_n). It is required to clarify if there is a functional or correlational connection between } and η.

6) Control of the process. One observes a stochastic process $\xi(t)$ where t might be discrete or continuous. Under some external influences the normal process can be changed and become, say, $\xi_1(t)$. One needs to find out the changing of the normal process and take some preventive measures for correction.

It should be noted that the problems listed above far from exhaust the basic problems of mathematical statistics. Industry and science are constantly posing new problems to mathematical statistics. In particular, the test planning itself is one of the main problems of modern mathematical statistics.

70 THE CLASSICAL PROCEDURE FOR ESTIMATING THE DISTRIBUTION PARAMETERS

The classical procedure for estimating the unknown parameters of the distribution function of a random variable ξ is based on the hypothesis that the form of the distribution is assumed to be known *apriori*. By assuming that fact and applying Bayes' Theorem, one is able to compute the *a posteriori* (post-experimental) distribution law for the parameters under the condition that the observed values of ξ are $x_1, x_2, \ldots, x_n$.

As was stated earlier, the entire discussion that follows will be concerned with determining the unknown parameters a and σ of the normal distribution

$$p(x|a, \sigma) = \frac{1}{\sigma\sqrt{2\pi}} e^{-(x-a)^2/2\sigma^2},$$

which random variable ξ obeys.

The density function

$$f(x_1, x_2, \ldots, x_n|a, \sigma) = \frac{1}{(\sigma\sqrt{2\pi})^n} e^{-S^2/2\sigma^2},$$

where

$$s^2 = \sum_{k=1}^{n} (x_k - a)^2$$

corresponds to the probability that $x_1, x_2, \ldots, x_n$ will be obtained as the result of n independent observations for the values of the random variable ξ, subject to the condition that the unknown parameters have the values a and σ. Let us introduce the notation

$$\bar{x} = \frac{1}{n} \sum_{k=1}^{n} x_k,$$

$$s_1^2 = \frac{1}{n} \sum_{k=1}^{n} (x_k - \bar{x})^2.$$

Then a simple computation shows that

$$f(x_1, \ldots, x_n|a, \sigma) = \frac{1}{(\sigma\sqrt{2\pi})^n} e^{-n/2\sigma^2[S_1^2 + (\bar{x}-a)^2]}. \qquad (1)$$

Let us recall the following three problems formulated in Section 60:

1) σ is known; it is required to estimate a;
2) a is known; it is required to estimate σ;
3) a and σ are unknown and are to be estimated.

Let us suppose that σ is known, and $\varphi_1(a)$ denotes the *a priori* density function of the quantity a; then for a given value of σ and the observed values $x_1, x_2, \ldots, x_n$ one obtains the following expression for the conditional density function of the variable a:

$$\varphi_1(a|x_1, x_2, \ldots, x_n; \sigma) = \frac{f(x_1, x_2, \ldots, x_n|a, \sigma)\,\varphi_1(a)}{\int f(x_1, x_2, \ldots, x_n|a, \sigma)\,\varphi_1(a)\,da}.$$

By substituting in this the value of the function f given by equation (1) and making obvious simplifications, one finds that

$$\varphi_1(a|x_1, x_2, \ldots, x_n; \sigma) = \frac{e^{-n(a-\bar{x})^2/2\sigma^2}\cdot\varphi_1(a)}{\int e^{-n(a-\bar{x})^2/2\sigma^2}\cdot\varphi_1(a)\,da}. \tag{2}$$

The corresponding expressions in the second and third problems are

$$\varphi_2(\sigma|x_1, x_2, \ldots, x_n; \sigma) = \frac{f(x_1, x_2, \ldots, x_n|a, \sigma)\,\varphi_2(\sigma)}{\int f(x_1, x_2, \ldots, x_n|a, \sigma)\,\varphi_2(\sigma)\,d\sigma}$$

and

$$\varphi_3(a, \sigma|x_1, x_2, \ldots, x_n) = \frac{f(x_1, x_2, \ldots, x_n|a, \sigma)\,\varphi_3(a, \sigma)}{\iint f(x_1, x_2, \ldots, x_n|a, \sigma)\,\varphi_3(a, \sigma)\,da\,d\sigma},$$

where the functions $\varphi_2(\sigma)$ and $\varphi_3(a, \sigma)$ denote the *a priori* density functions of the variable σ and the pair (a, σ).

By substituting in these formulas the value of f given by equation (1) and by making simple reductions, one finds

$$\varphi_2(\sigma|x_1, x_2, \ldots, x_n; a) = \frac{\sigma^{-n}e^{-(s^2/2\sigma^2)}\,\varphi_2(\sigma)}{\int_0 \sigma^{-n}e^{-(s^2/2\sigma^2)}\,\varphi_2(\sigma)\,d\sigma} \tag{3}$$

and

$$\varphi_3(a, \sigma \mid x_1, x_2, \ldots, x_n) = \frac{\sigma^{-n} e^{-(n/2\sigma^2)(s_1^2 + (a-\bar{x})^2)} \varphi_3(a, \sigma)}{\int_0^{\infty} \int \sigma^{-n} e^{-(n/2\sigma^2)(s_1^2 + (a-\bar{x})^2)} \varphi_3(a, \sigma) \, da \, d\sigma}. \quad (4)$$

The expressions obtained are not appropriate for practical application not only because of their complexity but also, and mainly, because the *a priori* probabilities appearing in them are usually unknown. Since these *a priori* densities are often not known, one makes more or less arbitrary assumptions about them, and on the basis of these obtains simple and practically applicable formulas. Let us choose a different path and, having made perfectly general assumptions about the character of the *a priori* distributions, deduce limit laws (as $n \to \infty$) for the *a posteriori* probabilities.

THEOREM 1. *If the a priori density function* $\varphi_1(a)$ *has a bounded first derivative and*

$$\varphi_1(\bar{x}) \neq 0,$$

then

$$\psi_1(\alpha \mid x_1, x_2, \ldots, x_n; \sigma) = \frac{1}{\sqrt{2\pi}} e^{-\frac{1}{2}\alpha^2} \left[1 + O\left(\frac{\sigma}{\sqrt{n}} \right)(1 + |\alpha|) \right], \quad (5)$$

uniformly to α *where*

$$\alpha = \frac{\sqrt{n}}{\sigma}(a - x), \quad (6)$$

and $\psi_1(\alpha \mid x_1, x_2, \ldots, x_n; \sigma)$ *denotes the a posteriori density function of the variable* a.

PROOF. From (6) one finds:

$$a = \bar{x} + \frac{\alpha\sigma}{\sqrt{n}} \quad (7)$$

and this means[1]

$$\frac{\sigma}{\sqrt{n}}\varphi_1(a|x_1,x_2,\ldots,x_n;\sigma)=\frac{e^{-\alpha^2/2}\varphi_1\left(\bar{x}+\frac{\alpha\sigma}{\sqrt{n}}\right)}{\int e^{-\alpha^2/2}\varphi_1\left(\bar{x}+\frac{\alpha\sigma}{\sqrt{n}}\right)d\alpha}.$$

By the formula of finite increments

$$\varphi_1\left(\bar{x}+\frac{\alpha\sigma}{\sqrt{n}}\right)=\varphi_1(\bar{x})+\frac{\alpha\sigma}{\sqrt{n}}\varphi_1'(z),$$

where

$$z=\bar{x}+\theta\frac{\alpha\sigma}{\sqrt{n}}\quad\text{and}\quad o<\theta<1.$$

By assumption of the theorem

$$|\varphi_1'(z)|\leqslant C<+\infty,$$

and therefore

$$\frac{\sigma}{\sqrt{n}}\varphi'(a|x_1,x_2,\ldots,x_n;\sigma)=\frac{e^{-\alpha^2/2}\left[\varphi_1(\bar{x})+\frac{\alpha\sigma}{\sqrt{n}}\varphi_1'(z)\right]}{\sqrt{2\pi}\,[\varphi_1(\bar{x})+r_n]},$$

where

$$r_n=\frac{\sigma}{\sqrt{2\pi n}}\int\alpha e^{-\alpha^2/2}\varphi_1'(z)d\alpha.$$

One can easily show that

$$|r_n|<\frac{2C\sigma}{\sqrt{2\pi n}}. \tag{8}$$

[1]Note that the density function of the variable α is given by $\psi_1(\alpha|X_1,X_2,\ldots,X_n;\sigma)=\frac{\sigma}{\sqrt{n}}\varphi_1(a|X_1,X_2,\ldots,X_n;\sigma)$.

Simple transformations lead to the expression

$$\psi_1(\alpha|x_1,x_2,\dots,x_n;\sigma)=\frac{1}{\sqrt{2\pi}}e^{-\alpha^2/2}\left[1+\frac{\sigma}{\sqrt{n}}\frac{\alpha\varphi'(z)-\frac{\sqrt{n}}{\sigma}r_n}{\varphi(\bar{x})+r_n}\right].$$

This expression, in conjunction with (8), proves the theorem.

THEOREM 2. *Under the conditions of the preceding theorem, the following is true*

$$\left.\begin{aligned}\mathbf{E}(a|x_1,x_2,\dots,x_n;\sigma)&=(\bar{x})+O\left(\frac{\sigma^2}{n}\right),\\ \mathbf{E}[(a-\bar{x})^2|x_1,x_2,\dots,x_n;\sigma)&=\frac{\sigma^2}{n}\left[1+O\left(\frac{\sigma}{\sqrt{n}}\right)\right]\end{aligned}\right\}\qquad(5')$$

PROOF: From (7), one finds:

$$\mathbf{E}(a|B)=\bar{x}+\frac{\alpha}{\sqrt{n}}M(\alpha|B)$$

and

$$\mathbf{E}[(a-\bar{x})^2|B]=\frac{\sigma^2}{n}M(\alpha^2|B),$$

where B stands for some event. Consequently,

$$\mathbf{E}(a|x_1,x_2,\dots,x_n;\sigma)=\bar{x}+\frac{\sigma}{\sqrt{n}}\int\alpha\psi_1(\alpha|x_1,\dots,x_n;\sigma)d\alpha$$

and

$$\mathbf{E}[(a-\bar{x})^2|x_1,x_2,\dots,x_n;\sigma]=\frac{\sigma^2}{n}\int\alpha^2\psi_1(\alpha|x_1,\dots,x_n;\sigma)\,d\alpha.$$

The substitution in these expressions of the value of the function ψ given in (5) followed by simple transformations yields the theorem.

The just-proved theorem allows one to write the following approximation:

$$a\sim\bar{x},$$

the mean-square error of which is approximately σ^2/n.

Theorem 1 enables one to obtain the probability that a will lie within given limits, under the condition that the quantities $\sigma, x_1, x_2, \ldots, x_n$ have taken on specific values. In fact,

$$\mathbf{P}\left\{|a-\bar{x}|<\frac{\sigma z}{\sqrt{n}}\middle| x_1,\ldots,x_n;\sigma\right\}=\mathbf{P}\{|\alpha|<z|x_1,\ldots,x_n;\sigma\}$$

and, consequently, due to (5),

$$\mathbf{P}\left\{|a-\bar{x}|<\frac{\sigma z}{\sqrt{n}}\middle| x_1,\ldots,x_n;\sigma\right\}=\frac{2}{\sqrt{2\pi}}\int_0^z e^{-t^2/2}dt+O\left(\frac{\sigma}{\sqrt{n}}\right).$$

Neglecting the remainder $O(\sigma/\sqrt{n})$ (which, in general, may be done only when σ is small or when n is sufficiently large), one can state that

$$\mathbf{P}\left\{|a-\bar{x}|<\frac{\sigma}{\sqrt{n}}z\middle| x_1,x_2,\ldots,x_n;\sigma\right\}=\frac{2}{\sqrt{2\pi}}\int_0^z e^{-t^2/2}dt.$$

THEOREM 3. *If the a priori density function $\varphi_2(\sigma)$ has a bounded first derivative and $\varphi_2(s-)\not\equiv 0$ then*

$$\psi_2(z|x_1,x_2,\ldots,x_n;a)=\frac{1}{\sqrt{\pi}}e^{-z^2}\left[1+O\left(\frac{1}{\sqrt{n}}\right)(1+|z|)\right],$$

uniformly in z, where ψ_2 is the a posteriori density of the variable

$$\beta=\frac{\sigma-\bar{s}}{\bar{s}}\sqrt{n} \tag{9}$$

and

$$\bar{s}=\frac{s}{\sqrt{n}}.$$

PROOF: From (9) one finds that

$$\sigma=\bar{s}\left(1+\frac{\beta}{\sqrt{n}}\right)$$

and

$$\psi_2(z|x_1, x_2, \ldots, x_n; a) = \frac{\bar{s}}{\sqrt{n}}\varphi\left(\bar{s}\left(1+\frac{z}{\sqrt{n}}\right)\middle| x_1, x_2, \ldots, x_n; a\right)$$

$$= \frac{\left(1+\dfrac{z}{\sqrt{n}}\right)^{-n} e^{-n/2(1+z/\sqrt{n})^2}\varphi_2\left(\bar{s}+\dfrac{z\bar{s}}{\sqrt{n}}\right)}{\displaystyle\int_{-\sqrt{n}}^{\infty}\left(1+\frac{\beta}{\sqrt{n}}\right)^{-n} e^{-s^2/2\sigma^2}\varphi_2(\sigma)d\sigma}.$$

By the formula of finite increments

$$\varphi_2\left(\bar{s}+\frac{z\bar{s}}{\sqrt{n}}\right) = \varphi_2(\bar{s}) + \frac{z\bar{s}}{\sqrt{n}}\varphi_2'(u),$$

where

$$u = \bar{s} + \theta\frac{z\bar{s}}{\sqrt{n}}, \text{ and } 0<\theta<1.$$

By assumption,

$$|\varphi_2'(u)| \leqslant C < +\infty,$$

and therefore

$$\psi_2(z|x_1, x_2, \ldots, x_n; a)$$

$$= \frac{\left(1+\dfrac{z}{\sqrt{n}}\right)^{-n} e^{-n/2(1+z/\sqrt{n})^2}\left[\varphi_2(\bar{s}) + \dfrac{z\bar{s}}{\sqrt{n}}O(1)\right]}{\varphi_2(\bar{s})\displaystyle\int_{-\sqrt{n}}^{\infty}\left(1+\frac{\beta}{\sqrt{n}}\right)^{-n} e^{-n/2(1+\beta/\sqrt{n})^2}d\beta\left[1+O\left(\frac{1}{\sqrt{n}}\right)\right]}$$

$$= \frac{\left(1+\dfrac{z}{\sqrt{n}}\right)^{-n} e^{-n/2(1+z/\sqrt{n})^2}\left[1+O\left(\dfrac{1}{\sqrt{n}}\right)(1+|z|)\right]}{\displaystyle\int_{-\sqrt{n}}^{\infty}\left(1+\frac{\beta}{\sqrt{n}}\right)^{-n} e^{-n/2(1+\beta/\sqrt{n})^2}d\beta}. \qquad (10)$$

However,

$$-n\ln\left(1+\frac{z}{\sqrt{n}}\right) - \frac{n}{2\left(1+\dfrac{z}{\sqrt{n}}\right)^2} = -n\left(\frac{z}{\sqrt{n}} - \frac{1}{2}\frac{z^2}{n} + \cdots\right)$$

$$-\frac{n}{2}\left(1-2\frac{z}{\sqrt{n}}+3\frac{z^2}{n}-\ldots\right) = -\frac{n}{2} - z^2 + O\left(\frac{1}{\sqrt{n}}\right) \qquad (11)$$

and

$$J=\int_{-\sqrt{n}}^{\infty}\left(1+\frac{\beta}{\sqrt{n}}\right)^{-n} e^{-n/2(1+\beta/\sqrt{n})^2}\,d\beta=n^{-n/2+1}2^{n-3/2}\int_0^{\infty}z^{n-3/2}e^{-z}\,dz$$

$$=n^{-n/2+1}2^{n-3/2}\Gamma\left(\frac{n-1}{2}\right).$$

By Stirling's formula,

$$\Gamma\left(\frac{n-1}{2}\right)=\sqrt{2\pi\frac{n-3}{2}}\left(\frac{n-3}{2}\right)^{n-3/2}e^{-n-3/2}[1+o(1)],$$

and therefore

$$J=\sqrt{2\pi}\left(\frac{n-3}{2}\right)^{n/2-1}2^{n-3/2}n^{-n/2+1}e^{-n/2+3/2}[1+o(1/)]$$

$$=\sqrt{\pi}\left(1-\frac{3}{n}\right)^{n/2}e^{-n/2+3/2}[1+o(1)]=\sqrt{\pi e}^{-n/2}[1+o(1)]. \qquad (12)$$

Equations (10), (11), and (12) prove the theorem.

Theorem 3 clearly implies the following result.

THEOREM 4. *Under the conditions of Theorem 3, the following relations hold*

$$\mathbf{E}(\sigma|x_1,x_2,\ldots,x_n;a)=\bar{s}+o(1/n)$$

and

$$\mathbf{E}[(\sigma-\bar{s})^2|x_1,x_2,\ldots,x_n;a]=\bar{s}^2/2n[1+o(1/\sqrt{n})].$$

This theorem allows one to conclude that for large values of n the approximate relation

$$\sigma\sim\sqrt{\frac{1}{n}\sum_{k=1}^{n}(x_k-a)^2},$$

holds and that its standard deviation is approximately $\bar{s}^2/2n$.

Theorem 3 can also be used to determine the probability that σ lies within given limits. Thus, by neglecting quantities of the order $1/\sqrt{n}$, one can state that for given values of $x_1,x_2,\ldots,x_n$, and a with probability

$$\frac{2}{\sqrt{\pi}}\int_0^z e^{-t^2}\,dt$$

σ will lie in the interval

$$\bar{s}\left(1-\frac{z}{\sqrt{n}}\right)<\sigma<\bar{s}\left(1+\frac{z}{\sqrt{n}}\right).$$

As to the third of the three problems formulated above, we shall restrict ourselves to the statement of results only, since their derivation is quite similar to the proof of Theorem 3. Let us introduce the notation

$$\alpha_1=\frac{a-\bar{x}}{\bar{s}_1}\sqrt{n}, \qquad \beta_1=\frac{\sigma-\bar{s}_1}{\bar{s}_1}\sqrt{n},$$

where $\bar{s}_1 = s_1\,(\sqrt{n}/n-1)$.

THEOREM 5. *If the a priori density function $\varphi_3(a,\sigma)$ has bounded first derivatives with respect to a and σ and $\varphi_3(\bar{x},\bar{s}_1)\neq 0$ then*

$$\psi_3(\alpha_1,\beta_1|x_1,x_2,\ldots,x_n)$$

$$=\frac{1}{\pi\sqrt{2}}e^{-\alpha_1^2/2-\beta_1^2}\left(1+O\left(\frac{1}{\sqrt{n}}\right)[1+|\alpha_1|+|\beta_1|]\right),$$

uniformly in α_1 and β_1 where ψ_3 stands for the a posteriori density function of the pair (α_1,β_1).

From Theorem 5 there follows the result:

THEOREM 6. *Under the conditions of Theorem 5,*

$$\mathbf{E}(a|x_1,x_2,\ldots,x_n)=\bar{x}+O\left(\frac{1}{n}\right),$$

$$\mathbf{E}\,[(a-\bar{x})^2|x_1,x_2,\ldots,x_n]=\frac{\bar{s}_1^2}{n}\left[1+O\left(\frac{1}{\sqrt{n}}\right)\right].$$

$$\mathbf{E}[\sigma|x_1, x_2, \ldots, x_n] = \bar{s}_1\left[1 + O\left(\frac{1}{n}\right)\right],$$

$$\mathbf{E}[(\sigma - \bar{s}_1)^2|x_1, x_2, \ldots, x_n) = \frac{\bar{s}_1^2}{2n}\left[1 + O\left(\frac{1}{\sqrt{n}}\right)\right].$$

Just as in the case of Theorems 1 and 3, Theorem 5 can be used for determination of the probability that a and σ will lie within given limits, subject to the condition that the observed values are x_1, $x_2, \ldots, x_n$.

The practical significance of Theorems 1, 3, and 5 is not the same. According to Theorem 1, the accuracy of the approximations (5) and (5′) not only increases with increasing n but also with decreasing σ. Therefore, one can determine a for given σ by formulas (5) and (5′) even for small values of n if sufficiently small σ is provided. In the case of Theorems 3 and 5, the remainder terms in the corresponding formulas decrease only with increasing n, and therefore they are inapplicable for small values of n.

The theorems just proved express, in a certain sense, the following elementary statements. If the random variable ξ is normally distributed, if the parameters a and σ are known, and if $x_1, x_2, \ldots, x_n$ are the outcomes of independent observations for ξ, then:

1. The density function of the random variable

$$\alpha = \frac{\sqrt{n}}{\sigma}(\bar{x} - a)$$

is

$$\psi(x|a, \sigma) = \frac{1}{\sqrt{2\pi}} e^{-x^2/2}.$$

2.
$$E(\bar{x}|a, \sigma) = \quad \text{and} \quad D(\bar{x}|a, \sigma) = \frac{\sigma^2}{n}.$$

3. The density function of the variable

$$\beta = \frac{\bar{s} - \sigma}{\sigma}\sqrt{2n}$$

is asymptotically equal to

$$\psi_2(x|a,\sigma)=\frac{1}{\sqrt{2\pi}}e^{-x^2/2}.$$

4.

$$\mathbf{E}(\bar{s}|a,\sigma)=\sigma\{1+O(1/n)\};\ \mathbf{D}(\bar{s}|a,\sigma)=\frac{\sigma^2}{2n}[1+O(1/n)].$$

5. The variables α and β are independent and the density function of the variable (α,β) asymptotically equals

$$\psi_3(x,y|a,\sigma)=\frac{1}{2\pi}e^{-(x^2+y^2/2)}.$$

6.

$$\mathbf{E}(\bar{x}|a,\sigma)=a;\ \mathbf{D}(\bar{x}|a,\sigma)=\frac{\sigma^2}{n};$$

$$\mathbf{E}(\bar{s}|a,\sigma)\approx\sigma;\ \mathbf{D}(\bar{s}|a,\sigma)\approx\frac{\sigma^2}{2n}.$$

Statements 1 and 2 require no proof.

Let us prove statement 3. In Section 20, we found that the density function for the random variable s^- is

$$\varphi(y)=\frac{\sqrt{2n}}{\sigma\Gamma(n/2)}\left(\frac{y\sqrt{n}}{\sigma\sqrt{2}}\right)^{n-1}e^{-ny^2/2\sigma^2}.$$

It is easy to verify that the density function for β is

$$\psi_2(x|a,\sigma)=\frac{\sigma}{\sqrt{n}}\varphi\left(\frac{\sigma x}{\sqrt{n}}+\sigma\right).$$

A set of simple transformations leads one to the expression

$$\psi_2(x|a,\sigma)=\frac{1}{\sqrt{2n}}e^{-x^2/2}.$$

To prove statement 4, one should notice that elementary computations lead to the relations

$$\mathbf{E}\bar{s}=\sqrt{\frac{2}{n}}\frac{\Gamma[n+(1/2)]}{\Gamma(n/2)}\sigma\sim e^{-1/4n}\sigma;\qquad \mathbf{E}\bar{s}^2=\frac{2}{n}\frac{\Gamma[n+(1/2)]}{\Gamma(n/2)}\sigma^2=\sigma^2.$$

It follows

$$\mathbf{E}\bar{s} \approx \sigma\left(1 - \frac{1}{4n}\right)$$

and

$$\mathbf{D}\bar{s} = \frac{\sigma^2}{2n}\left(1 - \frac{1}{8n}\right).$$

The independence of x-and s-will be proved later. Once this is done the remaining statements contained in 5 and 6 become obvious.

71 EXHAUSTIVE STATISTICS

The English statistician Fisher introduced a very important concept. Let us first explain it by a simple example. Assume that one has n observations of a random variable normally distributed and wishes to find a parameter a in the case where σ is known. If the prior distribution of the parameter a exists and equals $\varphi_1(a)$ then formula (2) from the preceding section shows that the conditional density function $\varphi_1(a/x_1, x_2, \ldots, x_n; \sigma)$ is completely defined by $\varphi(a)$, σ and the arithmetical mean of results of observations $x_1, x_2, \ldots, x_n$. Thus whatever may be is the prior distribution of parameter a, the only new information (in the case of the known variance) is contained in the single value $\bar{x}$. This is the reason why $\bar{x}$ is called exhaustive statistics for parameter a.

In an analogous way, under assumption that a and $\varphi_2(\sigma)$ are known, the only new information for the determination of the parameter σ is contained in a single value

$$\bar{s} = \sqrt{\frac{1}{n}\sum_{k=1}^{n}(x_k - \bar{x})^2}$$

(see (3), Section 61). Hence, in the problem of finding σ under the condition that a is known, the exhaustive statistics is the value of $\bar{s}$.

The general definition of the exhaustive statistics will be done following Kolmogorov.

Let an observed random variable have the distribution function depending on k parameters $\theta_1, \theta_2, \ldots, \theta_k$ whose meanings are unknown.

Any function $\chi(x_1, x_2, \ldots, x_n)$ of the results of observations and of parameters which are known is called statistics. The definition of the exhaustive statistics has the following natural generalization: a system of functions

$$\chi_i(x_1, x_2, \ldots, x_n) \qquad (i = 1, 2, \ldots, s)$$

is called an exhaustive system of statistics for the system of parameters $\theta_1, \theta_2, \ldots, \theta_k$ if the conditional k-dimensional distribution of these parameters for known $x_1, x_2, \ldots, x_n$ is completely defined by the prior distribution of parameters $\theta_1, \theta_2, \ldots \theta_k$ and by the values of sta tistics $\chi_1, \chi_2, \ldots, \chi_s$.

From formula (4) of Section 60, one concludes that for parameters a and σ the exhaustive statistics are $\chi_1 = \bar{x}$ and $\chi_2 = \bar{s}$. It is clear that for each of the parameters a and σ the system of statistics χ_1 and χ_2 is also exhaustive.

Without difficulties the reader can find that a random variable following the Poisson distribution

$$\mathsf{P}\{\xi = k\} = \frac{a^k e^{-a}}{k!} \qquad (k = 0, 1, 2, \ldots)$$

with unknown parameter a has the exhaustive statistic x^- for this parameter.

Analogously, if a two-dimensional random variable is normally distributed but parameters a, b, σ_1, σ_2 and r are unknown then the exhaustive system of statistics consists of the following five functions:

$$\chi_1(x_1, x_2, \ldots, x_n) = \frac{1}{n} \sum_{k=1}^{n} x_k = \bar{x},$$

$$\chi_2(y_1, y_2, \ldots, y_n) = \frac{1}{n} \sum_{k=1}^{n} y_k = \bar{y},$$

$$\chi_3(x_1, x_2, \ldots, x_n) = \sqrt{\frac{1}{n} \sum_{k=1}^{n} (x_k - \bar{x})^2} = s_1,$$

$$\chi_4(y_1, y_2, \ldots, y_n) = \sqrt{\frac{1}{n} \sum_{k=1}^{n} (y_k - \bar{y})^2} = s_2,$$

$$\chi_5(x_1, \ldots, x_n, y_1, \ldots, y_n) = \frac{1}{n} \sum_{k=1}^{n} (x_k - \bar{x})(y_k - \bar{y}) = \bar{r}.$$

Here $(x_1, y_1), (x_2, y_2), \ldots, (x_n, y_n)$ are the results of observations.

72 CONFIDENCE LIMITS AND CONFIDENCE PROBABILITIES

In the beginning of this chapter it was mentioned that the problem of determination of unknown parameters can be formulated as follows: it is required to find two such functions $\eta'(x_1, x_2, \ldots, x_n)$ and $\eta''(x_1, x_2, \ldots, x_n)$ based on the results of observations that one will have a practical assurance that the unknown parameter η lies between η' and η''. The functions η' and η'' are called the *confidence limits for parameter* η. For practical satisfaction, one should have the conditional probability (for the given $x_1, x_2, \ldots, x_n$)

$$\mathbf{P}\{\theta' \leqslant \theta < \theta'' | x_1, x_2, \ldots, x_n\}$$

close to 1. The measure of closeness is dictated by the practical needs. If one knows the *a priori* density function for the parameter η then for determining $\eta'(x_1, x_2, \ldots, x_n)$ and $\eta''(x_1, x_2, \ldots, x_n)$ it is natural to choose such η' and η'' which for a given ω close to 1 the equality

$$\omega = \mathbf{P}\{\theta' \leqslant \theta < \theta'' | x_1, x_2, \ldots, x_n\} = \frac{\int_{\theta'}^{\theta''} f(x_1, x_2, \ldots, x_n | \theta) \varphi(\theta)\, d\theta}{\int_{-\infty}^{x} f(x_1, x_2, \ldots, x_n | \theta) \varphi(\theta)\, d\theta},$$

holds and simultaneously the difference $\eta' - \eta''$ is minimal.

The problem formulated in such a form is difficult not only because it leads to sophisticated analytical operations but also because the *a priori* density $\varphi(\eta)$ is usually unknown. It was shown that the problem has a reasonable and simple solution independent on the *a priori* distribution of parameters only if the number of observation, n, is large and one can use the limit theorems.

One can use a different approach. Independently on the results of observations, $x_1, x_2, \ldots, x_n$, one can find such confidence limits $\eta'(x_1, x_2, \ldots, x_n)$ and $\eta''(x_1, x_2, \ldots, x_n)$ that with the given confidence probability the inequality

$$\theta'(x_1, x_2, \ldots, x_n) \leqslant \theta < \theta''(x_1, x_2, \ldots, x_n)$$

holds. Since the results of observations are unknown in advance, it is reasonable to consider not the conditional probabilities

$$\mathbf{P}\{\theta' \leqslant \theta < \theta'' | x_1, x_2, \dots, x_n\},$$

but unconditional probability

$$\mathbf{P}\{\theta' \leqslant \theta < \theta''\}$$

that no error appears if the rule will be applied.

For specified types of functions $\eta'(x_1, x_2, \dots, x_n)$ and $\eta''(x_1, x_2, \dots, x_n)$, the probability (1) of course depends on the distribution of random variables $x_1, x_2, \dots, x_n$. If the latter depends on k parameters $\eta_1, \eta_2, \dots, \eta_k$ and unconditional density function of these parameters is given by the function $\varphi(\eta_1, \eta_2, \dots, \eta_k)$, then

$$\mathbf{P}\{\theta' \leqslant \theta \leqslant \theta''\} = \int \cdots \int \mathbf{P}\{\theta' \leqslant \theta < \theta'' | \theta_1, \theta_2, \dots, \theta_k\} \, \varphi(\theta_1, \dots, \theta_k) \, d\theta_1, \dots, d\theta_k.$$

In practice, of great importance is the case where the conditional probability

$$\mathbf{P}\{\theta' \leqslant \theta \leqslant \theta'' | \theta_1, \theta_2, \dots, \theta_k\} \tag{2}$$

for any values of $\eta_1, \eta_2, \dots, \eta_k$ remains unchanged and equal to some number ω. In this case

$$\mathbf{P}\{\theta' \leqslant \theta \leqslant \theta''\} = \omega,$$

i.e., the unconditional probability (1) does not depend on the *a priori* unconditional distribution of parameters $\eta_1, \eta_2, \dots, \eta_k$. The hypothesis of the existence of the *a priori* distribution of parameters is far from always meaningful. Indeed, how is it possible to refer to the distribution of the parameter a of the Poisson distribution for any problem for which the random variable is characterized by the Poisson distribution? However if the conditional probability (2) does not depend on the values of parameters and equals the same value ω, then it is reasonable to assume the unconditional probability (1) existing and equaling ω even if the existence of the *a priori* distribution of parameters is not assumed.

Let us say that the *rule has the confidence probability* ω if for any possible values of parameters the conditional probability (2) equals ω.

Now let us return to the problems described in Section 60.

In the first of the problems considered, let us set

$$a' = \bar{x} + \frac{z_1}{\sqrt{n}}\sigma; \; a'' = \bar{x} + \frac{z_2}{\sqrt{n}}\sigma.$$

For any values of the parameters a and σ, one then obviously has

$$\mathbf{P}\{a' \leqslant a < a'' | a, \sigma\} = \mathbf{P}\left\{a - \frac{z_2\sigma}{\sqrt{n}} \leqslant \bar{x} \leqslant a - \frac{z_1\sigma}{\sqrt{n}} \middle| a, \sigma\right\}$$

$$= \frac{1}{\sqrt{2\pi}} \int_{-z_2}^{-z_1} e^{-t^2/2}\, dt = \frac{1}{\sqrt{2\pi}} \int_{z_1}^{z_2} e^{-t^2/2}\, dt.$$

From this, one concludes that the probability of the rule

$$\bar{x} + \frac{z_1\sigma}{\sqrt{n}} \leqslant a < \bar{x} + \frac{z_2\sigma}{\sqrt{n}}$$

is

$$\omega = \frac{1}{\sqrt{2\pi}} \int_{z_1}^{z_2} e^{-t^2/2}\, dt.$$

Therefore one has, in particular, the equality

$$\mathbf{P}\left\{|a - \bar{x}| \leqslant \frac{z}{\sqrt{n}}\sigma\right\} = \frac{2}{\sqrt{2\pi}} \int_0^z e^{-t^2/2}\, dt.$$

In the second problem, the parameter a is known, whereas the parameter σ is to be estimated. Let us set

$$\sigma' = \bar{s}\sqrt{n}/t_1, \quad \sigma'' = \bar{s}\sqrt{n}/t_2,$$

where

$$s = \sqrt{\frac{1}{n} \sum_{k=1}^{n} (x_k - a)^2}.$$

It is easy to see that

$$\mathbf{P}\{\sigma' \leqslant \sigma < \sigma'' | a, \sigma\} = \mathbf{P}\{\sigma t_2/\sqrt{n} < \bar{s}\, t_1\sigma/\sqrt{n} | a, \sigma\}.$$

In Section 21 (the χ^2-distribution) was found the density function of the variable s- under the assumption that a and σ are given. Namely,

$$\varphi(y \mid a, \sigma) = \frac{\sqrt{2n}}{\sigma\Gamma(n/2)} \left(\frac{y\sqrt{n}}{\sigma\sqrt{2}}\right)^{n-1} e^{-y^2 n/2\sigma^2}.$$

From this, one finds that

$$\mathbf{P}\{\sigma' \leqslant \sigma < \sigma'' \mid a, \sigma\} = \frac{\sqrt{2n}}{\sigma\Gamma(n/2)} \int_{\sigma t_2/\sqrt{n}}^{\sigma t_1/\sqrt{n}} \left(\frac{y\sqrt{n}}{\sigma\sqrt{2}}\right)^{n-1} e^{-y^2 n/2\sigma^2}\, dy$$

$$= \frac{1}{2^{n-2/2}\Gamma(n/2)} \int_{t_2}^{t_1} z^{n-1} e^{-z^2/2}\, dz.$$

We thus see that the conditional probability of the inequality

$$\sigma' \leqslant \sigma < \sigma''$$

under the assumption that the parameters a and σ are known, does not depend on the values of these parameters. Therefore, by the preceding discussion, the probability of the rule

$$\frac{1}{t_1}\sqrt{\sum_{k=1}^{n} (x_k - a)^2} \leqslant \sigma < \frac{1}{t_2}\sqrt{\sum_{k=1}^{n} (x_k - a)^2}$$

is

$$\omega = \frac{1}{2^{n-(2/2)}\Gamma(n/2)} \int_{t_2}^{t_1} z^{n-1} e^{-z^2/2}\, dz.$$

Finally we pass to a consideration of the last problem, where both of the parameters, a and σ, are unknown. Set

$$a_1' = \bar{x} + c_1\sqrt{n}\, s_1 \quad \text{and} \quad a_1'' = \bar{x} + c_2\sqrt{n}\, s_1,$$

$$\sigma_1' = \sqrt{n}\, s_1/t_1, \quad \sigma_1'' = \sqrt{n}\, s_1/t_2,$$

where

$$s_1 = \sqrt{\frac{1}{n} \sum_{k=1}^{n} (x_k - \bar{x})^2}.$$

Under the condition that a and σ are given, one has

$$\mathbf{P}\{a'_1 \leqslant a < a''_1 | a, \sigma\} = \mathbf{P}\left\{c_1 \leqslant \frac{a-\bar{x}}{s_1\sqrt{n}} < c_2 \middle| a, \sigma\right\}$$

and

$$\mathbf{P}\{\sigma'_1 \leqslant \sigma < \sigma'_2 | a, \sigma\} = \mathbf{P}\left\{t_2 \leqslant \frac{s_1\sqrt{n}}{\sigma} \leqslant t_1 | a, \sigma\right\}.$$

Let us now determine the conditional density functions for

$$\frac{a-\bar{x}}{s_1\sqrt{n}} \quad \text{and} \quad \frac{s_1\sqrt{n}}{\sigma}$$

under the condition that a and σ are known. Since

$$\frac{\bar{x}-a}{s_1\sqrt{n}} = \frac{\frac{1}{n}\sum_{k=1}^{n}(x_k-a)}{\sqrt{\sum_{k=1}^{n}[(x_k-a)-(\bar{x}-a)]^2}} = \frac{\bar{x}'}{\sqrt{\sum_{k=1}^{n}(x'_k-\bar{x}')^2}},$$

where $x'_k = x_k - a$ and $\bar{x}' = \frac{1}{n}\sum_{k=1}^{n} x'_k$ (the variables x'_k are independent and normally distributed with mathematical expectation 0 and variance σ^2). Let us now introduce a new system of orthogonal coordinates $(y_1, y_2, \ldots, y_n)$ into the n-dimensional space $(x'_1, x'_2, \ldots, x'_n)$, so that

$$y_1 = \sqrt{n}\bar{x}'_1$$

Then

$$ns_1^2 = \sum_{k=1}^{n}(x'_k - \bar{x}')^2 = \sum_{k=1}^{n} x'^2_k - n\bar{x}'^2 = \sum_{k=1}^{n} y_k^2 - y_1^2 = \sum_{k=2}^{n} y_k^2$$

and, consequently,

$$\frac{\bar{x}-a}{s_1\sqrt{n}} = \frac{y_1}{\sqrt{n\sum_{k=2}^{n} y_k^2}}.$$

Since

$$y_k = \sum_{i=1}^{n} \alpha_{ki} x'_i,$$

where the quantities α_{ki} satisfy the relations

$$\sum_{j=1}^{n} \alpha_{ij}\alpha_{jk} = \begin{cases} 1 & \text{for} \quad i = k, \\ 0 & \text{for} \quad i \neq k \end{cases}$$

and the variables x'_i are normally distributed, the variables y_k are also normally distributed. Further, $\mathbf{E}y_k = 0$ $(k = 1, 2, \ldots, n)$. Finally, from

$$\mathbf{E}y_i y_k = \sum_{j=1}^{n} \alpha_{ij}\alpha_{jk}\mathbf{E}x_j'^2 = \sigma^2 \sum_{j=1}^{n} \alpha_{ij}\alpha_{jk} = \begin{cases} \sigma^2 & \text{for} \quad i = k, \\ 0 & \text{for} \quad i \neq k, \end{cases}$$

independence of the variables y_k and the equality $Dy_k = \sigma^2$ $(k = 1, 2, \ldots, n)$ follow.

Further, since

$$\frac{y_1}{\sqrt{\sum_{k=2}^{n} y_k^2}} = \frac{y_1/\sqrt{n-1}}{\sqrt{\left(\sum_{k=2}^{n} y_k^2\right)\Big/(n-1)}}$$

and the numerator and denominator in this fraction are independent, and since, moreover, the density function of the numerator is

$$\frac{\sqrt{n-1}}{\sigma\sqrt{2\pi}} e^{-x^2/2\sigma^2(n-1)},$$

and that of the denominator (according to the χ^2-distribution, see Section 21) is

$$\frac{\sqrt{2(n-1)}}{\sigma\Gamma\left(\frac{n-1}{2}\right)} \left(\frac{y\sqrt{n-1}}{\sigma\sqrt{2}}\right)^{n-2} e^{-y^2/2\sigma^2(n-1)},$$

it follows from Section 21 (the Student distribution) that the density function of the quotient $y_1/\sqrt{\Sigma_{k=2}^{n} y_k^2}$ is $\Gamma(n/2)/\sqrt{\pi}\Gamma[(n-1)/2]$ $(1 + x^2)^{n/2}$. It is easy to verify that the density function of the variable

$y_1/\sqrt{n\Sigma_{k=2}^n y_k^2}$ is

$$\varphi(x|a,\sigma)=\frac{\sqrt{n}\,\Gamma(n/2)}{\sqrt{\pi}\,\Gamma\left(\frac{n-1}{2}\right)}(1+nx^2)^{-n/2},$$

and therefore

$$\mathsf{P}\left\{c_1 \leqslant \frac{a-\bar{x}}{s_1\sqrt{n}} < c_2\right\}=\frac{\sqrt{n}\,\Gamma(n/2)}{\sqrt{\pi}\,\Gamma\left(\frac{n-1}{2}\right)}\int_{c_1}^{c_2}(1+nx^2)^{-n/2}\,dx.$$

This probability does not depend on the values taken on by the parameters a and σ. Therefore one can say that in the third problem the confidential probability of the rule

$$c_1 s_1\sqrt{n} \leqslant a-\bar{x} < c_2 s_1\sqrt{n}$$

is

$$\omega=\frac{\sqrt{n}\,\Gamma(n/2)}{\sqrt{\pi}\,\Gamma\left(\frac{n-1}{2}\right)}\int_{c_1}^{c_2}(1+nx^2)^{-n/2}\,dx.$$

It still remains to determine the confidential probability of the rule which establishes limits for σ. Using the above transformations, one finds that

$$\mathsf{P}\{\sigma_1' \leqslant \sigma < \sigma_2' | a,\sigma\} = \mathsf{P}\left\{t_2 < \frac{s_1\sqrt{n}}{\sigma} \leqslant t_1 \middle| a,\sigma\right\}$$

$$= \mathsf{P}\left\{t_2\sigma < \sqrt{\sum_{k=2}^{n} y_k^2} \leqslant t_1\sigma \middle| a,\sigma\right\}.$$

By virtue of the results of Section 21 (χ^2-distribution), this yields

$$\mathsf{P}\{\sigma_1' \leqslant \sigma < \sigma_2' | a,\sigma\} = \frac{\sqrt{2(n-1)}}{\sigma\Gamma\left(\frac{n-1}{2}\right)}\int_{t_2\sigma/\sqrt{n-1}}^{t_1\sigma/\sqrt{n-1}}\left(\frac{y\sqrt{n-1}}{\sigma\sqrt{2}}\right)^{n-2} e^{-y^2(n-1)/2\sigma^2}\,dy$$

$$= \frac{1}{2^{(n-3)/2}\Gamma\left(\frac{n-1}{2}\right)}\int_{t_2}^{t_1} z^{n-2}e^{-(z^2/2)}\,dz.$$

Again, this probability does not depend on the values of the parameters a and σ. Consequently, the rule

$$\frac{\sqrt{n}s_1}{t_1} \leqslant \sigma < \frac{\sqrt{n}s_1}{t_2}$$

has the confidential probability

$$\omega = \frac{1}{2^{(n-3)/2}\Gamma\left(\frac{n-1}{2}\right)} \int_{t_2}^{t_1} z^{n-2} e^{-t^2/2}\, dt.$$

For the conclusion, notice that from the fact that

$$\mathbf{P}\{\theta' \leqslant \theta < \theta'' | \theta_1, \theta_2, \ldots, \theta_k\} = \omega$$

holds for any $\eta_1, \eta_2, \ldots, \eta_k$ it follows that

$$\mathbf{P}\{\theta' \leqslant \theta < \theta''\} = \omega,$$

and also that

$$\omega = \mathbf{P}\{\theta'(x_1, x_2, \ldots, x_n) \leqslant \theta < \theta''(x_1, x_2, \ldots, x_n) | x_1, x_2, \ldots, x_n\}.$$

73 TEST OF STATISTICAL HYPOTHESIS

Suppose that the functional form of the distribution of the random variable ξ is known but the values of the parameters $\eta_1, \eta_2, \ldots, \eta_k$ upon which it depends are unknown. There exists some basis for presuming that the parameters have certain specific values $\eta_1 = \eta_1^\circ$, $\eta_2 = \eta_2^\circ, \ldots$, $\eta_k = \eta_k^\circ$ (simple hypothesis) or belong to a certain set (composite hypothesis). It is required to determine whether or not the observed values of the variable ξ confirm this hypothesis.

In order to emphasize the practical importance of this problem, let us discuss some examples.

Example 1. Consider a large lot of the products of a certain manufacture. Each item of this product belongs to one of two categories: perfect or defective. Any lot of items is regarded as proper for shipment to a customer if the relative number p of defective items is not larger than a certain number $p_0 (0 < p_0 < 1)$. The quantity p is not known; its value is to be determined by investigating a small number (in comparison with the entire lot) of items. Let us consider the random variable ξ which equals 0 if a randomly sampled item occurs to be perfect and equals 1 if it is defective. The distribution function for ξ is

$$F(x)=\begin{cases}0 & \text{for } x\leqslant 0,\\ 1-p & \text{for } 0<x\leqslant 1,\\ 1 & \text{for } x>1.\end{cases}$$

The parameter p on which the distribution depends is unknown. The problem is to test the hypothesis that $p\leqslant p_0$.

Example 2. The random variable ξ is distributed normally:

$$p(x)=\frac{1}{\sigma\sqrt{2\pi}}e^{-(x-a)^2/2\sigma^2};$$

the parameters a and σ are unknown. It is required to test the hypothesis that

$$|a-a_0|\leqslant\alpha \quad \text{and} \quad \sigma<\sigma_0$$

where a_0, σ_0 and α are certain given quantities.

This, and analogous problems, are constantly arising in the theory of measurement as well as in the natural sciences and in industrial production.

Let n be the number of observations on the basis of which it must be decided whether a specified hypothesis is to be accepted or rejected. Let

$$x_1, x_2, \ldots, x_n \tag{1}$$

be the observed values. The testing process leading to the acceptance or rejection of the hypothesis is a certain rule according to which the set of all possible outcomes of the n observations can be divided into two disjoint sets R_{n1} and R_{n2}. If the quantities (1) belong to the set R_{n1} then this will correspond to the acceptance of the hypothesis in question, and if they belong to R_2 then the hypothesis is rejected. If numbers in (1) are represented as coordinates of a point in n-dimensional Euclidean space R_n then, obviously, every testing procedure corresponds to partitioning of the space R_n into two regions, R_{n1} and R_{n2}. If the point $(x_1, x_2, \ldots, x_n)$ is in the region R_{n1} then the hypothesis is accepted, and if the point $(x_1, x_2, \ldots, x_n)$ is in R_{n2} then the hypothesis is rejected. The region R_{n2} is referred to as the *critical region.* The selection of a testing rule for a hypothesis is equivalent to the selection of a critical region.

To illustrate this, let us return to Example 1. The space R_n in this case consists of all possible combinations of n quantities each of which takes on either the values 0 or 1. The critical region R_{n2} includes those elements of R_n for which

$$\frac{1}{n}(x_1 + x_2 + \cdots + x_n) > p_0.$$

Now let us turn to the following exceptional problem of testing a hypothesis for which there is a complete solution. Two simple hypotheses H_1 and H_2 are given. Hypothesis H_1 is that $\eta_i = \eta_i'(i = 1, 2, \ldots, k)$, and hypothesis H_2 is that $\eta_i = \eta_i''(i = 1, 2, \ldots, k)$. These hypotheses are the alternative hypotheses one which must be preferred on the basis of the observations.

Notice that accepting or rejecting of the hypothesis H_1 can lead to errors of two kinds. The first type of error is committed when the hypothesis H_1 is rejected when it is actually true. In the other words, an error of the first type occurs when the point $(x_1, x_2, \ldots, x_n)$ belongs to the region R_{n2} when the hypothesis H_1 is true. An error of the second type is made if the hypothesis H_1 is accepted when it is false. These two possible types of errors will be called errors of Type I and errors of Type II, respectively. If the critical region has been selected, the probabilities of the errors of both types can be computed. For specified n and R_{n2} these probabilities are denoted by α_1 and α_2, respectively.

It is clear that the smaller the values of α_1 and α_2 for a given critical region, the more successfully the critical region has been chosen. However, for a fixed number of trials n, it is impossible to make both quantities α_1 and α_2 simultaneously arbitrarily small for any choice of the critical region. However one can achieve arbitrarily small errors either of Type I or of Type II separately by corresponding changes of the critical region. Thus, if one set $R_{n2} = R_n$ then, obviously, $\alpha_2 = 0$. However, if one set $R_{n1} = R_n$ then $\alpha_1 = 0$. This hints at the following rational principle for the selection of the critical region. For given values of α_1 and n, one should choose that region R_{n2} for which α_2 attains its minimum. Of course, one should understand that the smaller the chosen value of α_1, the larger is the minimum value reached by α_2. It is impossible to say in advance what value of α_1 should be chosen so that the method of testing the hypothesis will be the most beneficial, because the practical side of the matter plays a main role in this situation.

For example, suppose that the rejection or acceptance of the hypothesis H_1 is concerned with some financial expenditure. Let the acceptance of the hypothesis H_1 when it is actually false lead to a large loss (for instance, it forces one to use a manual adjustment of certain details before assembling). Let the rejection of the hypothesis H_1 when it is true result in comparatively small expense. Then, obviously, it is necessary to select α_2 as small as possible and to accept as inevitable comparatively large values of α_1.

Suppose that practical considerations have been taken into account and that the quantity α_1 has been chosen. Then the following statement holds which we will prove, however, only for the case where the variable ξ has a finite density function both for the hypotheses H_1 and H_2.

THEOREM. *Among all of possible critical regions for which the probability of the error of Type I is α_1, the probability of the error of Type II reaches its least value for that critical region R^*_{n2} which consists of all those points $(x_1, x_2, \ldots, x_n)$ for which*[2]

$$\prod_{k=1}^{n} f(x_k | H_2) \geq c \prod_{k=1}^{n} f(x_k | H_1) \tag{2}$$

The number c is determined by the condition

$$\psi(c) = \mathbf{P}\{(x_1, x_2, \ldots, x_n) \subset R^*_{n2} | H_1\} = \alpha_1. \tag{3}$$

PROOF: Since for independent trials the probability of the point $(x_1, x_2, \ldots, x_n)$ being in some region S is

$$\mathbf{P}\{S | H_2\} = \int_S \cdots \int \prod_{k=1}^{n} f(x_k | H_1)\, dx_1\, dx_2 \ldots dx_n$$

[2]Thus the region R^*_{n2} is the best critical region. - *B. G.*

given that the hypothesis H_1 is true, and

$$\mathsf{P}\{S|H_2\} = \int_S \cdots \int \prod_{k=1}^{n} f(x_k|H_2)\,dx_1\,dx_2 \ldots dx_n$$

given that the hypothesis H_2 is true, then by the assumption

$$\mathsf{P}\{R^*_{n2}|H_1\} = \alpha_1$$

and for any other region R_{n2} in question

$$\mathsf{P}\{R_{n2}|H_1\} = \alpha_1.$$

By the axiom on the addition of probabilities,

$$\begin{aligned}\mathsf{P}\{R_{n2} - R_{n2}R^*_{n2}|H_1\} &= \mathsf{P}\{R_{n2}|H_1\} - \mathsf{P}\{R_{n2}R^*_{n2}|H_1\}\\ &= \alpha_1 - \mathsf{P}\{R_{n2}R^*_{n2}|H_1\}\end{aligned}$$

and

$$\mathsf{P}\{R^*_{n2} - R_{n2}R^*_{n2}|H_1\} = \alpha_1 - \mathsf{P}\{R_{n2}R^*_{n2}|H_1\},$$

i.e.,

$$\mathsf{P}\{R_{n2} - R^*_{n2}R_{n2}|H_1\} = \mathsf{P}\{R^*_{n2} - R_{n2}R^*_{n2}|H_1\}.$$

According to the definition of R^*_{n2} and the last equality,

$$\begin{aligned}\mathsf{P}\{R^*_{n2} - R_{n2}R^*_{n2}|H_2\} &\geqslant c\mathsf{P}\{R^*_{n2} - R_{n2}R^*_{n2}|H_1\}\\ &= c\mathsf{P}\{R_{n2} - R_{n2}R^*_{n2}|H_1\}.\end{aligned} \quad (4)$$

But for any point $(x_1, x_2, \ldots, x_n)$ not belonging to R^*_{n2},

$$\prod_{k=1}^{n} f(x_k|H_2) < c \prod_{k=1}^{n} f(x_k|H_1)$$

and, consequently, since the region $R_{n2} - R_{n2}R^*_{n2}$ lies entirely outside of R^*_{n2}, it must be true that

$$c\mathsf{P}\{R_{n2} - R_{n2}R^*_{n2}|H_1\} > \mathsf{P}\{R_{n2} - R_{n2}R^*_{n2}|H_1\}.$$

This inequality, together with (4), leads to the inequality

$$\mathsf{P}\{R^*_{n2} - R_{n2}R^*_{n2}|H_2\} > \mathsf{P}\{R_{n2} - R_{n2}R^*_{n2}|H_2\}.$$

Adding $\mathbf{P}\{R_{n2} - R^*_{n2} | H_2\}$ to both sides of this inequality, one finds that

$$\mathbf{P}\{R^*_{n2} | H_2\} > \mathbf{P}\{R_{n2} | H_2\}.$$

But since

$$\mathbf{P}\{R_n | H_2\} = 1$$

and $R^*_{n1} = R_n - R^*_{n2}$ and $R_{n1} = R_n - R_{n2}$, then it follows that

$$\mathbf{P}\{R^*_{n1} | H_2\} < \mathbf{P}\{R_{n1} | H_2\}.$$

Since $\mathbf{P}\{R_{n1} | H_2\}$ and $\mathbf{P}\{R^*_{n1} | H_2\}$ are the errors of Type II for the critical regions R_{n2} and R_{n2}, respectively, the theorem is completed.

One must still show that the selection of the constant c can actually be done in accordance with rule (3). Notice that the function

$$\psi(c) = \mathbf{P}\{R^*_{n2} | H_1\}$$

can only decrease as c increases, since the inequality (2) will be satisfied by smaller and smaller set of points $(x_1, x_2, \ldots, x_n)$ Moreover, it is clear that $\Psi(0) = 1$, since for each point $(x_1, x_2, \ldots, x_n)$

$$\prod_{k=1}^{n} f(x_k | H_2) \geqslant 0.$$

Further, from (2) it follows that

$$\mathbf{P}\{R^*_{n2} | H_2\} \geqslant c\mathbf{P}\{R^*_{n2} | H_1\}.$$

Substituting the left-hand side of this inequality by 1 and recalling the definition of $\Psi(c)$, one obtains the inequality

$$1 \geqslant c\psi(c).$$

Finally, one has

$$0 \leqslant \psi(c) \leqslant 1/c.$$

Thus, $\Psi(c) \to \infty$ as $c \to \infty$. Since the function $\Psi(c)$ does not increase, then one can find a value of c such that for any α_1 $(0 < \alpha_1 < 1)$

$$\psi(c-0) \geqslant \alpha_1 \geqslant \psi(c+0).$$

If the function $\Psi(c)$ is continuous at the point c then the selection of the constant c according to rule (3) is reasonable. If the function $\Psi(c)$ has a discontinuity at the point c the situation becomes somewhat more complicated: it requires a slight change in the definition of the region R^*_{n2}. One has to remove from this region a set of points $(x_1, x_2, \ldots, x_n)$ for which

$$\prod_{k=1}^{n} f(x_k|H_2) = c \prod_{k=1}^{n} f(x_k|H_1),$$

and add them to the region R^*_{n1}, so that the error of Type I would be equal to α_1.

Let us consider an example. Suppose that ξ is normally distributed and that the value of the variance, σ^2, is known. There are two hypotheses concerning the mathematical expectation a. Hypothesis H_1 consists in $a = a_1$ and hypothesis H_2 consists in $a = a_2$. It is required to find the best critical region.

In this example, relation (2) can be expressed in the following form:

$$e^{-1/2\sigma^2 \sum_{k=1}^{n} [(x_k - a)^2 - (x_k - a_1)^2]} \geqslant c.$$

Under the assumption that $a_2 > a_1$, this inequality is equivalent to the following one:

$$\sum_{k=1}^{n} x_k \geqslant \frac{\sigma^2 \ln c}{a_2 - a_1} + \frac{n}{2}(a_1 + a_2)$$

or, what is the same, to the inequality

$$\frac{1}{\sigma\sqrt{n}} \sum_{k=1}^{n} (x_k - a_1) \geqslant \frac{\sigma \ln c}{(a_2 - a_1)\sqrt{n}} + \frac{\sqrt{n}}{2\sigma}(a_2 - a_1) = k_1.$$

The inequality obtained defines the best critical region R^*_{n2}.

Since the quantity

$$\frac{1}{\sigma\sqrt{n}} \sum_{k=1}^{n} (x_k - a_1)$$

is normally distributed with the mathematical expectation 0 and variance 1, if the hypothesis H_1 takes place, one can determine k_1 (and hence c) for given α_1 by use of normal distribution tables. Suppose, for definiteness, that $\alpha_1 = 0.05$. Then $k_1 = 1.645$ and, consequently, the best critical region for $\alpha_1 = 0.05$ is given by the inequality

$$\sum_{k=1}^{n} (x_k - a_1) \geqslant 1.645\,\sigma\sqrt{n}.$$

It is interesting to note that the critical region in this example is not independent of the alternative value a_2. The region R^*_{n1} is defined by the inequality

$$\sum_{k=1}^{n} x_k < \frac{\sigma^2 \ln c}{a_2 - a_1} + \frac{n}{2}(a_2 + a_1),$$

which is expressible as follows:

$$\frac{1}{\sigma\sqrt{n}} \sum_{k=1}^{n} (x_k - a_2) < k_1 - \frac{\sqrt{n}}{\sigma}(a_2 - a_1).$$

Under the assumption that the hypothesis H_2 holds, the quantity on the left-hand side of the inequality is normally distributed with the mathematical expectation 0 and variance 1. It follows from this statement that the probability of an error of Type II equals

$$\Phi\left(k_1 - \frac{\sqrt{n}}{\sigma}(a_2 - a_1)\right) = \frac{1}{\sqrt{2\pi}} \int_{-\infty}^{k_1 - (\sqrt{n}/\sigma)(a_2 - a_1)} e^{-z^2/2}\, dz.$$

If α_1 and α_2 are specified, then there arises the problem of determining the minimal number of trials $n = n(\alpha_1, \alpha_2)$ that are necessary for the probabilities of getting erroneous conclusions to be not larger than given α_1 and α_2.

As n increases, the quantity $\alpha_{2_n} = \alpha_2(\alpha_1, n)$ is non-increasing and, in general, approaches 0. It is evident that $n(\alpha_1, \alpha_2)$ is the smallest of those values of n for which $\alpha_2(\alpha_1, n) \geq \alpha_2$.

In the example just considered, the number n can be found quite simply for given values of α_1 and α_2. In fact, from

$$1 - \Phi(k_1) = \alpha_1$$

and

$$\Phi\left(k_1 - \frac{\sqrt{n}}{\sigma}(a_2 - a_1)\right) = \alpha_2$$

one obtains the two equations:

$$k_1 = \psi(1 - \alpha_1)$$

and

$$k_1 - \frac{\sqrt{n}}{\sigma}(a_2 - a_1) = \psi(\alpha_2),$$

where Ψ is the inverse function to $\Phi(x)$. From here one has

$$n = \frac{\sigma^2}{(a_2 - a_1)}[\psi(1 - \alpha_1) - \psi(\alpha_2)]^2.$$

Let us consider a short numerical example. The following quantities are given:

$$a_1 = 135, \ a_2 = 150, \ \sigma = 25, \ \alpha_1 = 0.01, \ \alpha_2 = 0.03.$$

Since

$$\psi(0.99) = 2{,}33, \ \psi(0.03) = -1.88,$$

one easily finds

$$n = \frac{25^2}{15^2}[2.33 + 1.88]^2 = \frac{25}{9} \cdot 4.21^2 \sim 49.$$

Thus, the minimum number of observations that are needed to make a selection between the hypotheses H_1 and H_2 for the data specified above must be equal to 49. Only for this number of trials can one be certain that if the hypothesis H_1 is true, then the probability to reject it is not larger than 0.01. If the hypothesis H_2 is true, then the probability that it will be rejected is not larger than 0.03.

APPENDIX: THE HISTORY OF PROBABILITY THEORY

PART 1 MAIN CONCEPTS OF PROBABILITY AND RANDOM EVENT

A. First Steps

It is nearly impossible to identify the first person to measure (even imperfectly) the probability of reoccurrence of a random event. It is, however, clear that developing a model for measurement of random events preoccupied several generations of outstanding scientists. For years researchers restricted themselves to the consideration of games of chance such as tossing dice which allowed them to construct simple and transparent mathematical models. The conclusion drawn by many of these early researchers was eloquently expressed by Dutch physicist Christian Huygens (1629–1695) when he wrote, "... I suppose that an attentive reader will discover that he is dealing not only with a gambling game, but with the foundation of a very interesting and deep theory."

An intense probing of scientific theory and philosophical tenets played a major role in the development of probability theory. Even today, the relevance and need to further develop probability theory is apparent as new problems in science, technology, and defense emerge, challenging existing notions of probability and necessitating the widening of its arsenal of ideas, concepts, and methods of investigation.

Initially, scientists studying random phenomena focused on three problems:

Editing of the English translation of the Appendices was done by Ms. Susanne Willson.

(1) Accounting for the number of possible outcomes resulting from tossing dice several times;

(2) Distributing the winnings between players when a game is stopped before it is completed;

(3) Determining how many tosses of two or more dice are necessary in order to throw a roll in which all numbers on the dice are the same (for example all sixes). Or, ascertaining how many tosses are necessary to create a case in which the likelihood of the consecutive occurrence of a roll in which all numbers do not turn up the same is less than the likelihood of single roll of identical numbers. Bishop Vibold from Cambrais arrived at a number of conclusions employing three dice. He thought that there were 56 such outcomes. (He did not consider cases in which the same numbers turned up on all of the dice.) Vibold assigned religious principles to his finding: with the appearance of each three numbers he connected one of 56 virtues. A correct description of calculations was done in the eleventh century by chronicler Baldericus and was published in 1615.

An attempt to calculate the number of possible outcomes from the throwing of three dice, including potential transpositions, was offered in a poem by Richard de Fournival (1200–1250). In a segment of his poem "*De Vetula*" depicting gambling and games of chance he made the following consideration, "equal points on three dice can be achieved in six ways. If the number on two of the dice are the same, but differ from the third, we have 30 possibilities because there are six possible combinations of pairs (a die has six sides) and a third die might have five possible other values. There are 20 possible combinations resulting in a throw in which different points appear on all dice, because $30 \times 4 = 120$, but each possibility appears six times. Thus, there are a total 56 possibilities. Equal numbers on all dice can be obtained in one way only; equal numbers on two dice can be tossed in three ways."

Although this text affords a number of cases which illustrate Vibold's result, it was Richard de Fournival who wrote the following equation illustrating the entire number of equally probable cases when tossing three dice:

$$6 \times 1 + 30 \times 3 + 20 \times 6 = 216.$$

Furthermore, de Fournival devised a table in which a number of outcomes were calculated for different values of sums of numbers on all three dice. We offer a shortened version of this table.

Table 13

Sum		Number of ways
3	18	1
4	17	3
5	16	6
6	15	10
7	14	15
8	13	21
9	12	25
10	11	27

The first two columns illustrate the sums of the numbers thrown with three 3 dice. The third column depicts the numbers of possible combinations which result in such a sum. All calculations made by de Fournival are without errors and all arguments presented by him are logical and relatively contemporary. In 1477 Benvenuto d'Imola published Dante's "*Divine Comedy*" in Venice. D'Imola incorporated his own commentary into Dante's work. In part VI of "*In Purgatory*", d'Imola calculated the odds for a dice gambling game. He maintained that the sum of numbers thrown equaling 3, 4, 17, and 18 after three throws of the dice could be obtained only by one means. D'Imola's mistake is obvious and requires no further explanation.[1]

Special attention should be paid to the work appearing at the beginning of the Italian Renaissance by Luke Pacioli (Circa 1445–1614) entitled, "*Summa de Arithmetica, Geometria, Proportioni et Proportionalita*". This work, written in 1487 (but having been published in Venice only seven years later) played a significant role in generating interest in probability theory. Considering "uncommon problems", Pacioli presented the following two problems:

(1) Two teams play in a game the object of which is to score 60 points. The winner receives 22 ducats. For some reason, the game is stopped prematurely. When the game is called off one team has 50 points and the other 30 points. This raises the question: How should the prize money be distributed?

[1]Indeed, 3 can appear only in one combination: each of three dice shows 1 (same for 18: each die shows 6). 4 and 17 can appear in three different combinations. Possible outcomes delivering a sum of points equal to 4 are (2, 1, 1), (1, 2, 1), and (1, 1, 2). For 17 the outcomes are (5, 6, 6), (6, 5, 6), and (6, 6, 5).

(2) Three men compete in arbalest shooting. The one who hits the center of the target six times wins.The wager is six ducats. At the point when the first player hits four bull's eyes, the second hits three and the third hits two the game is stopped. How should the prize money be fairly distributed?

Pacioli suggested a solution to the problem which was later discredited as incorrect. He suggested that the prize money be divided up proportionally between the number of wins. His solution to the first problem would be to give 5/8 of the prize money, or 13.75 ducats, to the winning team and 3/8 of the prize money, or 8.25 ducats, to the second. Accordingly, Pacioli solved the second problem by awarding 4 and 4/9 ducats to the first place player, 3 and 3/9 ducats to the second, and 2 and 2/9 ducats to the third.

B. Investigations by Girolamo Cardano and Niccolo Tartaglia

The development of probability theory cannot be considered without studying the contributions of sixteenth century Italian scientists Girolamo Cardano and Niccolo Tartaglia. In an excerpt from Cardano's *De Lude Aleae* published in 1563 many problems associated with dice tossing and frequency questions raised from such tosses were solved. He correctly calculated a number of possible outcomes for cases involving tosses of two or three dice. Cardano's findings proved difficult to put into words. For instance, in Chapter XI "*On Tossing Two Dice*" he observed that, "If two dice are tossed there are 6 cases where each die shows an identical number and 15 tosses where each die shows a different result, i.e., including doubles, there are 30 tosses. Consequently, there are 36 possible tosses in all." In his writings the term "double" referred to tosses that resulted from permutations. For example, if the first die shows a 5 and the second die shows a 2, then a double occurs when a 2 and 5 are rolled correspondingly on those dice again.

Cardano calculated the number of possible rolls where a given number appeared at least on one of the dice. There were 11 such cases. The following observation by Cardano must be emphasized: "The frequency of such rolls is less than the frequency of rolls where such a number is absent. The frequency of such rolls relates to when the total number of tosses of two dice is larger than 1/6 and smaller than 1/4". Note Cardano's analysis is incorrect. He should have said smaller than 1/3, since 11/36 is not less but larger than 1/4.

Cardano's efforts are especially important because he twice prompts the consideration of a ratio that is now referred to as the classical determination of probability. Namely, 1/6 is the probability of rolling a particular number on one die, and 11/36 is the probability of rolling the same number on a roll of two dice. Does this mean that Cardano was considering corresponding probabilities rather than numbers of chance? Did he in fact introduce a classical definition of probability? Examples of Cardano's work cited indicate that he laid the foundation for the development of probability theory, rather than inventing it himself. Perhaps Cardano simply decided to clarify which event was more frequent: the appearance of a given number from the roll of a die or the appearance of that number at least once from a roll of two dice. Having satisfactorily answered his query he may have been satisfied, completely unaware that he was at the brink of discovering a concept that would provide for the development of a great branch of mathematics and all of quantitative science.

Cardano's lack of foresight is confirmed by the next chapter where he considers tossing three dice and fails to mention a ratio of the number of specified cases to the total number of cases. He simply provides a calculation for a number of possible chances.

In Chapter 13 "*On Complex Points, Up to Six and More and for two as well as for Three Dice,*" Cardano again considers a ratio of the frequency of cases with a specified sum of numbers to the total number of tosses. Again, Cardano does not find himself on the edge of an important scientific discovery. He wrote, "A sum of ten pips can be obtained by tossing two fives or from the toss of a six and a four. The latter can be obtained in two ways. In the same manner, nine can be obtained through the toss of a five and a four and from the toss of a six and a three. Thus it is 1/9 of the series[2] of tosses and 2/9 of its half. Eight pips can be rolled with a toss of two fours, with a toss of 3 and 5 and with a toss of 2 and 6. In total there are 5 cases or approximately 1/7 of all the series... 7 can be obtained by tossing a 1 and a 6, a 2 and a 5, and a 3 and a 4. In total, consequently, there are 6 cases which consist of 1/6 of the total series. 6 pips are obtained in a similar way as 8, 5 as 9, 4 as 10, 3 as 11, and 2 as 12."

---

[2]Cardano called a series of tossing all possible outcomes, i.e., 36 for tossing of two dice and 216 for tossing of three dice.

Cardano again actually operated with the classical concept of probability but he did not recognize its importance for the problems investigated. He considered these relations as arithmetical ones, as a portion of cases rather than measure of a possibility of occurrence of a random event.

In Chapter XI there is a statement which was interpreted in a very wide sense by some authors but we will see that it is rather uncertain. Cardano wrote: "A whole series of games does not give a deviation though a single game does. A large number of games shows that reality is very close to this assumption." Referring to this statement V. Bobynin[3] concluded that this law (meaning the law of large numbers – *B.G.*) with sufficient clearness was expressed in the sixteenth century by Cardano in his paper "*De Ludo Aleae*". Later O.Ore[4] in his book dedicated to Cardano wrote that the latter formulated and used the law of large numbers in rudimentary form. This opinion has met with some approval though we would like to emphasize that Cardano's statement is very uncertain.

In the same book Cardano came close to the definition of fair game. Let us cite: "So, there is one general rule for calculation: One needs to take into account the total number of outcomes and the number of outcomes presenting interest, then to find the ratio of the second and first numbers. The sizes of stacks should be related in the same manner for a fair game." But the opinion of some authors that Cardano here is close to the classical definition of probability seems to me erroneous.

Pacioli's problem of the stack sharing before the game terminated was also interesting for Cardano. In his book "*Practice of General Arithmetics*" published in 1539, Cardano gave some critical notices about Pacioli's solution. He wrote that the suggestion to share the stack proportionally to the numbers of games won by each gambler did not take into account how many games each of them should win to be a winner. In his opinion, Cardano suggested that if s is the number of games which should be won and p and q are the numbers of games actually won by the first and second players, then the stack

[3]V.V. Bobynin. *Jacob Bernoulli and Probability Theory.* Mathematical Education, 1914, No.4.

[4]Ore, O. (1953) Cardano. *The Gambling Scholar*. Princeton.

was to be shared as

$$[1+2+\cdots+(s-q)]:[1+2+\cdots+(s-p)].$$

We will show below that Cardano's suggestion in general is erroneous and leads to the correct results only in some particular cases.

Tartaglia returned to the problem of stack sharing in his "*La Prima Rarte del General Trattato di Numerix et Misure*" published in 1556. His approach was presented in Chapter 20 "*The Error of Brother Luca from Borgo.*" This critic is correct and has a strong basis: "This rule does not seem to me either beautiful or good. So, if one side has won 10 points and another none at all, then the stack should be given to the first gambler but this makes no sense."

For Pacioli's first problem (with slightly changed conditions), Tartaglia suggested the following solution. The first gambler who has won 10 points has to receive, first of all, a half of the whole stack and in addition (10 – 0)/60 of the whole stack, or 22/6 ducats, that is 14 and 2/3 in total. The second gambler should receive 7 and 1/3 ducats.

We will see that this solution is also incorrect. But everybody agrees that neither Tartaglia nor his predecessors could correctly solve this problem with no knowledge of concepts of probability and mathematical expectation. The following remark shows that Tartaglia himself has doubts about his own solution: "The solution of this problem is rather jurisdictional than mathematical because for any chosen rule one can find pro and contra. Anyway the following solution possesses some advantages..." Further he suggested to share the stack by the following rule: The deviation from the half of the stack share should be proportional to the difference of points which each gambler has won. This leads to the solution mentioned above

$$11+[(10-0)/60]\cdot 22=14 \quad \text{and} \quad 2/3,$$

and

$$11+[(0-10)/60]\cdot 22=7 \quad \text{and} \quad 1/3.$$

(Remember that the stack equals 22 ducats, the game continues to 60 points, at the stopping point the first gambler has won 10 points and the second has none.)

C. Investigations by Galileo Galilei

Purely probabilistic problems had already arisen by the sixteenth century. Researchers searched for a methodology for their solution. This inevitably necessitated the development of combinatorial methods and the construction of concepts which could be used for the analysis of phenomena. Mistakes made by one researcher were found by others who suggested their own solutions; the latter, in turn were critically analyzed by their followers. So, step by step methods were created that later became the foundation of a new theory or, at least, allowed for the solution of specific tasks.

Special attention should be paid to efforts made by outstanding philosopher and scientist of wide interests–Galileo Galilei (1564 – 1642). His work *On Point Outcome in Dice Tossing,* not published until 1718, was dedicated to the calculation of the number of all possible cases involving the tossing of three dice. Galileo calculated the entire number of all of possible cases through the simplest and most natural means. He took 6 (the number of equally likely outcomes for a single die toss) to the power of 3 and got $6^3 = 216$. This result was many times obtained before by direct calculation.

Then Galileo counted how many possible ways a specified sum of numbers could be rolled using three dice. It is clear this sum could be any value from 3 through 18. Galileo devised a powerful tool for calculating the sum. The dice were numbered (first, second, third) and all of the possible outcomes were written down as a triple figure that had the value of the pips corresponding to a die with the specified number. This simple idea was very useful for its time. Galileo wrote,"... *though* 9 *and* 12 *are obtained as result of the same number of combinations as* 10 *and* 11, *and consequently should be understood as equally valuable. After frequent observation we find that gamblers still believe rolls of* 10 *and* 11 *to be preferable to rolls of* 9 *and* 12. *It is absolutely obvious that* 9 *and* 10 (*we say this still taking into account* 12 *and* 11) *can be rolled in the same number of combinations*: 9 *from* {1,2,6}, {1,3,5}, {1,4,4}, {2,2,5}, {2,3,4} *and* {3,3,3}, i.e., *from* 6 *triples, and* 10 *from* {1,3,6}, {1,4,5}, {2,4,4}, {2,3,5}, {2,4,4} *and* {3,3,4} *and no other ways*". (G. Galilei, Opera, XIV, p.293, Fiorentina, 1855).

The question naturally arises–why does a sum equal to 10 appear to be preferable to one equal to 9? Galileo's answer is as follows: "(1) Triples with equal points on all three dice can be rolled by only one means. (2) Triples with equal points on two dice and different points on the third die can be realized through three different rolls. (3) Triples with different points on all three dice can be the result of six different rolls. From these arguments it is easy to construe all of the possible combinations of numbers a roll of three dice can yield." (*Ibid.*, p. 295).

At the end of his work, Galilei unveiled the following table:

Table 14

10		9		8		7		6		5		4		3	
631	6	621	6	611	3	511	3	411	3	311	3	211	3	111	1
622	3	531	6	521	6	421	6	321	6	221	3				
541	6	522	3	431	6	331	3	222	1						
532	6	441	3	422	3	322	3								
442	3	432	6	332	3										
433	3	333	1												
	27		25		21		15		10		6		3		1

In the upper row values are shown for the sum of points of three dice. The first three figures in each cell show how the sum in the corresponding columns is realized, the fourth figure corresponds to the number of different ways. For instance, next to the triple 631 6 ways are shown: 631, 136, 316, 613, 163, 361. Combination 361 means that the first die has 3 points, the second 6, and the third 1.

The table reveals only half of all the possible sums. The remaining half can be calculated similarly. Examination of the results reveals that the sum 11 can be achieved via 27 various possibilities–12 by 25, 13 by 21, 14 by 15, 15 by 10, 16 by 6, 17 by 3, and 18 by 1. Thus, the sum of all of possible variants of points on three dice equals $2(1+3+6+10+15+21+25+27)=216$.

Bear in mind that Galileo essentially repeated results obtained earlier by predecessors such as Bishop Vibold and Richard de Fournival among others. This simple problem, even for second year university students, was in its time a serious challenge for even the most advanced philosophers. Galileo's struggles were apparent when he wrote, "This problem has exhausted me. I am presenting my arguments with the hope that they will clarify some existing misconceptions and bring it closer to the fundamental truths which will shed light on all aspects of the game of chance". (*Ibid.*)

Note that Galileo and his predecessors argued not about the probability of random events, but about the number of chances corresponding to those events. For probability theory and mathematical statistics, Galileo's ideas on the theory of observational error are even more significant than the considered work above. He was the first to formulate and investigate this problem. At the time all that he wrote was revolutionary and is still important today. He divulged in detail his thoughts and inferences on observational problems in his masterpiece *Dialogue on Two Main Systems of the Universe: Ptolemaic's and Copernicus.*

According to Galileo, observational errors are inescapable when doing measurement and experimental investigation. He wrote, "With any combination of observations there will be some error; I think that it is inevitable..." (*Dialogue...*, p. 214). He listed two types of errors: (1)Systematic which are tied to the method of measurement and measurement tools; and (2) Random errors which unpredictably vary from one measurement to another. Galileo's classification of observational error is still standard today and widely used in manuals on the theory of measurement error.

Random errors possess some specific properties. Galileo very carefully analyzed them. First, smaller errors are encountered more often than larger ones, so correcting them is often unnecessary. Furthermore, negative values are met with the same frequency as positive ones. He maintained that, "One can make an error on one side or another." (*Ibid.*, p.125). Finally, Galileo noticed that the greatest number of measurements group around the true result. "Among all possible locations, the true value, we assume, is that which locates in the middle" (*Ibid.*, p.216).

Galileo's research is of a great importance because it is fundamental to a new scientific discipline–the theory of measurement errors. This theory, undoubtedly, played an important role in the development of probability theory, but was essential to the evolution of mathematical statistics. Recent random error theory of observation is considered to be a branch of mathematical statistics.

D. Contributions by B. Pascal and P. Fermat in the Development of Probability Theory

It is often assumed that probability theory began with the written correspondence between two great scientists, Blaise Pascal (1623 – 1662) and Pierre Fermat (1601 – 1665). In total only seven letters remain of the many letters exchanged between the two. Three letters survive by Pascal (of July 29, August 24, and October 27, 1654) and four by Fermat (one without date and others of August 9, August 29, and September 25, 1654). The first letter by Pascal was lost and speculation as to its contents can only be gleaned by referring to the letter Fermat wrote in response.

We would like to say that the evaluation of the contributions of Pascal and Fermat given by many historians of science is exaggerated. The correspondence between Pascal and Fermat does not explain the concept of probability. They restricted themselves to the analysis of the number of chances related to events. Of course, they were the first to correctly solve the stake sharing problem that had long perplexed their predecessors. Both understood that the stake should be shared proportionally to the probability of final victory. The solutions they proposed contained the seeds of the notion of mathematical expectation and in some imperfect form even theorems of summation and multiplication of probabilities. (More exactly, investigated events.) Though they did not invent probability theory, their efforts proved to be the catalysts necessary to arouse an interest in probabilistic problems and were fundamental to its creation. Pascal went a step further by developing combinatorial approaches and figuring out the role of combinatorics in the conception of probability theory.

The impetus for Pascal's interest in probabilistic problems came when he met and had talks with the aristocrat Chevalier de Mere (1607–1648). De Mere was interested in philosophy and literature and gambled compulsively. Gambling was the source of the problems Mere urged Pascal to solve. The problems were as follows:

(1) How many tosses of two dice were necessary to ensure that a roll of double sixes (at least once) was more likely than a roll without double sixes?

(2) How could the stakes be distributed fairly in a game which was ended before one of the players won the necessary number of rounds to emerge victorious?

De Mere pretended to solve the first problem. However, his arguments were flawed. In one of de Mere's letters to Pascal he maintained that, "If in one case there is one chance from N_0 in a single attempt to succeed, and in another case there is one chance from N_1, then the ratio of corresponding figures is $N_0:N_1$. Thus, $n_0:N_0 = n_1:N_1$".

The notation and meaning forwarded in his statement require explanation. De Mere argues that if a single die is tossed there are $N_0 = 6$ different outcomes, only one of which corresponds to the appearance of a six. Furthermore, if two dice are tossed only one outcome of $N_1 = 36$ possible outcomes corresponds to appearance of sixes on both dice. So, if a single die is tossed $n_0 = 4$ times, the number of cases where a six appears exceeds the number of cases where a six does not appear at all. Let, as de Mere's rule suggests, n_1 represent the number of tosses of two dice where the appearance of double sixes exceeds the number of cases where double sixes do not appear at all. According to that rule, 24 tosses of two dice would be necessary to ensure that the probability of rolling double sixes is greater than the probability of not rolling double sixes.

In fact, de Mere's rule is wrong because the probability that after four tosses of a single die a six will not appear equals

$$(5:6)^4 = 625:1296.$$

And consequently, the desired probability equals

$$1 - 625:1296 = 671:1296.$$

De Mere's calculations for tossing a single die were correct but, for 24 tosses of two dice, the probability of not throwing double sixes equals

$$(35:36)^{24} = 0.509.$$

And, the probability for 24 tosses of throwing double sixes equals

$$1 - (35:36)^{24} = 0.491.$$

Clearly, 24 tosses are insufficient. A minimum of 25 tosses of two dice are necessary for the probability of the appearance of double sixes to be greater than 0.5.

Our explanation of de Mere's blunder requires the use of modern language and applies concepts of probability that were unfamiliar in his time. De Mere's approach, although revolutionary for its time, was restricted to the calculation of cases related to his gambling.

The bulk of Pascal and Fermat's letters were devoted to questions revolving around the sharing of stakes. The following solution suggested by Pascal was expressed in detail in his letter of July 29:

"I make the following approximate determination of the price per game in a game involving two gamblers where three wins are required and the stake per player is 32 pistoles:

"Assume that the first gambler has won two games and the second has won one. The final game has yet to be played. If the first player wins the final game, then he receives the entire sum of 64 pistoles. If, however, the game is won by the second player, then each gambler having 2 won games, would receive half of the entire stake or 32 pistoles apiece.

"Take into account, Monseigneur, that if the first player had won the final game, he would have received 64 pistoles; had he lost he would have received 32. But, what if the gamblers did not risk playing the final game, opting instead to share the stake? In this case the first might say, 'I am entitled to 32 pistoles, because if I lose I will win that much anyway. The remaining 32 pistoles could be won by either of us. The risk for you is the same whether we continue to play or not. So, let us divide these remaining 32 pistoles evenly. The first 32 pistoles I am guaranteed anyway.'"

In addition, Pascal considered cases where the first gambler won two games and the second won none and where the first won a single game and the second won none. In both cases the arguments were similar to one mentioned above. The answers suggested by Pascal were as follows: In the first of the two additional cases cited, the first gambler should get 56 pistoles and the second 8 pistoles; in the second case the corresponding figures are 44 pistoles to 20 pistoles.

The solution suggested by Fermat to Pascal's problem comes only through a second-hand account in Pascal's letter of August 24. Fermat's original letter was lost. Pascal's version is as follows : If gambler A needs to win two games to complete the series, and gambler B still needs to win three games, then in order to complete the series, at most four additional games must be played. The possible outcomes of those games are represented in the following table:

In this table, A is used to denote a game won by gambler A, and B is used for gambler B. The order in which the games occur are presented in a row. In the first 11 cases gambler A has won, in the last five cases gambler B has won. Thus, the stake must be shared between gamblers A and B in the proportion 11:5. In other words, gambler A has 11/16 and gambler B has 5/16 of the entire stake. It is quite obvious that Fermat believed, as did Pascal, that the stake should be

Table 15

Game	possible outcomes			
1	AAAA	ABAA	ABBA	BBBA
2	AAAB	BAAB	BABA	BBAB
3	AABA	BAAA	BBAA	BABB
4	AABB	ABAB		ABBB
				BBBB
Games won by gambler	A after 2 games	A after 4 games	A after 3 games	B after 3 or 4 games

shared proportionally to the probability of the gambler's chances of winning the entire series. But, since such a notion of probability did not exist at that time, both were forced to use other means of expressing their ideas. As a result, neither Pascal nor Fermat could understand that their findings were in fact identical. In a letter dated October 27 Pascal wrote, "Monseigneur, I was deeply satisfied by your last letter and find your method of stake sharing admirable. Moreover, I understand completely that this method is entirely yours, has nothing common with mine, yet achieves the same result as mine."

In his letter of August 24 Pascal doubted that Fermat's method could be applied to cases involving more than two players. However, Fermat demonstrated that his method could indeed be used successfully to solve problems involving three gamblers. He applied his solution to a scenario where the first gambler (A) wins one point, and second and third gamblers (B and C) win two points each. This solution was also accompanied by the following table which needs no further explanation:

In his letter Pascal noted that Roberval (1602–1675) asked him why it was necessary to consider a continuation of the game until the

Table 16

A A A A A A A A A B B B C C C B C	A B B B C	C C C C B
A A A B B B C C C A A A A A A C B	B B B C B	C C C B C
A B C A B C A B C A B C A B C A A	B A C B B	C A B C C
A A A A A A A A A A A A A A A A A	B B B B B	C C C C C

one of the gamblers won four points if it was clear already which of the two would win the stake. Pascal clearly understood that this was necessary in order to consider all of possibilities. For instance, in the first four cases gambler A won the stake after two victories. Analogously, in the first nine cases in the second table gambler A won after only one attempt. Nevertheless, Fermat completed the table and considered all possible outcomes. Through their comprehensive methodology both Pascal and Fermat avoided a mistake Jean Batiste D'Alembert made a century later when he analyzed a number of equally likely outcomes for a coin tossing game. Careful inspection of the second table indicates that Pascal's reasoning was inaccurate. He thought that of all of the 27 possible outcomes only 13 were undoubtedly favorable for gambler A. However, the outcomes which appear in rows 5, 11, and 19 favor both gamblers A and B. Similarly the outcomes in rows 9, 15, and 24 favor both gamblers A and C. Thus, Pascal assigned these outcomes with a value equal to 1/2. As result, Pascal proposed to share stakes in a proportion of (16):(5.5):(5.5). Pascal's error is obvious to us.

Pascal carried on his correspondence on probabilistic approaches while simultaneously developing combinatorial methods. He published the culmination of his efforts "*Trait'e du Triangle Arithmetique*" in 1665. This tract proved to be a serious contribution to the development of combinatorics and included a section demonstrating a rule for the application of combinatorial results to stake sharing problems. The rule suggested by Pascal is the following: If gambler A wins m points, and gambler B wins n points, then the stake is shared between the gamblers in the following proportion:

$$\frac{\sum_{i=0}^{n-1}\binom{m+n-1}{i}}{\sum_{i=0}^{m-1}\binom{m+n-1}{i}}.$$

E. Work by C. Huygens

Christian Huygens (1629–1695) substantially influenced probability theory. His interest in the field was aroused during a trip to Paris in 1655. There he and other prominent scientists discussed stake sharing problems involving incomplete games of chance. Conceivably, it was these discussions which familiarized Huygens with the problem-solving techniques of Pascal and Fermat.

In his tractate *De Ratiocinics in Ludo Aleae*, Huygens observed that since neither Pascal nor Fermat had published their methods, he was forced to develop his own. In 1656 Huygens' teacher Frantz Van Schooten published a special edition of *Mathematishe Etuden* containing Huygens' methods. Van Schooten was so impressed by Huygens' work that he personally translated it into Latin.

Huygens' work consists of a brief preface and 14 statements. The probability problems raised in his statements are quite varied. The first three statements provide background for the remaining 11. Statements four through nine are dedicated to the problem of fair stack sharing. Statements 10 through 14 are devoted to dice throwing problems. At the end of his tractate, Huygens leaves the reader with five problems to solve. He did not publish the solution to those problems until 1665.

The first three statements are the most basic, so we will present them in their entirety with limited explanation.

STATEMENT 1. If my chance of obtaining a or b is equal, then my share equals $(a+b)/2$.

STATEMENT 2. If I have an equal chance of obtaining a, b or c, then my share equals $(a+b+c)/3$.

STATEMENT 3. If the number of times I win a equals p, and the number of times I win b equals q, then my expected gain equals $(ap+bq)/(p+q)$.

For us, it is clear that Huygens introduced the concept of the mathematical expectation for a random variable with two or three values. Huygens did not introduce the concept of the porbability, but instead focused on the ratio of the number of chances to the possible outcomes of a trial. Note than Van Schooten introduced the term "expectation" in his translation of Huygens' work.

Statements 1 and 2 provide a variation on the solution of problems involving stake sharing. The fragment of Huygens' text we chose to provide will illustrate how reminiscent Huygens' findings are to those of Pascal:

"Assume that I'm playing a game of stakes with someone which will be own by the first of us to win three times.. Assume I win twice and my opponent–only once. I would like to know what portion of the entire stake is mine, if we interrupt the game and decide to divide the winnings.... First, we must take into account the number of wins necessary for each of us to complete the game. For instance, if we divide the stakes based on the first to win twenty games and I have won 19 games and my opponent has won 18, the situation is the same as in the previous case where I won 2 games and my opponent only one: In both cases I need to win one game and he needs to win two. Then, consider what would happen if we were to continue play and on the basis of the outcome evaluate how to share the stake fairly. If we continue to play and I win the next game, I take the entire stake equals *a*. But, if my opponent wins the next game, both of us would have an equal chance of winning the entire game. This means that, according to Statement 1, each of us is entitled to $a/2$. Finally, I should win the sum of the halves, i.e., $(3/4)\ a$. So, my opponent would win the remainder equals $(1/4)\ a$".

Huygens also shared the stake among three gamblers in accordance with Statement 8. He considered a case where the first gambler won a single game and the remaining two won two games each. In Statement 9 he considered a case where three gamblers won games arbitrarily. He did not suggest a general solution, but indicated how to apply a general case to the particular cases he had solved before.

Statements 10 to 14 are confusing. They become clear only after analyzing the examples provided by Huygens. Let us consider Statement 10 in detail.

STATEMENT 10. Find how many times does one needs to toss a die to throw a six.

This statement is very unclear. Without the foreknowledge of the principles of probability Huygens was unable to ask the right questions or solve appropriate problems! More specifically, the problem he considered was: What is the probability that in n throws a person will throw a six at least once?

Huygens' solution was the following: The gambler has one chance of throwing a six and five chances of throwing another number. If the stake equals *a*, then according to Statement 3, the winnings equal $(1 \cdot a + 5 \cdot 0/6) = a/6$.

"A person who opts to throw the die once will get $(5/6)\ a$. So, a fair bid for a gambler who throws a die only once is 1 to 5."

For two throws of a die, Huygens argued the following: "If a six is rolled on the first toss, the gambler has *a*, but he has only 1 chance in favor and 5 chances against. But according to this theory, the second throw will result in less desirable odds or $a/6$. It follows that the game should cost the gambler $(1 \cdot a + (5a/6))/6 = 11a/36$."

In the same manner Huygens determined that for three throws of a die the considered probability equals $(91/216)\ a$. Further he found the corresponding probabilities for 4, 5, and 6 throws. They are $(671/1296)\ a$, $(4651/7776)\ a$ and $(31031/46656)\ a$.

In Statement 11, Huygens considered the likelihood of rolling a pair of sixes. Because of the archaic and therefore cumbersome language of Huygens, we chose to present the following loose interpretation of Statement 11:

A gambler can gain a sum *a* if he has thrown two sixes with two dice. What fair price he should pay for participating in the game? Huygens calculated that for one throw the gambler should pay $a/36$, for two throws $(71/1296)\ a$, and so on. Huygens remarked, "I found that he who agrees to play 24 games still has a slight disadvantage, but can assume the game will be profitable if he agrees to 25 games."

STATEMENT 12. Determine the number of dice necessary to roll a double six on the first throw.

STATEMENT 13. Determine what share of the stake the person deserves in the following case: I will win if the sum of numbers on two dice equals 7, my opponent will win if the sum equals 10, and we share the stack equally otherwise.

STATEMENT 14. Another gambler and I throw two dice sequentially. I win if the sum of numbers I roll is equal to 7, and my opponents wins if his equals 6. He throws first. The problem is to determine what his chances are and what my chances are of winning.

For completion of the discussion, we chose to present all five of the exercises Huygens presented to the reader for solution without assistance.

1. A and B play with two dice. A wins if he obtains 6 points, and B wins if he obtains 7 points. A starts with a throw and then B throws the dice twice, then A throws twice and they continue to throw twice in a sequence, until one emerges victorious. How do the chances of A relate to those of B?

2. Three gamblers take 12 tokens, 4 white and 8 black. The first to choose a white token is the winner. Gambler A chooses a token at random first, then B, then C, and then again A, and so on. How do their chances of winning relate to each other?

3. Gambler A bets with Gambler B that from a pack of 40 cards (ten of each suit) he will choose four cards each of a different suit.

4. As in Problem 2, there are 12 tokens, 4 white and 8 black. Gambler *A* bets Gambler B that if he randomly selects 7 tokens three of them will be white. How do *A*'s chances of winning relate to *B*'s chances of winning?

5. Gamblers A and B have 12 coins each. They throw three dice. If the sum of the points on all three dice is 11, then *A* gives *B* a coin. If, however, the sum of the points on all three dice is 14 then, B gives a coin to *A*. (This problem is a variation of the problem of the gambler ruin considered in the main text of the book.)

Ten years after the death of the outstanding philosopher, Benedict Spinoza (1632–1677) in Hague, an anonymous work was published in two parts. The two parts, *Investigation of a Rainbow* and *Notes on Mathematical Probability*, were topically unrelated. Historical research has confirmed that these works were written by Spinoza. The second part of the work provided a solution to the first of Huygens' problems and formulations for four others. It is important to emphasize that although the word probability appeared in the title of his work, Spinoza afforded no definition of probability and limited his arguments to problems involving games of chance.

It is useful to note that in 1692 D. Arbuthnot (1667–1735) arranged the English translation of Huygens' book and added to this edition a set of new problems including one of a quite different nature. The formulation of the problem is as follows: a parallelepiped is thrown on the plane. The edges of this body relate to each other as a: b: c. What is the probability that the parallelepiped will fall on sides *ab*, *bc* or *ac*?

At the end of the seventeenth century volumes of knowledge on random events and a number of correctly formulated probabilistic problems and their solutions were collected. Many outstanding scientists from different positions attempted the quantitative evaluation of the probability of the appearance of a random event. Fermat borrowed the concept of mathematical expectation which was developed by Huygens. Pascal, Fermat, and Huygens used some ideas introduced through the fields of summation and multiplication of probabilities coming very close to the concept of probability, but still they did not introduce it. In retrospect, going one step beyond accounting for probability of the occurrence of a random event to accounting the relationship between the frequency of the occurrence of a random event and the total number of events seems obvious. That crucial step, however, eluded all of the great philosophers mentioned. This lack of foresight impeded progress with overly sophisticated arguments and formulations that were often inaccurate. Had scientists of the time asked themselves which was more probable: to roll a six in 4 throws of a die or to roll double sixes in 25 throws of two dice?–they would have stumbled upon the concept of the classical probability. It was not until the next century, however, that the classical concept of probability was defined. The seventeenth century was a gestation period in the maturation of probability theory into a special branch of mathematics.

F. First Research in Demography

This section will demonstrate how the works of John Graunt (1620 – 1675) and William Petty (1623–1687) catalyzed the development of probability theory. These works were dedicated to demography, or–as it was then referred to–political arithmetics. These books were widely distributed and had a great influence on various scientists including mathematicians. The first work which established mathematical statistics as a branch of mathematical knowledge was a book by John Graunt published in 1662 under the long title: *Natural and Political Observations Listed in the Attached Contents and Performed with Death Bulletins. Related to Jurisdiction, Religion, Trade, Growth, Air, Ills and Various Changes of a Specified Town.* Rather than offering a description of the book, we have chosen to highlight some of its pertinent ideas.

Graunt was primarily interested in finding an accurate method for determining the average age of town's population based on an analysis of the average age at death. To achieve his aim, Graunt analyzed 229, 250 death registrations for deaths which had occurred over the previous 20 years in London. Among these he registered the deaths of 71,124 children under the age of six. The causes of death were carefully investigated by Graunt. He calculated that the number of these deaths compared to the total number of deaths equaled (71,124) :

(229,259), or approximately 1/3. In other words, Graunt introduced the concept of the frequency of an event. For the development of probability theory, it was a very important step. He also made the significant remark: "... we would like to emphasize that some randomness is inherent in the ratio of children's deaths to the total number of deaths" (page 32). Here Graunt came very close to the discovery of the statistical stability of averages.

He established that the relationship of the number of boys' births to the number of girls' births was 14:13, that of 11 families observed on average 3 members died in a given year, that one out of 2,000 women studied died during childbirth, and that on average 63 deaths per year were offset by only 52 births. (Graunt's analysis of the disproportionate ratio of deaths to births in London proved that London's population growth could only be attributed to the migration of people from the countryside.) Through death registrations he learned that in London out of any 100 men 34 were aged from 16 to 56, i.e., out of 199,112 males 67,694 had the mentioned age.

He composed the first mortality table which demonstrated that out every 100 of the migrants to London, 64 would survive to the age of six, 40 would survive to the age of 16, 25 would survive to the age of 26, 16 would survive to the age of 36, 10 would survive to the age of 46, 6 would survive to the age of 56, 3 would survive to the age of 66, 1 would survive to the age of 76, and none would live to be 86 years old.

We find a surprisingly high death-rate among children and teenagers: only 64% survived to the age of 6 years and only 40% survived to the age of 16 years.

Graunt completely understood that the more observations he made the more accurate his results would be. He noted that weekly mortality reports would not suffice to ensure precise results.

The concept of event frequency was so useful that it was immediately utilized by other researchers. Petty published his small book *Two Reviews on Political Arithmetic Related to People, Buildings and Hospitals in London and Paris* in 1682 in London. Two years later the book was translated in Paris. Petty issued a comparative analysis of mortality rates of charitable hospitals in Paris and London. He observed that in one Parisian hospital 338 patients died out of the total hospital patient population of 2,647, and in two London hospitals 461 out of 3,281 patients died. The frequencies of death were 0.136 and 0.140 for Parisian and London hospitals, respectively. Petty did not use decimal fractions; instead he referred to his frequency rates in terms of simple fractions. One Parisian hospital, "L'hotel Dieu" had a particularly high death rate: 5,360 deaths per 21,591 patients. So, for this hospital the mortality frequency equaled (5,630) : (21,491) $\approx$ 0.262 which Petty referred to as 1/4.

In the same book Petty found that every year in London one out of 30 people died as compared to the countryside where the figure was one out of 37. At the same time only one of 50 members of Parliament died. Petty maintained that mortality bulletins could be used to calculate a town's population. For instance, in London 22,331 deaths had been registered. It followed that since the mortality coefficient for London equaled 1/30, the total population of Londoners should be close to 669,930.

Without a doubt, the works by Graunt, Petty and their followers were the earliest examples of mathematical statistics.

A direct disciple of the research begun by Graunt and Petty was famous English astronomer Edmond Halley (1656–1742). In 1693 he published two papers in the London Royal Society. They were *An Estimate of the Degrees of the Mortality of Mankind, Drawn from Curious Tables of Births and Funerals of the City of Breslaw, with an Attempt to Ascertain the Price of Annuities upon Lives* and *Some More Remarks Concerning the Breslaw Tables of Funerals.* On the basis of these papers the author accumulated data on the changing population of Breslau in 1687–1691.

Halley paid attention to mortality tables in part because the numerical results published by Graunt and Petty were unsuitable. They investigated the general populations of London and Dublin where migration was a significant factor. In his first paper Halley wrote that this fact made those two cities "... inappropriate for the task at hand which requires that the populations under consideration be closed, i.e., the population must be such that everybody born in the town dies there as well, no immigrants, no emigrants at all." Halley remarked that the population data from Breslau was more fitting.

Based on his data, Halley constructed a mortality table which was considered by him to serve either as survival table or as table of the age distribution of the population. He introduced the concept of probable length of life which could be exceeded or not reached with the same frequency. In modern terms, it is the median length of the life. (Halley himself introduced neither the term median, nor the term probable life length.) The principles which laid the basis for the summation and multiplication theorems of probability theory as well as arguments which are very close to the law of large numbers can be traced to his work. Halley's works played a very significant role in the development of the scientific and applied methods of statistical research used in demography and insurance.

PART 2 FORMATION OF THE FOUNDATION OF PROBABILITY THEORY

G. Appearance of the Classical Definition of Probability

Up to the end of the seventeenth century, the classical definition of probability still had not been introduced and mathematicians continued to operate with the number of chances. However, in the eighteenth century familiarity with the classical definition of probability became widespread and the simple calculation of chance obsolete. Who was the first to introduce this concept? Remember that improving on existing notions of probability was a slow process as was the transition from focusing on the particular cases to general concepts.

Huygens' *On Calculations in Games of Chances* (1637), lacked the understanding of probability as a number within the interval [0, 1] and equal to the ratio of the number of successes to the total number of trials. It was J. Bernoulli's tractate *Ars Conjectandi* (1713) that introduced the concept (though imperfectly) and led to its extensive use. What happened in the half century between the publication of these two books? What moved J. Bernoulli to embrace this new concept?

Certainly the formulation of the law of large numbers by J. Bernoulli compelled him to think in terms of probabilities. However, there was another factor which cannot be ignored–the tremendous influence that the works of Graunt and Petty had on all significant mathematicians of the time.

Works by Graunt and Petty certainly proved the advantages of including frequency in comparisons with the absolute number of chances. Petty and Graunt's research was only one step shy of defining classical probability. In addition, the fact that their work mentioned stability of frequency indicates that Bernoulli's introduction of the law of large numbers was also a continuation of the efforts of Petty and Graunt.

An imperfect definition of classical probability appeared in Chapter 1 of Part 4 of J. Bernoulli's tractate *The Art of Assumptions.* In that work he maintained, "A probability is like a measure of confidence and differs from it as a part of something differs from the whole." The following statement by Bernoulli confirms he understood the concept of probability much as we do today: "If a complete and unconditional confidence is denoted by α or 1, (it will be assumed to consist of five probabilities–or five parts–three of which are desirable for some event, existence or occurrence and two of which are not), then one says that this event has $3/5\alpha$ or 3/5 of confidence."

In Chapter 5 of Part 4 of *The Art of Assumptions,* J. Bernoulli again wrote about the ratio of the number of desired events to the total number of events. Though he did not expressly write that the occurrence of desirable events was equally probable for each trial, it is assumed that Bernoulli believed this to be the case. He also suggested a ratio for the number of desirable to undesirable cases. This measure has not survived the test of time, because it is a non-additive measure which could assume values from 0 to infinity.

J. Bernoulli expanded further upon and probably borrowed from the efforts of Graunt and Petty in Chapter 4 of Part 4 of *The Art of Assumption.* In Chapter 4 he asked and attempted to answer how to find the probability of the occurrence of a random event when it is impossible to count all of the desirable chances relating to that event. The solution he afforded was as follows: "Here another route is available to reach the desired result. What cannot be found *a priori* can be obtained *a posteriori,* i.e., by multiple observation of results of similar trials.... For instance, from observing three hundred people of the same age and of the same bodily constitution of Titus, it was determined that two hundred of them died after 10 years and the remainder survived. From this reliable data one can conclude that it is twice as likely for Titus to die within the next ten years as it is for him to survive.... This experimental method of determination based on the number of cases observed is not new or unusual." The fragments of Bernoulli's arguments cited introduce the concept of statistical probability.

J. Bernoulli's work reflected upon statistical and classical probability. Although neither concept was thoroughly investigated, the introduction and awareness of both were crucial to the future development of science: In addition, Bernoulli proved that there are two methods of determining probability: (1) by counting the number of chances, and/or (2) by performing large numbers of trials. Though the "prehistory period" laid the foundation, it is at this moment that the history of the probability theory as a mathematical discipline began!

According to J. Bernoulli it took him some 20 years to complete *The Art of Assumption.* It was not published until eight years after his death in 1731. The scientific community, however, was familiar with Bernoulli's manuscript prior to its publication, because Bernoulli continually shared his work with his contemporaries. Not surprisingly, Fontel (1705) and Soren (1706) paid homage to the great service of J. Bernoulli. P. Montmort (1687–1719) praised Bernoulli's publications in his book *Review of Hazard Game Analysis* (first edition in 1706 and second edition in 1713). In addition, Montmort subsequently published an analysis of the *Ars Conjectandi.* Thus, Bernoulli's tractate influenced the development of the probability theory long before its publication.

Montmort's *Review of Hazard Game Analysis* applied the concept of probability to difficult problems. In particular, Montmort considered and correctly solved the following problem: Let there be n things numbered from 1 to n. The question arises: What is the probability that through sequential random choice without substitution, at least one thing will be chosen in such a way that the number of trials coincides with the number of things? The probability of this occurring is equal to:

$$1-\frac{1}{2!}+\frac{1}{3!}-\cdots+(-1)^{n-1}\frac{1}{n!}.$$

We know that now this problem has various formulations.

DeMoivre accepted the classical definition of the porbability given by J. Bernoulli. He defined the probability of the occurrence of an event in almost modern terms. He wrote: "Hence we construct a fraction where the numerator is equal to the number of appearances of the event and the denominator is equal to the total number of cases when the event can or cannot happen. Such a fraction expresses the probability of the event happening."

DeMoivre supported his definition by succeeding it with Bernoulli's example mentioned above. He wrote: "If some event has 3 favorable chances and 2 undesirable chances of outcomes then the fraction 3/5 will characterize the probability of the event's occurrence and might be considered as its measure (*The Doctrine of Chances*). DeMoivre, like J. Bernoulli before him, did not emphasize that the chance of the desirable event occurring in different trials should be equally possible. It was Laplace who first drew attention to this phenomenon in his *Theorie Analytique des probabilities*. Laplace was probably influenced by D'Alembert's earlier suggestion that when tossing two coins the probability of them turning up heads on one and tail on another simultaneously equals 1/3. D'Alembert assumed that there are three possibilities: (1) both coins turn up heads; (2) both coins turn up tails; (3) one coin turns up heads and the other tails. His mistake is obvious for us because there are four outcomes for two coins: head-head, tail-tail, head-tail, and tail-head, i.e., the probability mentioned equals 1/2.

H. Formation of Geometric Probability

During the first half of the eighteenth century it was clear that the applications of the classical definition of probability were very limited. In its fledgling state it was inappropriate for many of the problems it attempted to address. The definition of classical probability needed to evolve beyond existing notions of the early eighteenth century. It is commonly believed that French researcher George Buffon (1707–1788) took the lead in expanding the role of probability theory through his exploration of geometric probability. He is often credited as the first to consider a now famous problem which involved throwing a needle on a lined plane. Perhaps this problem needs to be expanded upon. Buffon was trying to determine the probability of the needle landing on the lines (or between them) for varied proportions between length of the needle and the distance between the lines. Actually Buffon's work was not as revolutionary as many mathematical histories suggest. Long before Buffon, the analogous problem involving geometric probability was considered but at the time, however, concepts of probability had not been formulated.

In 1692 in London Huygens' *On Calculations for Games of Chances*" was translated by J. Arbuthnot (1667–1735). At the end of the book Arbuthnot added several new problems including one that was of a completely different nature than those of the author. (We briefly described this problem at the end of Section 5.) Arbuthnot formulated the following problem: One throws a parallelepiped (a solid with six faces, each of which is a parallelogram) with edges, a, b and c at a right angle, onto a plane. What is the probability that it will fall on the plane on one of its side's with edges ab, bc or ca?

Arbuthnot never attempted to solve his own problem. It was solved much later by T. Simpson (1710–1761) in his book *Nature and Laws of Random* (1740).

The solution suggested by Simpson was the following: Let us surround a parallelepiped with a sphere in such a way that all eight vertexes are placed on the surface of the sphere. Then let us project all edges from the center of the sphere. Then let us project all adges from the center of the sphere (the center of weight). As a result, the sphere will be divided into six non-intersected areas. Then Simpson wrote: "It is obvious that the part of the sphere specified by the mentioned way will relate to the total surface of the sphere as the probability that the parallelepiped falling on the corresponding side will relate to 1".

This statement includes all of the principles for determining geometrical probabilities: The measure of a set of desirable outcomes is introduced and then the ratio of this measure to the measure of the whole set is taken. In this case the total measure coincides with the square of the sphere's surface. Notice that Simpson said nothing about the physical interpretation of his solution.

Let us introduce for further explanation the notation that: $R^2 = a^2 + b^2 + c^2$, P_{ab}, P_{bc} and P_{ca} are the probabilities of the parallelepiped falling down on side ab, bc or ca respectively. These probabilities must be increased two times because of symmetry. The formulas for the notion are:

$$P_{ab} = \frac{1}{\pi}\,\text{arctg}\,\frac{ab}{cR} \quad P_{bc} = \frac{1}{\pi}\,\text{arctg}\,\frac{bc}{aR} \quad P_{ca} = \frac{1}{\pi}\,\text{arctg}\,\frac{ac}{bR}$$

Buffon twice published works dedicated to geometrical probabilities. His first publication in 1733 was based on his report at the Paris Academy of Sciences titled *Memoire sur le Jeu Franc-Careau.* Later this paper, *Essay d'une Arithmetique Morale,* was published in its entirety as a supplement to Volume 4 of Buffon's *History of Nature.*

He formulated his goal to demonstrate that, "Geometry can be used as an analytical tool to solve probability problems." Previously he noted, "Geometry was rarely applied to such problems, because everybody relied exclusively on arithmetic."

The Franc-Careau game can be described as follows: The floor is divided into identically shaped figures. A coin is thrown with a diameter $2r$, which is less than that of the smallest of the small figures on the floor, i.e., the coin could land just within the parameters of one of the figures. What is the probability that the coin will intersect at least one of the figure's sides?

For accuracy, consider a rectangle on a plane with edges a and b such that min $(a, b) < 2r$. Inside each rectangle let us construct a smaller inner rectangle such that, if the center of a coin is within the inner rectangle, then the coin will not touch the edges of the initial rectangle. The square of this separating "strip" equals $2r(a+b-2r)$. If the center of a coin is located within this strip, it crosses the edge of the initial rectangular. So, the probability that the coin will intersect at least one of the above-mentioned edges equals $2r(a+b-2r)/ab$.

The second of Buffon's problems was: A plane is divided by parallel lines of equal distance. A needle is thrown randomly onto the plane. Two players bet on where the needle will fall. Player A bets that when the needle falls it will intersect the lines of the plane. Player B bets that it will land between the lines. Find the probability of either player winning. The solution to this problem is well known and no requires no further explanation. Less known is Buffon's problem involving throwing a needle onto a plane divided into squares. Buffon himself erroneously solved this problem only to be corrected later by Laplace. Buffon maintained that the desired probability equaled $2r(a-r)/\pi a$ but the correct answer is $4r(2a-r)/\pi a^2$. Thanks to Buffon's efforts, problems concerning geometrical probabilities began systematically to appear in tractates and textbooks on probability theory. Laplace in his famous book *Theorie Analytique of probabilities* included all of Buffon's problems and considered each in detail. Unfortunately, he never cited Buffon as the source.

It is important to remember Laplace's terminology was far from perfect. For instance, he wrote that "$8r$ is equal to the sum of all cases where the needle crosses one or more parallel lines" and that "$2a\pi$ is equal to the number of all possible combinations." Here $2r$ represents the length of the needle and a the distance between the parallel lines.

In the second problem considered by Laplace, a plane is divided through two systems of parallel lines. In the first system the distance between lines is a and in the second, b. A needle of length $2r$, $2r <$ min (a, b), is thrown onto the plane. The question arises: What is the probability that the needle will intersect at least one line? The solution is derived through the assumption that lines of one system are perpendicular to lines of the other. (Laplace did not stipulate this assumption). As a result of calculating the "number of desirable cases" to the "total number of all possible cases," Laplace concluded that the probability considered was $4r(a+b-r)/\pi ab$.

Outstanding for its time was the textbook by V. Bounyakovsky (1804–1889) *Fundamentals of the Mathematical Theory of Probabilities* (1846). It included a large section dedicated to geometrical probabilities. Included in that section was Buffon's problem concerning needle throwing and a particular variant of the Franc-Careau game where a plane is divided by isosceles triangles. In retrospect, the terminology used by Bounyakovsky is flawed. Misused language is apparent in the following excerpt from this book: "Situations are encountered where the numbers of desirable cases and possibilities are infinite. In such cases the investigated probability is defined as the ratio of these two infinite numbers. The resulting ratio is a finite and completely defined number."

Significant progress in development of geometric probabilities was made by G. Lame (1795–1870), J. Barbier (1797–1856), D. Silvester (1814–1897), and M. Crofton (1814–1874). These men not only solved new problems but employed the concept of a measured set at an intuitive level. Later, on the basis of their analysis integral geometry was founded.

In 1860 while on the faculty of the Paris Normal School, Lame presented a series of lectures on geometry. His lectures considered Buffon's problem and applied it to a case where the needle is thrown randomly at the center of an ellipse or a symmetrical polyhedron. One of Lame's students, Barbier, applied this approach to an arbitrary convex contour. He brought nothing new to the problem other than demonstrating that Lame's method had no specific restrictions.

Silvester was first to expand upon Buffon's problems. He suggested a problem involving four points which was later referred to as Silvester's problem. Its formulation is the following: Four points are taken randomly from a plane. What is the probability that if all of these points are used as vertexes a convex quadrangle will result?

Silvester suggested the following solution: Let A denote the area of a convex figure. Throw four points at random at the convex figure, choose three of them arbitrarily, and construct a triangle. Let the area of the triangle equal M. Now, consider the remaining fourth point. If this point is located inside the triangle (the probability of this equals M/A), then it is impossible to construct a convex quadrangle using those four points. Three of the four points used to construct the triangle, however, can be chosen in four different ways. So, if one throws four points the probability of creating a non-convex quadrangle equals $p = 4M/A$. Thus, the probability of creating a convex quadrangle by the same method equals $1 - 4M/A$. The mean size of M depends on the shape of the figure into which the points are thrown. For some convex figures this value is calculated. M. Crofton in his paper *Probability* published in the Encyclopedia Britannica (9th edition, vol. 19, p. 786, Edinburgh, 1885) constructed a table referring to the work of Wallhouse.

Silvester was acutely aware that for computation of geometrical probabilities it was necessary to compare the ratio of the areas or volumes (common measures) of the shapes which corresponded to the occurrence of a desirable event to the measure of the entire shape where all of possible events might happen. The problem Silvester solved was not unique, however, he conveyed his findings with a clarity that had eluded his predecessors. Silvester suggested utilizing his procedure to find those shapes that were most likely to result in the formation of a convex quadrangle. The first to utilize Silvester's suggestion successfully was Crofton. He published his findings in his paper *Probability*. He proved that the circle was an optimal shape and assumed the same to be true for an ellipse. His assumption was later proven to be true by V. Blaschke (*Vorlesungen uber Differential Geometrie*, Berlin, 1923). Deltail proved that a triangle was the optimal shape for forming a quadrangle according to Silvester's method. Problems calculating the likelihood of rendezvous are common in mathematics. My pupil M. T. Lorino Peres traced the origin of such problems to Whiteworth's book *Choice and Chance* (London, 1886). Whiteworth presented the following problem: Persons A and B go separately to a reception. Each of them must attend the reception for an hour. Person A arbitrarily chooses his arrival time to be between 3

Table 17

prob.	triangle	rectangular	hexagon	circle
p	$1/3 \approx 0.3333$	$11/36 \approx 0.3056$	$189/972 \approx 0.2971$	$35/(12\varsigma^2) \approx 0.2955$

and 5 p.m. Person B randomly selects his arrival time to be between 4 and 7 p.m. What is the probability that the two will bump into each other for at least a moment?

Whiteworth solved the problem in a common way. It is easy to find that the answer is 1/3. This problem often resurfaced in other later books.

Doubtlessly, Crofton influenced the development of the geometric probability in the nineteenth century. He began to study intersections of randomly positioned lines with different specified convex figures. We will not recall his results because they have already been widely published in books on integral geometry and geometry probability.

The book *Calcul de Probabilite* (Paris, 1899) by Joseph L. F. Bertrand (1822–1900) distinctly illustrated the deficiencies of geometrical probabilities. Speculating with vague and uncertain language, he showed that a slight variation in the understanding of conditions surrounding a problem could lead to different results to the same problem. He focused on the well-known problem of randomly charting a chord in a circle. Bertrand's criticism of the reliability of results forced mathematicians to question the logical foundation upon which probability theory was based.

In the twentieth century, interest in geometrical probabilities intensified, because of their applicability to physics, biology, medicine, engineering, and other areas of human activity.

I. Basic Theorems of Probability Theory

Now we will investigate who conceived of the essential theorems of probability theory and the time frame during which these developments took place. It is the theorems of summation and multiplication and the theorem of complete probability which made possible the development of probability as we know it today.

These theorems were not conceived of in the correspondences between Fermat and Pascal, nor in Huygens' tractate. Yet, embryonic forms of these theorems can be traced to the early evolutionary stages of probability theory.

Pascal's works clearly demonstrate that he understood implicitly how to compute the number of desirable chances of event A happening, if there were known chances of disjoint events A_j occurring which when combined would result in the creation of the entire event A. Though he had yet to formulate summation theory, Pascal was moving in the right direction. Pascal's stake sharing problems were argued in the following manner: Gambler A needs to win 3 games to win the competition and gambler B 4 games. So, six games are sufficient to complete the competition. Thus, A's chances of winning equal

$$\binom{6}{6}+\binom{6}{5}+\binom{6}{4}+\binom{6}{3}=1+6+15+20=42$$

Pascal applied the same arguments to a general case where one gambler required n games to win and another needed m games.

In works by Jacob and Nicholas Bernoulli a clear formulation for the computation of the probability of complimentary events was derived for cases where the probability of a primary event is known.

Developing formulas which were later referred to as Bernoulli's Formulas, Jacob Bernoulli effectively employed the rules of summation and multiplication of probabilities, but he did not formulate them himself. He assumed their formulation was too obvious.

Proof that J. Bernoulli clearly understood the specifics of the theorem of summation of probabilities for disjointed events is evident from his remark: "Two men sentenced to the death are forced to throw dice. The one who rolls less points will be executed. The one who rolls more points will be set free. If the number of points rolled are equal, both of them will be set free. We can derive that the likelihood of survival for one of them equals: 7/12.... But based on this statement it is erroneous to assume that the likelihood of survival for the other prisoner is 5/12, since it is obvious that both fates are similar. If the chance of survival for both prisoners is 7/12 respectively, then their combined survival rates would be 7/6, i.e., more than one life. The explanation of this erroneous statement lies in the fact that there are several situations where both of them will survive, but there is no situation where both will die." J. Bernoulli was very close to formulating the important equation which is expressed in modern terms as:

$$P\{A\cup B\}=P\{A\}+P\{B\}-P\{A\cap B\}.$$

In Cardano's *De Lude Aleae* in Chapter 14 "*On Joining Points*" he expressed a similar result in terms of chances rather than probabilities. He considered the number of cases where at least one die shows 1 point. He proved that this number equals 11, because the first die shows one point in 6 cases: (1, 1), (1, 2), (1, 3), (1, 4), (1, 5), (1, 6). The number would be the same for the second die in the same number of cases with the exception of (1, 1) which happens twice.

Though their efforts were important it was not Cardano, Pascal or Fermat who wrote the first and final formulation of the theorem of summation of probabilities. It was Thomas Bayes (1702–1761) in his long titled work *Experience of Solution of Problems of the Probability Theory by Deceased Honorable Mister Bayes, Member of the Royal Society. Reported by Mister Price in His Letter to John Kenton, Magister of Art, Member of the Royal Society.* Bayes' work was read at a meeting of the London Royal Society on December 27, 1763, two years after the author's death. Definition 1 of the work contains the definition of disjointed events. Bayes used the term "inconsistent events." Bayes wrote, "If several events are inconsistent, then the occurrence of one of them excludes the possibility of the occurrence of others." His formulation of the theorem of summation of probabilities in Statement 1 read, "If several events are inconsistent, then the probability of the occurrence of one of them is equal to the sum of the probabilities of occurrence of each of them." This formulation is almost indistinguishable from its contemporary equivalent.

In the same way, the theorem of multiplication of probabilities was derived on the basis of multiple examples of particular cases and on the computation of the number of chances of the simultaneous occurrence of two or more events. Almost all of J. Bernoulli's predecessors had written similar computations. He relied on their computations to deduce his famous formulas. Montmort widely employed the rules of summation and multiplication for probabilities. However, it was DeMoivre who first formulated the authoritative definition of multiplication for probabilities. In his introduction to *Doctrine of Chances* DeMoivre presented the important independent property of random events. He worte: "We will refer to two events as independent, if one of them has no relation to the other, and occurrence of one of them has no influence on occurrence of the other." He went on to provide a definition for dependent events: "Two events are dependent if they are tied to each other and if the probability of the occurrence of one of them influences the probability of the occurrence of the other."

DeMoivre supported his statements with this simple example: Let there be two packs of playing cards of a single suit two through ace. Then, the probability that an ace is picked at random first from one of the packs and then from the other equals $1/13 \times 1/13 = 1/169$. In this case each time an ace is picked it is an independent event. If one uses only one pack of cards and randomly picks up an ace followed by a two, then the probability of the first event occurring equals 1/13 and the probability of the second event occurring equals 1/12. Thus, the probability of picking an ace followed by a two equals $1/13 \times 1/12 = 1/156$.

The following statement by DeMoivre is of particular importance: "The probability of the occurrence of two dependent events equals the product of the probability of occurrence of one of them divided by the probability that the other must occur if the first one has already occurred. This rule can be generalized for several events."

Thus, DeMoivre deserves credit for the formulation of the multiplication theorem. The theorem was published for the first time in 1718 in his *Doctrine of Chances.*

Of the occurrence of several events DeMoivre wrote, "One needs to choose one of them as the first, another as the second, and so on. Then, the occurrence of the first event must be considered as independent from all others; the second must be considered under the condition that the first one has happened; the third one under the condition that both first and second have occurred, and so on. Hence, the probability of the occurrences of all of events is equal to the product of all of the probabilities." DeMoivre soon discovered that identifying conditional probabilities, as a rule, was no simple task.

DeMoivre demonstrated his statement by solving some problems. Let us consider one of them. Let events A, B and C be totally independent of one another. Let x, y and z be the probabilities of their occurrence. Then, xyz is the probability of the occurrence of all three events and $1-(1-x)(1-y)(1-z)$ is the probability of occurrence at least one of them.

Bayes' work is reminiscent of DeMoivre's formulation of the theorem of multiplication. Bayes wrote, "The probability that two mutually tied events will occur can be determined by multiplying the probability of the first event occurring by the probability of the second event occurring with the stipulation that the first event has already occurred." Since DeMoivre's work was widely known, Bayes must have borrowed his ideas from DeMoivre. Bayes did go beyond DeMoivre in one area, however, namely the formulation of the condition probability $P\{B|A\}$ using probabilities $P\{AB\}$ and $P\{A\}$. On the basis of that statement many scientists credit him as the creator of the theorem which bears his name, Bayes Theorem, the theorem for the formulation of probabilities. The honor is, however, not completely his, because Bayes was still not able to formulate complete probabilities.

It was, however, Laplace who developed the majority of what is to this day referred to as Bayes Theorem. In the chapter "General Principals of Probability Theory" of his *Experience in Philosophy of the Probability Theory*, Laplace formulated Principle VI. Principle VI relates to the probability of hypothesis or, as Laplace wrote, the probability of causes. Let event A happen in conjunction with one of n disjointed events B_1, B_2, or ..., B_n and only in conjunction with that disjointed event. (Laplace called the latter events causes). The question arises: If event A is known to have occurred, what is the probability that B_j has also occurred? Laplace presented the answer as follows: "The probability of the existence of one of these causes equals a fraction with a numerator equal to the probability of the event which follows this cause and denominator which is the sum of similar probabilities related to all of the possible causes. If these different causes *a priori* are not equally probable, then instead of taking the probability of the event following each cause, take the product of this probability times the probability of the cause." In essence LaPlace verbally expressed the formula for "Bayes Rule."

$$P\{B_j - A\} = \frac{P\{B_j\}.P\{A - B_j\}}{\sum_{1 \leq j \leq n} P\{B_j\}.P\{A - B_j\}}.$$

In addition, Laplace's principle contains the formula of complete probability widely used by researchers in the field of probability theory from the beginning of the eighteenth century. The principle of the formula of complete probability was used, though never formulated.

Perfection of the principles of probability theory was a long process. Principles were correctly applied for years, but it took almost a century to introduce a definitive rule governing probabilities. Some useful accompanying concepts–such as an understanding of independent and disjoint events–were formulated during this long difficult process of development.

J. A Problem of the Gambler's Ruin

The problem of the gambler ruin played a key role in probability theory evolution. It was extraordinarily complex for its time and depended on the invention of new approaches for its solution. This problem's importance was not limited to probability theory alone, but also served as a springboard for the development of stochastic processes. The gambler ruin problem was first formulated by Huygens in his book *On Calculations in Gambling* (see Section E of Chapter 1 of this essay). It was a problem that intrigued and challenged many outstanding scientists of the past including J. Bernoulli, N. Bernoulli, DeMoivre, and Laplace. J. Bernoulli even criticized Huygens' numerical formulation and solution of the problem, arguing that it restricted the possibility of finding a general rule.

First attempts to solve the problem of gambler ruin were undertaken almost simultaneously by three mathematicians: P. Montmort (1687–1719), A. DeMoivre and N. Bernoulli (1687–1759). They obtained their results independently in 1710–1711. This slight variation of Huygens' solution is what is commonly accepted today: Gamblers A and B have a and b francs, respectively. After each game one of them wins 1 franc from the other. The probability of gambler A winning equals p, and similarly the probability of gambler B winning equals $q = 1 - p$. The question arises: What is the probability p_A of A's win and p_B of B's win, i.e., the probability of one of the gamblers totally ruining the other?

DeMoivre published his results in the journal "*Philosophical Transaction*" in 1711. He found that

$$p_A = \frac{\left(\dfrac{q}{p}\right)^a - 1}{\left(\dfrac{q}{p}\right)^{a+b} - 1},$$

and

$$p_B = \frac{\left(\dfrac{p}{q}\right)^b - 1}{\left(\dfrac{p}{q}\right)^{a+b} - 1},$$

Furthermore, the mathematical expectation of the number of games N which are necessary for the complete ruin of one of the gamblers equals

$$E\{N\} = \frac{bp_A - ap_B}{p - q}.$$

He also found the probabilities $p_{A,n}$ and $p_{B,n}$ of gambler A (or respectively, gambler B) winning. In modern terms the results are expressed as

$$p_{a,n} = \sum_t \Big\{ p^{ts-b} q^{ts} \sum_i \binom{n}{i} (p^{n-b-2ts-i} q^i - q^{n-s-2ts-i} p^i) \Big\} - \sum_t \Big\{ p^{ts-i} q^{ts+i} \times \sum_i \binom{n}{i} (p^{n-b-2ts-2a-i} q^i - q^{n-b-2ts-2a-i} p^i) \Big\}.$$

Here the notation $s = a + b$ is used and summation is taken for all non-negatives t.

In addition, he considered a case where a is infinite.

In 1710 Montmort discovered formulas for $p_{A,n}$ and $p_{B,n}$ with $p = q$. He sent his findings to Johann Bernoulli[5] who forwarded the letter to his nephew, Nicholas. Nicholas's response of February 26, 1711, considered the solution for $p \neq q$. This letter was published by Montmort in 1713 in his tractate *Essay D'Analyse Jeux Asard.*

J. Bernoulli also considered the problem of the gambler ruin both in a particular case ($a = b = 2$) and a general one. While solving the problem, he followed Huygens' approach and obtained far-reaching results (for probabilities p_A and p_B).

The results of J. Bernoulli, N. Bernoulli, Montmort, and DeMoivre clearly demonstrate that they had at their disposal methods which enabled them to manipulate the probabilities of complex events. With few exceptions they correctly applied the theorems of summation and multiplication of probabilities and the formula of complete probability, in spite of the fact that those theorems had yet to be formulated.

K. Appearance of Limit Theorems

J. Bernoulli's law of large numbers generated a number of improvements to the development of probability theory in and beyond the seventeenth century. He chose to go beyond analyzing only exact

[5]In this text the reader meets a number of famous Bernoullis: Daniel, Jacob, Johann and Nicholas. Johann Bernoulli' is mentioned only once. In other contexts J. Bernoulli refers to Jacob Bernoulli.–*I.U* and *S.W*.

solutions to probabilistic problems. In developing his theorem, J. Bernoulli questioned the asymptotic behavior of solutions subjected to the infinite growth of some parameter.

J. Bernoulli wrote a formula for his theorem which differs slightly from what is now referred to as the law of large numbers. Because his calculations were based on demography, he used the term "fertile" to indicate the occurrence of an event and the term sterile to indicate an event had not taken place. The original theorem which J. Bernoulli spent 20 years advancing is the following:

"Let the number of fertile cases relate to the number of sterile cases as r/s, and this number relate to the entire number of cases[6] as $r/(s+r)$, or as r/t. Hence the latter ratio lies between $(r-1)/t$ and $(r-1)/t$. It is necessary to prove that it is possible to produce a large enough number of trials to ensure that the ratio of the number of occurred fertile cases to the total number of cases is larger than $(r-1)/t$ and smaller than $(r-1)/t$".

It is quite clear that J. Bernoulli's formulation differs from the modern one only in terms of choice of terminology.

We already mentioned that J. Bernoulli's book *Art of Assumptions* was well known to his contemporaries long before it was published. It was studied with particular care by N. Bernoulli. N. Bernoulli defended his own dissertation titled *On Application of the Art of Assumptions to Juridical Problems* in 1709. In Chapter 2 *"On the Method of Calculation of the Probability of Human Life"*, based on Graunt's tables, he analyzed the probability of life span exceeding some limit. While conducting his research N. Bernoulli examined large numbers of records. This led to his unexpected discovery that boys were born more frequently than girls. He found that the number of male births related to the number of female births as 18:17, i.e., $p_{boy} \approx 0.514$ and $p_{girl} \approx 0.486$.

N. Bernoulli considered an example with 14,000 births. According to J. Bernoulli's formulas the following inequality was valid (here m is the actual number of boys birth):

$$P\{1\mu - 7200/ < 163\} = P\{7037 < \mu < 7363\}$$
$$= \sum_{i=7038}^{7362} \binom{14000}{i} p^i q^{14000-i}.$$

[6] J. Bernoulli found it redundant to mention that $t = r + s$. Notice that r, s, and t are not fixed, the only condition taken into consideration is that r/t equals a given value. From here it follows that $1/t$ can be assumed to be arbitrarily small. – *B.G.*

The actual number of boys' births was the result of random causes. The formula proved the solution, but the calculations required were very complex.

Exactly the same example was considered by Laplace in his *Analytical Probability Theory*. The researched value of the probability of inequality $7037 < \mu < 7363$ he found was 0.994303.

In the two last editions of his book *Doctrine of Chances*, DeMoivre added a translation of his paper *Approximation ad summum terminonum Binomii* $(a+b)^n$ *in serien-expansis.* Of his translation he wrote: "I added a translation of my work written on November 12, 1733. At that time the work was extended to some my friends but has never been published." In a brief introduction DeMoivre cautioned that for some probabilistic problems it was necessary to calculate the following types of sums:

$$\sum_{l \leqslant m \leqslant k} P_n(m)$$

where $P_n(m)$ are components of a binomial distribution. These numerical calculations proved to be very difficult if the number of trials was large. So, DeMoivre constructed an asymptotic expression, and successfully solved the problem. The main calculation difficulty was estimating factorial $m!$ for large m. DeMoivre proved that

$$m! \approx B\sqrt{m}e^{-m}m^m$$

where B is a constant. To find this constant, the following set was constructed

$$\ln B = 1 - \frac{1}{12} + \frac{1}{360} - \frac{1}{1260} + \frac{1}{1680} - \ldots.$$

DeMoivre found that this constant B approximately equaled 2.5074. But he was unsatisfied with this, because it failed to tie his constant to some constant previously introduced in mathematics. He solicited help from James Stirling (1692–1770). Stirling successfully solved the problem and demonstrated that $B = \sqrt{2\pi} \approx 2.506628\ldots$. It should be stressed that the Stirling formula should be more aptly referred to as the DeMoivre formula, or at least the DeMoivre-Stirling formula. In addition, DeMoivre was the first to calculate and publish a table for the function $\ln n!$ for n from 10 through 900. Using the "Stirling" formula, DeMoivre began by proving that for $p=q=0.5$ the middle term of binomial $(1/2+1/2)^n$ asymptotically equals $1/\sqrt{2n\pi pq}$. Then, in his *Doctrine of Chances* DeMoivre went on to prove the Local Theorem (later named after him). DeMoivre naturally began with the case $p=q=1/2$, because this case is the most important to simple problems of demography. Later DeMoivre developed the more contemporary and accurate Local Theorem for $p \neq 0.5$.

Employing the Local Theorem, DeMoivre proved without difficulty the Integral Theorem, though only within symmetrical bounds. However, the Integral Theorem for symmetrical bounds is easily generalized for use with cases involving asymmetrical bounds. He emphasized the importance of the expression $\sqrt{npq}$ for his new theory and even suggested a special name for it–module.

Using the Newton-Cotes method of approximate integration, DeMoivre calculated $p=q=0.5$ as the probability

$$p\left\{\frac{1}{2}n-\sqrt{n}<\mu<\frac{1}{2}n+\sqrt{n}\right\}.$$

According to his calculations this value equals 0.95428. A comparison of his results to the accurate contemporary tables provided reveals that his error was negligible. (tabulated value equals 0.95450). In addition he calculated the probability

$$p\left\{\frac{1}{2}n-\frac{3}{2}\sqrt{n}<\mu<\frac{1}{2}n+\frac{3}{2}\sqrt{n}\right\}.$$

His result is 0.99874 (accurate one is 0.99731).

DeMoivre noticed that the Integral Theorem could be used to estimate an unknown probability p, i.e., to solve an inverse problem relating to mathematical statistics.

L. Quality Control of Mass Production

With the development of mass production, interest in problems of quality control strongly increased during the last 50–60 years. The theory of statistical methods of quality control, deep in its theory and very important in its applications, appeared. This applied direction is strongly related to probability theory.

The first step in this direction was taken by T. Simpson in his book "*Nature and Laws of Chance*" (1740). The formulation of the problem is the following. There are a given number of things of different quality: n_1 of the first type, n_2 of the second type, and so on. At random one chooses m things. Find the probability that one picks m_1 things of the first type, m_2 things of the second type, and so on. Now this problem is hardly difficult for a student, but at the time it was a subject for a serious scientific tractate.

A century later, M. V. Ostrogradsky (1801–1862) returned to this problem in his work "*On a Problem Related to Probability Theory*" (1846). This work is not interesting in a mathematical sense but its deep understanding of a practical problem attracts our attention. Probably, in this aspect Ostrogradsky is more accomplished than other researchers. Simpson never made practical conclusions from his work but Ostrogradsky even calculated tables for practical applications. Let us cite his original words.

"There are white and black balls in the bowl. We know the total number of balls but don't know how many of them are of each color. We extract several of them at random then return them to the bowl. Find the probability that the total number of white balls does not exceed a previously specified level. Stated more clearly, try to find the dependence between this probability and the limits mentioned above.

"To understand the importance of the problem, let us put ourselves into the position of a person who must get a large number of the items while satisfying some conditions and who must spend some time checking these conditions. Such problems always arise before military providers. For them, the balls in the bowl represent delivered things–white balls, for example, the things with satisfactory properties, and black balls, inappropriate things.

"Thus if the problem formulated above would be solved, the provider could use this to decrease the volume of tiresome and difficult mechanical work, for instance, checking a large number of flour sacks."

Thus one needs to estimate the content of the bowl by a random sample. For this purpose Ostrogradsky used the Bayes' Formula. However his arguments have a serious deficiency: He suggested that all of n possible numbers of balls of one color are equally probable, i.e., there are $n+1$ situations: 0 white balls, 1 white ball,..., n white balls and all of them happen with the probability $1/n+1$. This assumption hardly could be met in practice. More realistic is the assumption which is used in problems of quality control: each item might fail with probability p. For a well-organized production, this probability should be very small and practically stable over time. In this case the probability to find m failed items from among n is described by the Bernoulli Formula

$$P_n(m) = \binom{n}{m} p^m (1-p)^{n-m}.$$

Methods of statistical quality control became especially important during World War II because many mass production items cannot be checked without destroying them (for instance, shells and detonators). And even in common cases total control was impossible because it demanded too much labor under wartime restrictions.

It is almost impossible to list those who have worked in this area. Among Soviet scientists we must mention A. N. Kolmogorov, V. I. Romanovsky (1879–1954), S. H. Sirazhdinov (1921–1988), and Yu. K. Belyaev (1932–).

PART 3 FORMATION OF THE CONCEPT OF RANDOM VARIABLE

M. Development of the Theory of Errors of Measurement

We already know from Chapter 1 that Galileo Galilei put forth the fundamentals of the theory of errors of measurement and introduced a set of concepts which are important up to our day.

Later, due to the needs of astronomy and geodesy, interest in the theory of measurement errors increased even more. The famous astronomer Ticho Brahe (1546–1601) found that each measurement has a possible error and the accuracy of the measurement could increase if one took several measurements and calculated their arithmetic average. Indeed, the idea of using the arithmetical average to make measurement more exact had been in practice long before Brahe's time. By some sources (see L. E. Maistrov "*Development of the Concept of Probability*"), this idea was used in ancient India for measurement of earthworks.

It seems that we should expect more attention to the theory of measurement errors from Johann Kepler (1571–1630) who contributed to much to the formation of the laws of the planets' movement. He mentioned once that a good observer made measurements with small errors: "Thank God Who gave us in the person of Ticho such an accurate observer that an error in eight minutes is impossible. Let us thank God and use this gain. These eight minutes which we cannot neglect will give us the possibility to reformat entire astronomy."

The first attempts to construct the mathematical theory of measurement errors were made by R. Cotes (1682–1716), T. Simpson (1710–1761), and D. Bernoulli (1700–1782).

Assumptions about the laws of measurement errors distribution made by different authors were quite different. Cotes believed that errors were distributed uniformly within the interval $(-a, a)$. Simpson thought that small errors happened more frequently than large ones which, in turn, were restricted by some specified number a. Simpson thought that errors distributed by a triangle distribution. Its density equals 0 on $[-\infty, -a]$ and $[a, \infty]$. The equation describing the density on $(-a, 0)$ is $x - 2a^2y = -a$ and, finally, on $(0, a)$ the equation is $x + 2a^2y = a$.

Simpson for his distribution proved that the average gives a better estimate than each single measurement. He evaluated this result and published it in his work "*On The Advantage of Averaging of Several Observations in Practical Astronomy*" (1755).

D. Bernoulli's work, "*Most Probable Determination by Several Unequal Observations and Inferences From This the Most Likely Conclusion*" published in 1778 by the Sanct Petersburg Academy of Sciences, attracts significant interest. This work is interesting because it was the first attempt to estimate an unknown parameter by the method of maximal likelihood. D. Bernoulli began his paper with doubts about the expediency of the commonly used method of arithmetical average. As a density function he took $y = \sqrt{R^2 - (x - \bar{x})^2}$ where R is known and $\bar{x}$ must be determined by results of observations. Notice that D. Bernoulli did not pay attention to the fact that the integral from the chosen function equals $(\pi/2)R^2$, and, consequently, can represent the probability density function only for one particular value of R.

For this work of D. Bernoulli, the commentary was written by L. Euler (1707–1783) in which the latter, first, criticized the method of maximal likelihood (of course, there was no such term at the time) and, second, suggested excluding observations far from the average because they are slightly probable.

It is necessary to mention works by I. Lambert (1728–1777) who in two papers of 1760 and 1765 exposed goals of the theory of measurement errors (incidently, the term itself belongs to him), pro- perties of errors, estimating the accuracy the observations and rules of approximation of discrete observations by continuous curves. Later the work by J. Lagrange (1736–1813) dedicated to the role of the mean for estimating the real value of a measured value was published.

P. Laplace (1749–1827) considered probability theory as a discipline of natural science rather than of mathematics. In connection with astronomy, he inevitably had to pay attention to the theory of measurement errors and then became interested in probability theory. Indeed, Laplace got some important results in the theory of measurement errors which became a solid part of practical approaches. We will not describe them because our goal is to emphasize his influence on the development of probability theory. In this context two of Laplace's works are of great interest. One of them generated an interest in the limit theorems on summation of independent random variables. Namely, Laplace claimed that an observed error of measurement is a consequence of the summation of a very large number of elementary errors. If these errors are uniformly small then their sum should have a normal distribution.

His second idea concerns estimating observed value by the results of measurement $x_1, x_2, \ldots, x_n$. As the estimate of an unknown observed value of a, Laplace suggested taking $\hat{a} = a(x_1, \ldots, x_n)$ which minimizes the sum $\sum_{1 \leq k \leq n} |x_k - a|$. In this case $\hat{a}$ equals the empirical median, i.e., such x that equal numbers of observed values lie on the left and on the right of this value. This approach was not widely distributed at the time because very soon another method was suggested which led to simpler results. The new method was developed by K. Gauss (1777–1855), Legendre (1752–1833) and American mathematician R. Adrain (1775–1843). Their works formed a real epoch in the theory of measurement errors. Gauss and Legendre suggested and developed the Least Square Method. Gauss gave it in the second volume of his long tractate "*Theory of Movement of the Heavenly Bodies Rolling Around the Sun by Cone Sections*" (1809). Legendre exposed his ideas in the work "*New Methods of Determination of Comet's Orbits*" (1806) where he put a special addition "*On the Least Square Method.*" Gauss more than once wrote that he used this method beginning in 1795. Gauss and Adrain showed that for some wide conditions the density of errors of measurement has the form

$$\varphi(\Delta)=\frac{h}{\sqrt{\pi}}e^{-h^2\Delta^2}.$$

We should notice that the influences of the works by Gauss and Adrain on the development of science were quite different. Adrain published his paper in a little-known American journal and this made its appearance almost invisible. The works by Gauss and Legendre became well known almost immediately to all in the scientific world. Scientists accepted their method and began to use it in their practical research.

A big contribution was made by S. Poisson (1781–1840). In particular, he put the question: Is it correct that the arithmetic average is always better than a single observation? The answer was-no. He found the distribution

$$p(x)=\frac{1}{\pi(1+x^2)}, \quad -\infty<x<\infty$$

and proved that for this distribution the statement above is wrong. Poisson found that sum of two random variables with this distribution has the same distribution but with another scale. Then he found that the arithmetical average of independent random variables of this type has exactly the same distribution as a single one. A. Cauchy (1789–1857) repeated the same result 20 years later and afterwards this distribution was called by his name. Poisson, who discovered this distribution, was forgotten.

Later the theory of measurement errors attracted attention of practically all prominent scientists in the area of the probability theory. P. L. Chebyshev and A. A. Markov (1856–1922) paid attention to the Least Squares Method and other problems of the theory of measurement errors. This theory became an important branch of mathematical statistics.

N. Formation of a Concept of a Random Variable

We have mentioned that formation of scientific concepts happens before they are completely understood. This was also true for the concept of a random variable which is a baseline of modern probability theory. Introduction of this term is connected with the names of many scientists who factually investigated its properties though they never used the term itself.

We already know that beginning with Cotes, Simpson, and D. Bernoulli in the eighteenth century the theory of measurement errors began to develop. At first, this theory served for astronomy. Long before them, Galileo mentioned that measurement error might take different values depending on random chance. He also introduced into usage the terms "random error" and "systematic error" of measurement. The latter depends on the quality of the tool, the skill of the observer, and the conditions of observation. But the first depends on a number of influences which cannot be predicted and which change from trial to trial. Now we clearly understand that a measurement error represents a random variable with some unknown distribution of probabilities.

But J. Bernoulli, N. Bernoulli, Montmort, and DeMoivre had already worked with the concept of a random variable. Indeed, J. Bernoulli considered a number of observations of the appearance of event A in n trials. For us now, this is a random variable taking values 1,2,...,n with probabilities determined by the Bernoulli formulas. N. Bernoulli, Montmort, and DeMoivre, analyzing the problem of the gambler's ruin, also dealt with a random variable–a number of games up to the gambler's ruin. DeMoivre came even further: He introduced the normal distribution. However, none of them realized that a powerful new concept needed to be introduced. J. Bernoulli kept the position of a sequence of random events, and the remaining three mathematicians were restricted by the concepts of the problem they were solving. DeMoivre found only that the normal distribution is a very convenient approximation for the distribution of random variables under consideration.

We said that at the beginning that scientists thought that possible values of measurement errors formed an arithmetical set with unknown but very small differences. Later this assumption was rejected and they began to think that possible values of measurement errors were filling an interval with some probability density. If D. Bernoulli allowed some liberty with the probability density, Laplace, Gauss and Legendre took it with all strictness. For them it was a non-negative function with the integral taken over the entire numerical axis equal to 1. Laplace already knew how to find a density of the sum with the help of density of separate random variables. In his famous book

"*Theorie Analitique of Probabilites*", he proficiently operated with distribution densities and solved interesting related problems but nowhere introduced the concept of a random variable. He used either terminology of the theory of measurement errors or the language of mathematical calculus and really did not feel a necessity to introduce a new concept into probability theory.

The first half of the nineteenth century brought new problems which needed a concept of a random variable. First of all, they were investigations of Belgian naturalist A. Quetelet (1796–1874) who found that the sizes of organs of animals of the same age are normally distributed. Many scientists analyzed the deviation of shells in artillery and also found that the normal distribution is appropriate to describe this phenomenon. In the middle of the nineteenth century the outstanding works by J. C. Maxwell (1831–1879) and some other scientists appeared in the mathematical theory of molecular physics of gas. And again the normal distribution took its honorable place.

We should mention one problem formulated by Gauss. In his diary of October 25, 1800, under the number 113 this problem was recorded. Twelve years later he wrote about this problem to Laplace in the letter of January 30, 1812. This problem was concerned with on interesting branch of mathematics, namely, the Metric Theory of Numbers, and simultaneously it related to investigation of uniformly distributed random variables. In his letter to Laplace, Gauss wrote: "... I recollect a curious problem which I dealt with around 12 years ago but had no satisfactory solution to at the time. Should you permit yourself to spend several minutes solving it and I'm sure that you could find more complete solution. Here it is. Let M be an unknown value between 0 and 1 for which all its meanings are equally possible or adhere more or less to the same law. Assume that it can be represented by continuous fraction

$$M = \frac{1}{a^{(1)}} + \frac{1}{a^{(2)}} + \cdots.$$

What is the probability that after exclusion of a finite number of terms up to $a^{(n)}$ one will have the remaining part

$$\frac{1}{a^{(n)}} + \frac{1}{a^{(n+1)}} + \cdots.$$

belonging to the interval from 0 to x? I denote this via P(n, x) and assume that for M all its meanings are equally probable: $P(0, x) = x$."

Gauss's hypothesis was that

$$\lim_{n\to\infty} P(n, x) = \frac{\ln(1+x)}{\ln 2}.$$

He wrote then that all his "attempts to solve the problem were in vain." The solution of the problem appeared only in 1928 thanks to R. O. Kuzmin (1981–1949). Within a year, P. Levy (1886–1971) obtained the same result by pure probabilistic methods and gave a better estimate of the quickness of convergence to the limit than Kuzmin. Later it was proven that the result is valid for any random variable M for which P(0, x) has the restricted derivative. This fact makes clear what Gauss meant by his uncertain words concerning "all its meanings are equally possible or adhere more or less to the same law."

It is interesting to notice that function P(0, x) as well as P(n, x) are distribution functions.

We see that research of many outstanding mathematicians provided an excellent background for introducing the concept of a random variable. It seems that the first step in this direction was taken by Poisson in 1832 in his work "*On Probability of Average Results of Observations.*" There is no term "a random variable" in this work but he wrote about "some thing" which one can take unclear as meaning from a set $a_1, a_2, \ldots, a_\lambda$ with corresponding probabilities $p_1, p_2, \ldots, p_\lambda$. He considered also continuous random variables and their densities.

So, Poisson took an important step in probability theory: He introduced in scientific usage a new concept–a random variable. His original term "thing" did not survive. Chebyshev in his works used the term "variable" and tacitly assumed that all random variables he dealt with were independent. Liapounov in his works systematically used the term "random variable" and, everywhere necessary, accentuated the variables' independence.

We emphasize that at the very beginning of his work "*On One Statement in the Probability Theory*", Liapounov defined the distribution function in the same manner as we do today. He gave the formula which is widely used now:

$$\Pr\{a < \xi \leqslant b\} = F(b) - F(a).$$

Notice that in tractates on probability theory published before 1912–"*Des Probabilite Calculus*" by Poincare, "*Probabilities Calculation*", "*Theory of Probability and Mathematical Statistics*" by Czuber–there were no concepts of the distribution function.

The Poisson definition of a random variable cannot now be accepted as a mathematical one. It is rather intuitive, based on the practical experience of dealing with real objects. But this description is very useful even now for teaching the beginning of probability theory. Even simple analysis of this definition shows that from it not follow the rules of operations with random variables: summation, extraction, multiplication, etc. To be a real mathematical object, the concept of a random variable should be strongly formalized. This was done at the end of the 1920's in a brief paper of Kolmogorov dedicated to axiomatics of the theory of probabilities and later with all details represented in his famous book "*Fundamental Concepts of the Probability Theory.*" Kolmogorov's approach is now common because it allowed the inclusion of probability theory in modern mathematics.

O. The Law of Large Numbers

Bernoulli's famous theorem about convergence of the frequency of event A to its probability when the number of observations increases was the first time generalized only by S. Poisson in 1837 in his work "*Investigation of Probabilities in a Court Decision Concerning Criminal and Civil Cases*". Just in this work he introduced the term "the law of large numbers".

Poisson considered a sequence of n independent trials each of which can generate event A with the probability p_k depending on the trial ordering number. If μ_n is the number of appearance of event A in n sequential observations, then for any ε for $n \to \infty$

$$P\left\{\left|\frac{\mu_n}{n} - \frac{p_1 + p_2 + \cdots + p_n}{n}\right| < \varepsilon\right\} \to 0.$$

On the occasion on this Poisson's Theorem, in 1843 Chebyshev in his brief note wrote: "... however artfully the technique was used by the great geometrician, it did not deliver a limit of error which connected with this approximate analysis, and because of this unknown value this proof has no necessary strictness" (P. L. Chebyshev, *Works Collection*, Vol. 2, p.14, The Academy of Sciences of the USSR, 1947).

Chebyshev estimates number n for which given ε and η for the inequality

$$P\left\{\left|\frac{\mu_n}{n}-\frac{p_1+p_2+\cdots+p_n}{n}\right|<\varepsilon\right\}>1-\eta$$

in his note.

However interesting these results were, they did not contribute substantial progress to probability theory because conceptually they did not came out of the frame of Bernoulli's original conception. An essential jump in this direction is connected with the work by Chebyshev "*On Average Values*" (1867) published simultaneously in Russian and French. In this work he considered random variables instead of random events and so relocated the accent in probability theory to analysis of random variables. It is necessary to notice that Chebyshev did not mention independence of random variables because in accordance with the traditions of that time he did not consider dependence. Now Chebyshev's Theorem is considered in all textbooks on probability theory. Not once it was a subject for the future generalizations.

In 1909 E. Borel showed that the Bernoulli Trials led to a stronger statement than the law of large numbers for $p=0.5$. In 1917 Italian mathematician Cantelli extended this fact to an arbitrary p that

$$\mathrm{P}\left\{\lim_{n\to\infty}\frac{\mu_n}{n}=\mathrm{p}\right\}=1.$$

This statement became to be called the Strengthened Law of Large Numbers.

A wide generalization of the Strengthened Law of Large Numbers was given by A. N. Kolmogorov in 1930 and was put in his famous book "*Fundamental Concepts of the Probability Theory*" (1932).

Necessary and sufficient conditions for the Strengthened Law of Large Numbers are found in the works by Yu.V. Prokhorov from 1958 and 1959 (see "*On the Strengthened Law of Large Numbers*", Proceedings of the Academy of Sciences of the USSR, Issue Mathematics, No. 14, p.6, 1958; "*Some Comments on the Strengthened Law of Large Numbers*", Theory of Probabilities and Its Applications, Vol. 4, No. 2, pp. 215–220, 1959).

In 1935 A.Ya. Khinchine introduced a new concept of the relative stability of sums which had to give the extremely general form of the Strengthened Law of Large Numbers for positive random variables. The sum $S_n = \xi_1 + \xi_2 + \cdots + \xi_n$ is said to be relatively stable if there exist such constants $A_n > 0$ that for any $\varepsilon > 0$ and $n \to \infty$ the following relation is valid

$$P\left\{\lim_{n\to\infty} \frac{\mu_n}{n} = p\right\} = 1.$$

For identically distributed random variables ξ_n, Khinchine found necessary and sufficient conditions of relative stability of sums S_n. Khinchine's pupil, A. A. Bobrov (born in 1912) extended this result on differently distributed variables.

The law of large numbers up to 1939 was considered as a special limit theorem separate from the remaining limit theorems for sums of independent random variables. In the work by B.V. Gnedenko which is considered in Section 17, the law of large numbers was included in the common theory of limit theorems where the limit distribution has the only point of growth in zero. In the same manner, theorems on relative stability of sums are limit theorems for the case where the limit distribution has the only point of growth in $x = 1$.

Substantial extension of the area concerned with the law of large numbers was made by V. I. Glivenko in 1929–1933 when he began to consider the limit theorems for random variables taken from functional spaces. Probably the climax of his results is his famous theorem on convergence of the empirical distribution to the genuine distribution function of observed random variables. Glivenko's Theorem just after its publication was named by Cantelli as the Fundamental Theorem of Mathematical Statistics.

Glivenko's Theorem was not once generalized.

P. Central Limit Theorem

The DeMoivre Theorem on the binomial distribution convergence to the normal distribution for a long time was a subject for future generalization. Probably, the first generalization was made by Laplace. This approach was formulated as a limit theorem for a sum of random variables ξ_k each of which uniformly distributed on the interval $(-h, h)$. It was presented in the work "Memoire sur les Approximations? des Formules Qui Sont Functions de Tres Grand Nombre et sur Leur Application aux Probabilites" (1809). Laplace there considered discrete random variables with increasing number of their meanings. This gave an approximation of a continuous distribution by discrete one. Laplace proved that the following is valid for any s

$$\lim_{n\to\infty} Pr\left\{-s\leqslant\frac{S_n}{\sqrt{n}}\leqslant s\right\}=\frac{2\sqrt{3}}{h\sqrt{2\pi}}\int_0^s e^{x^2/2\sigma}\,dx,$$

or

$$S_n=\sum_{1\leqslant k\leqslant n}\xi_k,\quad \sigma^2=\frac{h^2}{3}.$$

It is interesting to notice that Laplace used the method of characterization function which, naturally, did not have this name at the time.

A substantial approach to limit theorems belongs to Poisson. In his famous memoir of 1837 "Recherches sur la Probabilite des Jugements..." he considered the scheme of sequential trials with different probabilities of appearance of event A in each trial. Let the probability of the appearance of event A in trial k equal p_k. Poisson proved for this case the following local theorem. If a set $\sum_{1\leqslant k<\infty} p_k(1-p_k)$ diverges, then the probability that in n trials the event will appear m times equals

$$Pr\left\{m=np-\theta c\sqrt{n}\right\}=\frac{1}{c\sqrt{n\pi}}e^{-\theta^2}-\frac{h\theta}{2c^4n\sqrt{\pi}}(3+2\theta^2)e^{-\theta^2}$$

where

$$p=\frac{1}{n}\sum_{1\leqslant k\leqslant n}p_k,\ c^2=2\sum_{1\leqslant k\leqslant n}p_k(1-p_k),\ h=\frac{4}{3n}\sum_{1\leqslant k\leqslant n}(2p_k-1)p_k(1-p_k).$$

In the same book Poisson made a mistaken generalization of this theorem for the sum of arbitrary independent random variables with finite variances under the condition of their centering by sums of the means and norming by the square root of the sum of their variances. Justice forces us to mention that this theorem is absolutely correct if random variables are identically distributed. But this proof was obtained much later, by our contemporaries, Lindeberg, Feller and Khinchine.

Notice that the works by Laplace and Poisson and their followers dealing with the Central Limit Theorem were stimulated by the theory of measurement errors. It seems that the first to change this tradition was Chebyshev.

Interest in the normal distribution increased in the beginning of the nineteenth century in connection with the famous research by Adrien Legendre and Karl Gauss dedicated to the method of smallest squares. F.V. Bessel in 1818 in his work "*Fundamenta Astronomiae pro Anno 1755 Deducta ex Observationibus Viri Incomportabilis James Bradley in Specula Grenovicensi pro Annis 1750–1762 Institutis*" noticed that observations made by Greenwich astronomer Bradley excellently confirm the appearance of the normal distribution. Bessel's explanation of this fact coincides with Laplace's idea which the latter had been maturing thirty years: The sum of a large number s of random influences each of which is small in comparison with the sum of the remaining has the common distribution and this distribution is normal. This idea was clearly repeated by Bessel in 1838 in "*Untersuchungen uber die Wahrscheinlichkeit der Beobachtungsfehler*". Bessel noticed that this rule is not universal and other errors distributions might appear in particular cases. For instance, if during angles measurement one source of errors is prevailing in comparison with all of the remaining, the density of the resulting error can be $1/\pi\sqrt{a^2-x^2}$.

The same concept of the appearance of the normal distribution as the distribution of measurement errors one can meet in the famous book by Henry Poicare "*Calcul des Probabilites*" (Paris, 1912). He wrote that "the error of the measurement tool is determined by a very large number of independent errors each of which brings only a small influence into the resulting error. This resulting error has the Gaussian distribution law."

It is necessary to emphasize that all these arguments have a qualitative character and need mathematical formulation and strict proofs.

The second stimulus for the increasing interest in limit theorems in probability theory was the development of statistical physics the basis for which began to be built in the middle of the nineteenth century. The first result in this direction was formulated in 1887 by P.L. Chebyshev in his work "*About two theorems in probability.*" The theorem formulated by Chebyshev reads as follows:

Let the mathematical expectations of random variables $u_1, u_2, \ldots$ equal 0, and the mathematical expectations of all their powers be values lower than some threshold. The probability that their sum $u_1 + u_2 \cdots + u_n$ divided by the square root of the doubled sum of expectation of their squares is bounded by some t and t′ is equal to $1/\sqrt{\pi}\int_t^{t'} e^{-z^2}\,dz$ if n increases to infinity.

To prove this statement, Chebyshev developed an excellent method which was called the Method of Moments and which was one of the strongest achievements of science at the time. However the proof had some deficiencies which Chebyshev's pupil, A. A. Markov, began to correct almost immediately. The criticism of Chebyshev's paper is contained in the letters which A. A. Markov sent to Professor of the Kazan University A.V. Vasilyev. The latter found it necessary to publish these letters of Markov's in 1898 with the title "*The Law of Large Numbers and Method of Least Squares*". In these letters Markov absolutely strongly proved a slightly corrected Chebyshev's theorem: If S_n is a sequence of sums $u_1 + u_2 + \cdots + u_n$ and $\phi_n(x)$ is the distribution of u_n, then from the assumption that for any integer k the following condition is valid

$$\int_{-\infty}^{\infty} x^k d\Phi_n(x) \to \frac{1}{\sqrt{2\pi}} \int_{-\infty}^{\infty} x^k e^{-x^2/2} dx$$

follows that for any a and b

$$Pr\left\{a < \frac{S_n}{\operatorname{Var} S_n} < b\right\} \to \frac{1}{\sqrt{2\pi}} \int_a^b e^{-x^2/2} dx.$$

Chebyshev's Method of Moments celebrated a success. It seemed that the final theorem had been proven. Some dissatisfaction arose only from the fact that for a simple result one needed a counting number of conditions. Quite unexpectedly in several publications by A. M. Liapounov in 1900 and 1901 it was shown that the final result needed to satisfy a very simple condition. In addition this condition made clear. The meaning of those assumptions which led to the normal distribution of normed and centered sums.

At first Liapounov proved that if the random variables have finite third moments, $c_k = E\{|\xi_k - a_k|^3\}$, and, $C_n = \sum_{1<k<n} c_k$, $B_n^2 = \sum_{1<k<n} \operatorname{Var}\{\xi_k\}$, and ratio $C_n/B_n^3 \to 0$ when $n \to \infty$, then distribution functions of S_n converge to the normal distribution.

The next year, Liapounov found that for the final result the existence of the third moments is not necessary. It is enough if moments of the order $2+\delta$ exist where $\delta>0$. Liapounov showed that the sum of independent random variables normed by the square root of the sum of their variances converges to the normal distribution if the following conditions take place: let $c_k = E\{|\xi_k - a_k|^{2+\delta}\}$ *and* $C_n \sum_{1\leqslant k\leqslant n} c_k$ then with $n\to\infty$ the ratio $C_n/B_n^{2+\delta}$ must go to 0.

Liapounov offered even more: he estimated how fast the distribution of the sum converges to the normal one. The order of this estimate equals $n^{-1/2}\ln n$.

The generality of Liapounov's results impressed contemporaries. It seems that just then the term "*Central Limit Theorem*" had appeared for the convergence of the distribution of centered and normed sums of random variables to the normal distribution.

Markov came to the same results from a quite different position. In this connection, it would be interesting to cite Markov's original words: "Generality of results of the latest Liapounov work seriously overcame the results obtained with the help of the method of mathematical expectations. It seemed even impossible to reach such general results by that method because it assumed the consideration of an infinite number of these mathematical expectations. Liapounov's method does not use such an assumption. To rehabilitate the method of mathematical expectation it was necessary to show that this method has not been exhausted by these works". In 1908 Markov suggested the excellent idea of truncation of random variables. This approach allowed one to prove the limit theorem in Liapounov's conditions by the method of moments or, as Markov said, by the method of mathematical expectation. The idea of truncation resolutely entered into probability theory.

The later history of the Central Limit Theorem is the following. In 1922 Finnish mathematician Lindeberg went further than Liapounov–he denied the assumption of the existence of any moments but the second. Namely, he proved that if for any $\tau>0$

$$\lim_{n\to\infty}\frac{1}{B_n}\sum_{1\leqslant k\leqslant n}\int_{|x-a_k|>\tau B_n}(x-a_k)^2\,dF_k(x)=0$$

then the distribution of the sum of random variables, centered by its mathematical expectations and normed by the square root of the sum of its variances, converges to the standard normal distribution.

Twelve years later W. Feller showed that Lindeberg's condition is also necessary under the assumption of uniformly small terms of the sum.

It is clear that as the corollary of Lindeberg's Theorem one obtains a long-expected result: If random variables are independent, identically distributed and have finite variance differing from 0 then the Central Limit Theorem can be applied to the sum of such variables.

In 1927 S. N. Bernshtein considered a more general problem. Let $\xi_1, \xi_2, \ldots, \xi_n, \ldots$ be a sequence of independent random variables for which no assumptions about the existence of the mathematical expectations or variances have been made. The question is when can such constants $B_n > 0$ and A_n can be found that the distribution function of sums $(S_n - A_n/B_n)$ converges to the normal distribution?

Sufficient conditions for this problem were found by Bernshtein. Some eight years later W. Feller showed that the same conditions are not only sufficient but also necessary under the condition that all terms in the sum are uniformly small.

In the same year 1935 A.Ya. Khinchine and P. Levy found independently from each other the necessary and sufficient conditions of convergence of the distribution of the sum of independent and identically distributed random variables to the normal distribution (in the Bernshtein formulation of the problem).

In 1926 in his special course on limit theorems, A.Ya. Khinchine formulated the question: Is there a connection between the law of large numbers and the Central Limit Theorem? The answer for this question has been found by D.A. Raikov and A.A. Bobrov who proved the following theorem. For a distribution function of sums

$$\frac{\xi_1 + \cdots + \xi_n - A_n}{B_n}$$

to converge (under a corresponding choice of real constant $B_n > 0$ and A_n) to the normal distribution, it is necessary and sufficient that sums

$$(\xi_1 - a_1)^2 + (\xi_2 - a_2)^2 + \cdots + (\xi_n - a_n)^2$$

be relatively stable. Here $a_n = \int_{-\varepsilon}^{\varepsilon} x\,dF_n(x)$, $\varepsilon > 0$ *is arbitrary.*

Investigation of the convergence of distributions to the normal one continues today. But now different problems are under consideration: how fast distributions converge to normal, how to converge the distribution of the sum of a random number of random variables, how to converge the distribution of the sum of non-uniformly small random variables.

Q. General Limit Distributions for Sum of Random Variables

A natural question arises: What kind of limit distributions could be possible for the sum of independent random variables under the condition that they are "almost equal" by their values? This question arose only in 20s–30s of this century, though some particular results on the subject were known. In this context, Poisson, in his paper "*On Probabilities of Average Results of Observations*", where he used in his paper the methods of characteristic functions, inferred the distribution of the sum of a large number of independent errors and considered the distribution which later was called the Cauchy distribution. For this distribution, Poisson found the density

$$f(t) = \frac{1}{\pi(1+x^2)}$$

and proved that this distribution possessed the two properties:

(1) The arithmetical mean of the observation errors distributed by the Cauchy Law has also Cauchy distribution;
(2) For this distribution the accuracy is not improved if the arithmetical mean is taken from several series of observations.

This paper was published in 1832.

Thirty years later, in 1853, in the paper "*On the Average Results of Observations of the Same Nature and on Most Probable Ones*" O. Cauchy obtained the characterization function for all of those distributions for which the distribution of the sum differs from the distribution of a single random variable only by a multiplier of the argument (scale factor). Cauchy found that all of these functions have the form $f(x) = \exp(-t^{\mu})$ where μ is a positive number. Later it was found that $f(t)$ then and only then is a characterization function when $0 < \mu \leqslant 2$.

P. Levy in the book "*Calcul des Probabilite*" (1925) in Chapter 6 "*Exponential Distributions*", constructed the first theory of stable distributions. This theory, naturally continuing Cauchy's research, went far ahead: Let $F(x)$ be a distribution function and $f(t)$ its characterization function. The distribution $F(x)$ is called *stable* if for any positive a_1 and a_2 there can be found such constant a that the equality

$$f(a_1 t) \cdot f(a_2 t) = f(at)$$

is true.

In terms of random variables, this class of distributions possesses the following characterization property: If ξ_1 and ξ_2 are independent random variables with the same distribution, and a_1 and a_2 are arbitrary positive numbers, then for any pair a_1 and a_2 there can be found such a positive number a that the sum $a_1\xi_1 + a_1\xi_1$ has the same distribution as $a\xi_1$ has.

P. Levy showed that for stable distributions, the function f(t) has the form

$$e^{-c(1+i\beta t\,|t|)|t|^{\alpha}}$$

where $0 < \alpha \leqslant 2$. P. Levy also introduced the concept of the *area of attraction* of the stable distributions. This is a set of all such distributions F(x) for which the distribution functions under some certain normalization converge to the given stable distribution.

In 1935 A.Ya. Khinchine added the concept of the stable distribution introduced by Levy. He suggested celling those distributions 'stable' for which the linear form $a_1\xi_1 + a_1\xi_1$ for arbitrary a_1 and a_2 has the same distribution as $a\xi_1 + b$ where a is an arbitrary positive number and b is a material constant. The class of Khinchin stable distributions appeared wider than that of Cauchy.

In 1939, independently, B. V. Gnedenko and W. Döbline found the areas of attraction of stable distributions. The conditions of belonging to the attraction area of the stable distributions are very simple. They are reduced to the behavior of the "tails" of distribution functions.

The main result obtained by P. Levy and A.Ya. Khinchin can be formulated in the following form: if $\xi_1, \xi_2, \ldots$ are a sequence of identically distributed and independent random variables, then sums

$$S_n = \frac{\xi_1 + \cdots + \xi_n - A_n}{B_n} \qquad (1)$$

under the corresponding choice of constant $B_n > 0$ and material A_n can converge only to the stable distributions. Each stable distribution law is a limit distribution for the distribution of sum (1).

In 1936 P. Levy and A. Ya. Khinchin gave the final presentation of stable distributions via logarithm of characterization function. Function $\varphi(t)$ is a characterization function if and only if it can be presented in the form

$$\ln \phi(t) = i\gamma t - c|t|^{\alpha}\left[1 + i\beta\frac{t}{|t|}\omega(t,\alpha)\right]$$

where α, β, γ, and c are material constants $(-1 \leqslant \beta \leqslant 1, c \geqslant 0, 0 < \alpha \leqslant 2)$ and

$$\omega(t,\alpha) = \begin{cases} tg\frac{\pi}{2}\alpha & \text{if } \alpha \neq 1 \\ \frac{2}{\pi}\ln|t|, & \text{if } \alpha = 1. \end{cases}$$

This result finally completed the research begun by Poisson and Cauchy.

A natural question about a class of limit distributions for sums (1) where the terms might be differently distributed was formulated in the letter from A.Ya. Khinchine to P. Levy. Khinchin suggested celling this class the L-class. The restriction on terms ξ_k/B_n of the sum is that each of them influence has a negligible on the entire sum. This requirement can be represented in the following form: All values ξ_k/B_n are constant in limit, i.e., there exists such a sequence of constants m_{nk} that uniformly relatively any k $(1 \leqslant k \leqslant n)$ and for any $\varepsilon > 0$ the following relation is valid:

$$Pr\left\{|\frac{\xi_k}{B_n} - m_{nk}| > \varepsilon\right\} \to 0 \quad \text{if} \quad n \to \infty.$$

The characterization property of the laws of class L founded by P. Levy consists of the following. For function $\varphi(t)$ to be a characterization function of class L, it is necessary and sufficient that for any $\alpha(0 < \alpha \leqslant 1)$ the equality $\varphi = \varphi(\alpha t)f_\alpha(t)$ where $f_\alpha(t)$ is some characterization function holds.

The history was not yet complete. One needed to answer the question raised by B.V. Gnedenko. What are the classes of possible limit distributions if random variables $\xi_1, \xi_2, \ldots, \xi_n, \ldots$ can be distributed only by $k(1 \leqslant k < \infty)$ various distributions $F_1(x), F_2(x), \ldots, F_k(x)$? The completed solution of the problem was offered by A.A. Zinger in 1971.

In the frame of this research one more special problem was considered: What happens if sums (1) of independent and identically distributed random variables are considered not for all possible n but for some subsequence? What kind of limit distributions can be met in this case? This question was formulated by A. Ya. Khinchine, who gave this answer: The class of limit distributions in the definition above coincides with the class of infinitely divisible distributions which were introduced in 1930 by Italian mathematician Bruno de Finetti and investigated in detail by A. N. Kolmogorov, P. Levy, and A. YA. Khinchine. A random variable is called infinitely divisible if for any integer n it can be represented as a sum of n independent and identically distributed random variables. This property gives its name to this distribution.

In 1933 A. N. Kolmogorov formulated the hypothesis that if almost equally powerful random variables are summed, then with an increase of the number of terms their distributions will converge to infinitely divisible distributions. Consequently, if distributions of consequent sums converge to a limit one, this limit distribution must with necessity be infinitely divisible. Kolmogorov's pupil G. M. Bavli (1908–1941) proved this hypothesis in 1934 under the assumption that all terms have finite variances and the variances of consequent sums are restricted. This hypothesis was completely proved in three years by A.Ya. Khinchine with the use of cumbersome analytical methods. Based on this work, B. V. Gnedenko constructed the theory of summation of independent random variables using a simply proved fact: If a sum consists of constant in limit independent variables and distribution functions of centered sums converge to some limit distribution, then it is possible to construct a sequence of infinitely divisible random variables whose distribution functions converge to the distribution functions of the sum. From this statement the theorems proven by Bavli and Khinchine follow as particular cases. In addition, this approach gave in a very clear form the possibility of finding conditions of existence of limit distributions and conditions of function convergence to any limit distribution. In particular, the necessary and sufficient conditions for the law of large numbers, for convergence to the normal distribution, the Poisson distribution, and to stable distributions were found. All of these aspects are reflected in a monograph by A. N. Kolmogorov and B. V. Gnedenko "*Limit Theorems for Sum of Random Variables*" (1949).

In recent years many researchers have begun to analyze limiting behavior of sums of random numbers of random variables. At the beginning only conditions of convergence to the normal distribution were investigated and the accomplishments of requirements of the law of large numbers were analyzed. Then the problem of the existence of limit distributions and their classes were analyzed. This problem was solved under the condition of independence and identity of distributions of random variables, and also under the requirement of independence of a number of summed variables on these variables. Notice that these problems arose in reliability theory and physics. The main theorem concerning this direction of research was called the theorem of transposition.

R. Law of Repetitive Logarithm

The law of large numbers gave a beginning to the new limit law which is called the Law of Double Logarithm. This theorem does not have as a goal to find a limit distribution but moves the problem of sequential sums into an entirely new area, namely, into the area of studying the behavior of these sums together. First let us consider this problem for the simplest case, the Bernoulli trials. This is very natural additionally because it coincides with the historical development of the subject.

Denote μ_n to mean a number of appearances of event A in n independent trials and consider the difference $S_n = \mu_n - np$. In 1909 E. Borel gave a generalized formulation of the law of large numbers by proving that there exists the stronger formulation, namely,

$$Pr\left\{\frac{S_n}{n} \to 0\right\} = 1.$$

With in four years F. Hausdorff (1868–1942) proved that there was an even stronger statement: for any $\varepsilon > 0$

$$Pr\left\{\frac{S_n}{\sqrt{n^{1+\varepsilon}}} \to 0\right\} = 1.$$

The next year, G. Hardy (1877–1947) and J. Littlewood (1885–1977) found the new stronger statement which says that with the probability equal to 1 the ratio $|S_n|/\sqrt{n \ln n}$ remains restricted. In 1922 A.Ya. Khinchine gave an estimation for the growth of sum S_n: $S_n = O(\sqrt{n \ln \ln n})$. In two years A. Ya. Khinchine found the final result:

$$Pr\left\{\limsup_{n\to\infty}\frac{|S_n|}{\sqrt{2npq\ \ln\ \ln n}}=1\right\}=1.$$

In 1926 A. Ya. Khinchine generalized this result for the Poisson scheme, i.e., the scheme of sequential trials with the varying probability of the occurrence of event A.

The work by A. N. Kolmogorov (1929) essentially overlapped the results obtained by Khinchine. Khinchine's results were corollaries of it. In saying this, we are not trying to decrease the role of Khinchin's works, because the discovery of a new law, even for a particular case, is a great advancement.

Let there be a sequence $\xi_1, \xi_2, \ldots$ of mutually independent random variables with the means $a_k = E\{\xi_k\}$ and variances $b_k = Var\{\xi_k\}$; $B_n = \sum_{1 \leq k \leq n} b_k$, $S_n = \sum_{1 \leq k \leq n}(\xi_k - a_k)$. If the sequence ξ_k satisfies two more conditions for $n \to \infty$: (1) $B_n \to \infty$, and (2) $|\xi_n| < m_n = o\left(\sqrt{B_n/(\ln\ \ln\ B_n)}\right)$, then this sequence satisfies the Law of Double Logarithm, i.e., the following statement is true

$$Pr\left\{\limsup_{n\to\infty}\frac{|S_n|}{\sqrt{2B_n \ln \ln B_n}}=1\right\}=1.$$

In other words, the following statement was formulated: under the given conditions for any positive ε and δ an arbitrary large integer number N can be found that

(1) the probability that at least for one $n > N$ the inequality

$$|S_n| > (1+\delta)\sqrt{2B_n \ln \ln B_n}$$

fulfills is less than ε, and

(2) the probability that at least for one $n > N$ the inequality

$$|S_n| > (1-\delta)\sqrt{2B_n \ln \ln B_n}$$

fulfills is larger than $1 - \varepsilon$.

Later the problem of the double logarithm attracted the attention of many probabilists: P. Levy, W. Feller, Zygmund and Marcinkiewicz, Hartmann, T. A. Sarymsakov, V. V. Petrov, B. V. Gnedenko, and others. Among a numbers of excellent results we emphasize one: If random variables ξ_k are identically distributed and have the finite variance (of course, non-zero), then the Law of the Double Logarithm takes place. As was shown by A. I. Martikainen, this result allows the inversion.[7]

An analogous problem for stable distributions differing from normal was also formulated and solved. As it became clear (B.V. Gnedenko) for any non-decreasing function u(n) and for any stable distribution with the power $\alpha (0 < \alpha < 2)$ with the probability 1 the ratio $\lim_{n \to \infty} \sup \{|S_n|/u(n)\}$ equals to 0 or infinity.

S. Formation of a Concept of Mean and Variance

The term "mathematical expectation" was introduced into probability theory during the early stages of the development of the theory. It was mentioned first in the correspondence between Pascal and Fermat described. In earlier more determined form it was introduced by Huygens. These three determinations deal with the mathematical expectation of random variables which takes two or three meanings. As we mentioned above in Chapter 1, the term "expectation" itself was introduced by Schooten, Huygens' teacher. This term survived and has been kept until our day. In those days, this term was understood as an expectation of a gain for the game which gave x_1 with probability p_1, x_2 with probability $p_2, \ldots, x_n$ with probability p_n.

This idea is a dominant in "*On Application of the Art of Guessing in Law Problems*" by Nicholas Bernoulli. He wrote there that "this rule (*calculation of the expectation—B.G.*) is equivalent to that which is usually used for finding the arithmetic average of several values and also with the rule of mixture my uncle found appropriate to refer." Further he cited an example taken from "*Art of Guessing*" by Jacob Bernoulli: "If three glasses of beer with the price of 13 each are mixed with two glasses of beer with the price of 8 each, then after mixture we have the total price of all volume equals $3 \times 13 + 2 \times 8 = 55$. It gives for each of 5 glasses of mixture a price equals 55:5 = 11. The same rule must be used for calculation of expectation of something: if there are three possible cases with value of 13 and two cases with value of 8." Notice that this is a direct repetition of Huygens' rule. It should be emphasized not only that N. Bernoulli considered the calculation of

[7]Inversion of the Law of the Double Logarithm for Random Walking. Theory of Probabilities and Its Applications, 1980, vol. 25, No. 2, pp. 364–366.—*B. G.*

expectation in the case where the number of random values take more than two or three meanings, but compared the expectation calculation with the calculation of the center of weight of a system of material points. Below is the citation from the same book by N. Bernoulli.

"Even more remarkable is the exclusive coincidence which take place for this rule and the rule of calculation of the center of weight of several loads. Indeed, the sum of moments, or the sum of products of weights by the distance from some point divided by the sum of weights shows the distance of this particular point from the center of weight, i.e., such a point t at which all loads are in the state of equilibrium. The expectation found above is, so to speak, the center of weight of all probabilities which balance them. Nothing can overweight another load. To keep the same equilibrium in suspicious and unclear cases our layers usually keep the middle."

An eighteenth century reference to the mathematical expectation was not typical. All attention was focused on the term of the probability of a random event. In an encyclopedia of the science of probability, the famous book "*Analytical Theory of Probabilities*" by P. Laplace, there is no mention of mathematical expectation and, naturally, no discussion of it. Probably this was connected with the fact that Laplace did not consider a random variable. Instead he wrote about errors of observations and density functions of their distributions. He even got the formula for the density of two independent errors. (He did not mention that the variables are independent because others were not considered at all at the time.)

It seems natural that the development of the theory of errors of observations should stimulate the development of understanding of numerical characteristics of random variables (they were called errors of observations). However this did not happen. Nevertheless for the normal distribution, the concepts of true value and accuracy of measurement were introduced and the rules for how to find them on the basis of the known density function were given. Thus for this particular case, the formulas for calculation of the mathematical expectation and variance were known.

Notice that in the beginning of the nineteenth century the normal distribution overshadowed all other concepts because it related to the important practical application, the theory of measurement. The proofs of normality of errors distribution in most of the measurements by Gauss and Legendre strengthened the position of the normal distribution. The same distribution appeared in the theory of artillery shelling. The Belgian biologist A. Quetelet gave much evidence of a dominant role of the normal distribution in biology. Nobody thought about another distribution. This led to the situation in which nobody thought how to prove theorems concerning the mathematical expectation and variance, because for the normal distribution all this was clear. In this connection it is very interesting to notice that in Chebyshev's "*Experience of Elementary Analysis of the Probability Theory*" (1845) the concepts of random variable, mathematical expectation, and variance are not even mentioned. However in his lectures on probability theory systematically read by him in the Sanct Petersburg University, Chebyshev did discuss variables (assuming random variables), and their mathematical expectation and variance.

We should emphasize that in these lectures one can find formulations and proofs of the theorems about mathematical expectation and variance of a sum of random variables. In the same lectures he gave an inference of his famous inequality. He assumed as obvious that independent random variables were considered. Notice also that the fact that the variance of the sum of random variables equals the sum of the variances was used by Chebyshev in his paper "*On the average Values.*" In this paper he published his inequality for the first time. Notice that in the popular textbooks of the beginning of this century there were no theorems at all about mathematical expectation and variance.

Naturally, the question arises: Is the fact that the mathematical expectation of the sum of random variables is equal to the sum of mathematical expectations always valid, not only for independent random variables? Now we only can confirm that there is no such statement in Czuber's textbook (1908) and in the second edition of the Poincare's book (1912). But in the popular textbook at the time "*Probabilities Calculus*" (1913 and 1924) there is the strong proof of the theorem of the mathematical expectation of the product and sum of random variables with the note that this theorem is valid not only for independent values.

In conclusion we should say that the history of the concept of mathematical expectation and variance has been investigated absolutely insufficiently. We see that the concepts of mathematical expectation appeared simultaneously with the concepts of probability but its main properties were formulated only in the middle of the nineteenth and at the beginning of the twentieth centuries. It is not clear how the concept of the moment of inertia existing at the time influenced the concept of variance.

Everything exposed in this chapter is only a rough approximation of the history of this important branch of scientific knowledge.

PART 4 HISTORY OF STOCHASTIC PROCESSES

The concepts of stochastic processes were formulated in this century and are connected with the names of Andrei Kolmogorov, Alexander Khinchine, Evgenii Slutsky, and Norbert Wiener (1894–1965). The field of stochastic process is one of the most important not only in probability theory but also in engineering, economics, management and telecommunication. This is one of the fastest developing mathematical theories today. Doubtlessly this is because of its deep connections with practical problems.

Modern science could not be satisfied only by the heritage left from the past. Indeed, engineers, physicists, biologists were interested in processes developing over time but probability theory delivered only tools for the analysis of stationary states. For time-developing processes there were neither particular tools nor general methods. At the same time the necessity of such approaches grew quickly. Analysis of Brown's processes in physics led mathematics to the threshold of the theory of stochastic processes. Investigations by Dutch scientist A. K. Erlang formed an important area of applications connected with traffic of telephone networks. The number of callers changes in time stochastically and the duration of each call is essentially individual–these are conditions of a "double randomness" for calculation of telephone network capacity, commutation equipment and traffic control systems. Doubtlessly these works by Erlang influenced not only the solution of problems in telecommunication but led to the formation of elements of the theory of stochastic processes, in particular, Birth and Death Processes.

In the second decade of the twentieth century the study of biological populations began. Italian mathematician Vito Volterra developed a mathematical theory of these processes on the basis of pure deterministic concepts. Later several mathematicians and statisticians developed these ideas with a stochastic background. Initially in this theory Birth and Death Processes were exclusive tools. In particular, the name of this process has its roots in biology.

Let us imagine that we decided to observe the track of a particle of gas or liquid. This particle at random moments collides with other particles and changes direction and speed of its movement. Thus the state of such a particle is subjected to random influences and, so, represents a stochastic process. This process is defined by six parameters: three coordinates and three components of the speed. Many physical applications need to calculate probabilities that a specified portion of particles will move from one area of space into another. For instance, if two liquids are interacting the mutual penetrating of molecules takes place between them. This effect is known as diffusion. With what speed is the diffusion running and by what laws? When does the mixture of two liquids become practically homogeneous? These and other questions are answered in the statistical theory of diffusion which is based on stochastic processes theory. Obviously very similar problems appear in chemistry. What portion of molecules is under active reaction? What specific type of the interaction process takes place? When can the reaction be considered as completed?

One very important class of problems concerns the nature of radioactive emanation. The nature of this process consists in transformation of one substance into another. Each atom has a blast type emanation, and some energy is freed. Multiple experiments and observations showed that emanation of individual atoms appears at random moments of time and these moments are independent under the condition that the quantity of the freed energy does not exceed some level. For analysis of radioactive emanation, it is very important to calculate the probability that there will be not more than some specified quantity of emanated atoms. Formally, from the purely mathematical side, many other processes develop in a similar way: threads break in textile production, a number of Brown's particles exist in a specified area of space, a number of calls take place in a switch of a telephone network,etc.

The theory of Brown's processes based on probabilistic arguments was developed in 1905 by two famous physicists, M. Smoluchowsky (1872–1917) and A. Einstein (1955). Later their idea was used for different physical problems and in engineering. In particular, after this work (as well as after Erlang's work) a wide interest in the Poisson process arose. Incidentally, Poisson himself introduced only the Poisson distribution, and he had not even a dream about the Poisson Process. But nevertheless, he doubtlessly earned the right to be mentioned every time one considers a stochastic process connected with his distribution. It sometimes happens that a new phenomenon is called after a researcher though the researcher himself had not the slightest idea about the related phenomenon. For instance, now Gaussian Processes are widely known though Gauss himself had no understanding of such a process. Even more, the Gauss distribution was discovered long before his birth by DeMoivre, Laplace and others. In the Theory of Errors, Legendre described this distribution simultaneously with Gauss.

Attempts of studying the phenomenon of diffusion by means of probability theory was undertaken in 1914 by two prominent physicists, M. Planck (1958 – 1947) and Fokker.

N.Wiener in the middle of the 1920s, studying the Brown Moving of particles, introduced processes satisfying the following conditions:

1) distribution of the difference $\xi(t_0 + t) - \xi(t_0)$ does not depend on the initial moment t_0 (homogeneousity in time);

2) increments of the process $\xi(t)$ during non-intersected time intervals in finite number are mutually independent (independence of increments);

3) values $\xi(t_0 + t) - \xi(t_0)$ are normally distributed with the mean 0 and variance $\sigma^2 t$.

We should mention two important groups of research begun at different times and for different reasons. First, there are the works by A. A. Markov (1856 – 1922) on chain dependence.

Secondly, there are the works by E. E. Slutsky (1880 – 1948) on the theory of stochastic functions. Both of these directions played a substantial role in the formation of the general theory of stochastic processes. The time to formulate such a theory had come and all necessary material was accumulated. What was needed was to analyze the existing works and on their basis perform the synthesis.

In 1931 A. N. Kolmogorov published a big paper "*On Analytical Methods in Theory of Probabilities*" and within three years A. Ya. Khinchine published. "*The Theory of correlation of stochastic processes.*" These two papers we can consider to be the beginning of construction of the general theory of stochastic processes. The first of these works contains fundamentals of Markov processes, and the second is dedicated to stationary processes. These works were a basis for many others in this direction. Among them one should select the paper by W. Feller "*On the theory of stochastic processes*" (1936) which presented integral-differential equations for the jumping type of Markov processes.

The two fundamental works mentioned above contain not only mathematical results but also a deep philosophical analysis of what led to the construction of the theory of stochastic processes. To introduce the reader to this aspect of research we present a big piece from the introduction to Kolmogorov's work.

If one is willing to infer mathematically natural or social phenomena, one should first of all formalize them. The matter is that one can apply mathematical analysis to the process of a system state changing only if any possible state of the system might be determined with the help of known mathematical methods, for instance, by means of values of parameters of the system. Of course, such a mathematically determined system is not a reality itself but a scheme usable for the description of the reality.

Classical mechanics uses only such schemes for which system state y at moment t is uniquely determined via its state x at any previous moment t_0; mathematically it is expressed as the formula $y = f(x, t, t_0)$.

If such a unique function exists, as is always assumed in classical mechanics, one says that the scheme is the scheme of a *completely determined* process. Processes for which state y is not completely determined by the state of a single time moment t but essentially depends on the changing of the state before t, could also be joined to completely determined processes. However one usually prefers to avoid this type of dependence on the previous system behavior and widens the concept of the system state at the moment t by introducing new parameters.[8]

Beyond the area of classical mechanics with completely deterministic processes, one often considers schemes where system state x at some moment t_0 is characterized only the known probability of occurrence of some possible state y in some coming moment $t > t_0$. If for all of given

[8]We have a well-known example of the application of this method for description of the state of a mechanical system with the help not only of coordinates of material points but also of components of its speeds. – *B. G.*

t_0, $t > t_0$, and x there exists the distribution function for state y, we say that this scheme is the scheme of a stochastically determined process. In general, this distribution function is represented as $P(t_0, x, t, A)$ where A is some set of states and P is the probability that at the moment t one of the states belonging to this set will be realized."

But general philosophical aspects of this work of Kolmogorov are not the main issue. This work laid a foundation for the theory of memoriless stochastic processes and differential equations (direct and inverse) controlling the transition processes were obtained. In this work were also given general ideas of memoriless jumping processes which were detailed and analyzed later by W. Feller and V.M. Dubrovsky.

Now the theory of Markov processes has come to power and forms a branch of mathematical science which has a number of applications in physics, engineering, geophysics, chemistry, and other areas of human knowledge.

The fundamentals of another class of stochastic processes on the basis of problems of physics have been developed by A.Ya. Khinchine. He introduced the aspect of the stochastic processes in wide and narrow senses and has a famous formula for autocorrelation. This work was a generating point for many works including that by H. Cramer, A. Wald, A.N. Kolmogorov, and others.

During the development of the theory of stochastic processes, a close aspects were notably divided. If a random variable $\xi(t)$ or a vector $\{\xi_1(t), \ldots, \xi_n(t)\}$ with values taken from numerical axes depends on a single real parameter t then one refers to stochastic process $\xi(t)$. The parameter, t, usually is called "time." If time takes a discrete sequence of meanings $t_1, y_2, \ldots$, then one refers to stochastic sequence rather than a stochastic process. If a random variable (or a vector) depends on not a single parameter but several of them, one refers to stochastic field.

Stochastic fields appeared first of all in biology and geophysics. Later it became clear that such situations are observed in almost all applications. One must consider not only stochastic processes but stochastic fields as well. Here are several examples.

Let $\rho(t, x, y, z)$ be the density of the water of an ocean. This value changes from one point to another and from one moment in time to another. Numerous observations show that ρ might be considered as a stochastic field.

Consider the changing of the strength and direction of the wind. For each point of space and for each moment of time the strength of the wind is a scalar value and the wind direction $\xi(t, x, y, z)$, $\eta(t, x, y, z)$, $\zeta(t, x, y, z)$ is a stoachatic vector. This is a typical example of scalar and vector fields.

The number of examples can be easily continued.

In the history of science one often meets a situation when some branch of the science has not yet been created but separate particular problems related to this still non-existing branch of science are beginning to be solved. Such a situation we observed in arithmetics and geometry, algebra and the theory of numbers. The same situation we meet in the theory of stochastic processes . There were no theories, no related concepts, not even an idea of time dependence of random variables, but particular problems were successfully analyzed and solved. This is an appropriate place to mention Nicholas Bernoulli, Montmort, and DeMoivre who dealt with the problem of a gambler ruin and the state of gamblers after N parties before the game's finish. These were typical problems of the theory of stochastic processes where the role of time was played by the number of completed parties. The same situation arises in the Laplace problem concerning one-by-one replacement of balls from one box to another where the boxes' contents after N replacements are computed. The new is always born from the old and as time passes the necessity of a new theory becomes a demand of life. Sometimes it takes years and years to appear.

The theory of probabilities has a long and didactic history. It visually shows how different main concepts and methods were developed in accordance with the needs of practice. The history of the theory of probabilities is far from completed. One needs a systematic work to restore the traversed path and to render homage to its creators. This will show us how initial guesses became knowledge, and how the creation of the probability theory led from a strong deterministic understanding to wider probabilistic concepts which allowed thinkers to penetrate deeper into the essence of the nature of the world.

The theory of probabilities dynamically continues to develop. New branches appear: optimal control of stochastic processes, the theory of martingales, the theory of diffusion, stochastic operators, probabilistic concepts for algebraic and topological structures. All of these directions are of great interest in both theoretical and applied senses. This historical essay is restricted by the beginning of the 40s of the twentieth century. Only some remarks concern more recent years. I hope that the problem of the history of the theory of probabilities will be a challenge to some of the readers and they will develop and add to this essay in various directions.

TABLES OF FUNCTION VALUES

Table 1 *Table of values of the function* $\varphi(x) = (1/\sqrt{2\pi})\, e^{-x^2/2}$

x	0	1	2	3	4	5	6	7	8	9
0.0	0.3989	0.3989	0.3989	0.3988	0.3986	0.3984	0.3982	0.3980	0.3977	0.3973
0.1	0.3970	0.3965	0.3961	0.3956	0.3951	0.3945	0.3939	0.3932	0.3925	0.3918
0.2	0.3910	0.3902	0.3894	0.3885	0.3876	0.3867	0.3857	0.3847	0.3836	0.3825
0.3	0.3814	0.3802	0.3790	0.3778	0.3765	0.3752	0.3739	0.3726	0.3712	0.3697
0.4	0.3683	0.3668	0.3653	0.3637	0.3621	0.3605	0.3589	0.3572	0.3555	0.3538
0.5	0.3521	0.3503	0.3485	0.3467	0.3448	0.3429	0.3410	0.3391	0.3372	0.3352
0.6	0.3332	0.3312	0.3292	0.3271	0.3251	0.3230	0.3209	0.3187	0.3166	0.3144
0.7	0.3123	0.3101	0.3079	0.3056	0.3034	0.3011	0.2989	0.2966	0.2943	0.2920
0.8	0.2897	0.2874	0.2850	0.2827	0.2803	0.2780	0.2756	0.2732	0.2709	0.2685
0.9	0.2661	0.2637	0.2613	0.2589	0.2565	0.2541	0.2516	0.2492	0.2468	0.2444
1.0	0.2420	0.2396	0.2371	0.2347	0.2323	0.2299	0.2275	0.2251	0.2227	0.2203
1.1	0.2179	0.2155	0.2131	0.2107	0.2083	0.2059	0.2036	0.2012	0.1989	0.1965
1.2	0.1942	0.1919	0.1895	0.1872	0.1849	0.1826	0.1804	0.1781	0.1758	0.1736
1.3	0.1714	0.1691	0.1669	0.1647	0.1626	0.1604	0.1582	0.1561	0.1539	0.1518
1.4	0.1497	0.1476	0.1456	0.1435	0.1415	0.1394	0.1374	0.1354	0.1334	0.1315
1.5	0.1295	0.1276	0.1257	0.1238	0.1219	0.1200	0.1182	0.1163	0.1145	0.1127
1.6	0.1109	0.1092	0.1074	0.1057	0.1040	0.1023	0.1006	0.0989	0.0973	0.0957
1.7	0.0940	0.0925	0.0909	0.0893	0.0878	0.0863	0.0848	0.0833	0.0818	0.0804
1.8	0.0790	0.0775	0.0761	0.0748	0.0734	0.0721	0.0707	0.0694	0.0681	0.0669
1.9	0.0656	0.0644	0.0632	0.0620	0.0608	0.0596	0.0584	0.0573	0.0562	0.0551
2.0	0.0540	0.0529	0.0519	0.0508	0.0498	0.0488	0.0478	0.0468	0.0459	0.0449
2.1	0.0440	0.0431	0.0422	0.0413	0.0404	0.0396	0.0387	0.0379	0.0371	0.0363
2.2	0.0355	0.0347	0.0339	0.0332	0.0325	0.0317	0.0310	0.0303	0.0297	0.0290
2.3	0.0283	0.0277	0.0270	0.0264	0.0258	0.0262	0.0246	0.0241	0.0235	0.0229
2.4	0.0224	0.0219	0.0213	0.0208	0.0203	0.0198	0.0194	0.0189	0.0184	0.0180
2.5	0.0175	0.0171	0.0167	0.0163	0.0158	0.0154	0.0151	0.0147	0.0143	0.0139
2.6	0.0136	0.0132	0.0129	0.0126	0.0122	0.0119	0.0116	0.0113.	0.0110	0.0107
2.7	0.0104	0.0101	0.0099	0.0096	0.0093	0.0091	0.0088	0.0086	0.0084	0.0081
2.8	0.0079	0.0077	0.0075	0.0073	0.0071	0.0069	0.0067	0.0065	0.0063	0.0061
2.9	0.0060	0.0058	0.0056	0.0055	0.0053	0.0051	0.0050	0.0048	0.0047	0.0046
3.0	0.0044	0.0043	0.0042	0.0040	0.0039	0.0038	0.0037	0.0036	0.0035	0.0034
3.1	0.0033	0.0032	0.0031	0.0030	0.0029	0.0028	0.0027	0.0026	0.0025	0.0025
3.2	0.0024	0.0023	0.0022	0.0022	0.0021	0.0020	0.0020	0.0019	0.0018	0.0018
3.3	0.0017	0.0017	0.0016	0.0016	0.0015	0.0015	0.0014	0.0014	0.0013	0.0013
3.4	0.0012	0.0012	0.0012	0.0011	0.0011	0.0010	0.0010	0.0010	0.0009	0.0009
3.5	0.0009	0.0008	0.0008	0.0008	0.0008	0.0007	0.0007	0.0007	0.0007	0.0006
3.6	0.0006	0.0006	0.0006	0.0005	0.0005	0.0005	0.0005	0.0005	0.0005	0.0004

Table 1 (Continued).

x	0	1	2	3	4	5	6	7	8	9
3.7	0.0004	0.0004	0.0004	0.0004	0.0004	0.0004	0.0003	0.0003	0.0003	0.0003
3.8	0.0003	0.0003	0.0003	0.0003	0.0003	0.0002	0.0002	0.0002	0.0002	0.0002
3.9	0.0002	0.0002	0.0002	0.0002	0.0002	0.0002	0.0002	0.0002	0.0001	0.0001

Table 2 Table of the function $\Phi(x) = (1/\sqrt{2\pi})\int_{-\infty}^{x} e^{-z^2/2}\,dz$

x	0	1	2	3	4	5	6	7	8	9
0.0	0.0000	0.0040	0.0080	0.0120	0.0159	0.0199	0.0239	0.0279	0.0319	0.0359
0.1	0.0398	0.0438	0.0478	0.0517	0.0557	0.0596	0.0636	0.0675	0.0714	0.0753
0.2	0.0793	0.0832	0.0871	0.0909	0.0948	0.0987	0.1026	0.1103	0.1064	0.1141
0.3	0.1179	0.1217	0.1255	0.1293	0.1331	0.1368	0.1406	0.1443	0.1480	0.1517
0.4	0.1554	0.1591	0.1628	0.1664	0.1700	0.1736	0.1772	0.1808	0.1844	0.1879
0.5	0.1915	0.1950	0.1985	0.2019	0.2054	0.2088	0.2123	0.2157	0.2190	0.2224
0.6	0.2257	0.2291	0.2324	0.2356	0.2389	0.2421	0.2454	0.2486	0.2517	0.2549
0.7	0.2580	0.2611	0.2642	0.2673	0.2703	0.2734	0.2764	0.2793	0.2823	0.2852
0.8	0.2881	0.2910	0.2939	0.2967	0.2995	0.3023	0.3051	0.3078	0.3106	0.3133
0.9	0.3159	0.3186	0.3212	0.3238	0.3264	0.3289	0.3315	0.3340	0.3365	0.3389
1.0	0.3413	0.3437	0.3461	0.3485	0.3508	0.3531	0.3554	0.3577	0.3599	0.3621
1.1	0.3643	0.3665	0.3686	0.3708	0.3728	0.3749	0.3770	0.3790	0.3810	0.3830
1.2	0.3849	0.3869	0.3888	0.3906	0.3925	0.3943	0.3962	0.3980	0.3997	0.4015
1.3	0.4032	0.4049	0.4066	0.4082	0.4099	0.4115	0.4131	0.4147	0.4162	0.4177
1.4	0.4192	0.4207	0.4222	0.4236	0.4251	0.4265	0.4279	0.4292	0.4306	0.4319
1.5	0.4332	0.4345	0.4357	0.4370	0.4382	0.4394	0.4406	0.4418	0.4429	0.4441
1.6	0.4452	0.4463	0.4474	0.4484	0.4495	0.4505	0.4515	0.4525	0.4535	0.4545
1.7	0.4554	0.4564	0.4573	0.4582	0.4591	0.4599	0.4608	0.4616	0.4625	0.4633
1.8	0.4641	0.4648	0.4656	0.4664	0.4671	0.4678	0.4686	0.4692	0.4659	0.4706
1.9	0.4713	0.4719	0.4726	0.4732	0.4738	0.4744	0.4750	0.4756	0.4761	0.4767
2.0	0.4772	0.4778	0.4783	0.4788	0.4793	0.4798	0.4803	0.4808	0.4812	0.4817
2.1	0.4821	0.4826	0.4830	0.4834	0.4838	0.4842	0.4846	0.4850	0.4854	0.4857
2.2	0.4861	0.4864	0.4868	0.4871	0.4874	0.4878	0.4881	0.4884	0.4887	0.4890
2.3	0.4893	0.4896	0.4898	0.4901	0.4904	0.4906	0.4909	0.4911	0.4913	0.4916
2.4	0.4918	0.4920	0.4922	0.4924	0.4927	0.4929	0.4930	0.4932	0.4934	0.4936
2.5	0.4938	0.4940	0.4941	0.4943	0.4945	0.4946	0.4948	0.4949	0.4951	0.4952
2.6	0.4953	0.4955	0.4956	0.4957	0.4958	0.4960	0.4961	0.4962	0.4963	0.4964
2.7	0.4965	0.4966	0.4967	0.4968	0.4969	0.4970	0.4971	0.4972	0.4973	0.4974
2.8	0.4974	0.4975	0.4976	0.4977	0.4977	0.4978	0.4979	0.4979	0.4980	0.4981
2.9	0.4981	0.4982	0.4982	0.4983	0.4984	0.4984	0.4985	0.4985	0.4986	0.4986
3.0	0.4986									
3.1	0.4990									
3.2	0.49931									
3.3	0.49952									
3.4	0.49966									
3.5	0.4998									
3.6	0.4998									
3.7	0.49989									
3.8	0.49993									
3.9	0.49995									
4.0	0.499968									
4.5	0.499997									
5.0	0.49999997									

Table 3 Table of values of the function $P_k = a^k/k!\,e^{-a}$

$k\backslash a$	0.1	0.2	0.3	0.4	0.5	0.6
0	0.904837	0.818731	0.740818	0.670320	0.606531	0.548812
1	0.090484	0.163746	0.222245	0.268128	0.303265	0.329287
2	0.004524	0.016375	0.033337	0.053626	0.075816	0.098786
3	0.000151	0.001091	0.003334	0.007150	0.012636	0.019757
4	0.000004	0.000055	0.000250	0.000715	0.001580	0.002964
5		0.000002	0.000015	0.000057	0.000158	0.000356
6			0.000001	0.000004	0.000013	0.000035
7					0.000001	0.000003

$k\backslash a$	0.7	0.8	0.9	1.0	2.0	3.0
0	0.496585	0.449329	0.406570	0.367879	0.135335	0.049787
1	0.347610	0.359463	0.365913	0.367879	0.270671	0.149361
2	0.121663	0.143785	0.164661	0.183940	0.270671	0.224042
3	0.028388	0.038343	0.049398	0.061313	0.180447	0.224042
4	0.004968	0.007669	0.011115	0.015328	0.090224	0.168031
5	0.000695	0.001227	0.002001	0.003066	0.036089	0.100819
6	0.000081	0.000164	0.000300	0.000511	0.012030	0.050409
7	0.000008	0.000019	0.000039	0.000073	0.003437	0.021604
8		0.000002	0.000004	0.000009	0.000859	0.008101
9				0.000001	0.000191	0.002701
10					0.000038	0.000810
11					0.000007	0.000221
12					0.000001	0.000055
13						0.000013
14						0.000003
15						0.000001

$k\backslash a$	4.0	5.0	6.0	7.0	8.0	9.0
0	0.018316	0.006738	0.002479	0.000912	0.000335	0.000123
1	0.073263	0.033690	0.014873	0.006383	0.002684	0.001111
2	0.146525	0.084224	0.044618	0.022341	0.010735	0.004998
3	0.195367	0.140374	0.089235	0.052129	0.028626	0.014994
4	0.195367	0.175467	0.133853	0.091226	0.057252	0.033737
5	0.156293	0.175467	0.160623	0.127717	0.091604	0.060727
6	0.104194	0.146223	0.160623	0.149003	0.122138	0.091090
7	0.059540	0.104445	0.137677	0.149003	0.139587	0.117116
8	0.029770	0.065278	0.103258	0.130377	0.139587	0.131756
9	0.013231	0.036266	0.068838	0.101405	0.124077	0.131756
10	0.005292	0.018133	0.041303	0.070983	0.099262	0.118580
11	0.001925	0.008242	0.022529	0.045171	0.072190	0.097020
12	0.000642	0.003434	0.011262	0.026350	0.048127	0.072765

Table 3 (Contd.)

$k\backslash a$	4.0	5.0	6.0	7.0	8.0	9.0
13	0.000197	0.001321	0.005199	0.014188	0.029616	0.050376
14	0.000056	0.000472	0.002228	0.007094	0.016924	0.032384
15	0.000015	0.000157	0.000891	0.003311	0.009026	0.019431
16	0.000004	0.000049	0.000334	0.001448	0.004513	0.010930
17	0.000001	0.000014	0.000118	0.000596	0.002124	0.005786
18		0.000004	0.000039	0.000232	0.000944	0.002893
19		0.000001	0.000012	0.000085	0.000397	0.001370
20			0.000004	0.000030	0.000159	0.000617
21			0.000001	0.000010	0.000061	0.000264
22				0.000003	0.000022	0.000108
23				0.000001	0.000008	0.000042
24					0.000003	0.000016
25					0.000001	0.000006
26						0.000002
27						0.000001

Table 4 Table of values of the function $\sum_{m=0}^{k} (a^m/m!)e^{-a}$

$k\backslash a$	0.1	0.2	0.3 ·	0.4	0.5	0.6
0	0.904837	0.818731	0.740818	0.670320	0.606531	0.548812
1	0.995321	0.982477	0.963063	0.938448	0.909796	0.878099
2	0.999845	0.998852	0.996390	0.992074	0.985612	0.977885
3	0.999996	0.999943	0.999724	0.999224	0.998248	0.997642
4	1.000000	0.999998	0.999974	0.999939	0.999828	0.999606
5	1.000000	1.000000	0.999999	0.999996	0.999986	0.999962
6	1.000000	1.000000	1.000000	1.000000	0.999999	0.999997
7	1.000000	1.000000	1.000000	1.000000	1.000000	1.000000

$k\backslash a$	0.7	0.8	0.9	1.0	2.0	3.0
0	0.496585	0.449329	0.406570	0.367879	0.135335	0.049787
1	0.844195	0.808792	0.772483	0.735759	0.406006	0.199148
2	0.965858	0.952577	0.937144	0.919699	0.676677	0.423190
3	0.994246	0.990920	0.988542	0.981012	0.857124	0.647232
4	0.999214	0.998589	0.997657	0.996340	0.947348	0.815263
5	0.999909	0.999816	0.999658	0.999406	0.983437	0.916082
6	0.999990	0.999980	0.999958	0.999917	0.995467	0.966491
7	0.999998	0.999999	0.999997	0.999990	0.998904	0.988095
8	1.000000	1.000000	1.000000	0.999999	0.999763	0.996196
9				1.000000	0.999954	0.998897
10					0.999992	0.999707
11					0.999999	0.999928
12					1.000000	0.999983
13						0.999996
14						0.999999
15						0.100000

REFERENCES

Borovkov, A. A. Probability Theory (Russian). 2nd ed., Moscow, Nauka, 1986

Doob, J. L. Stochastic Processes, New York, Wiley, 1953

Feller, W. An Introduction to Probability Theory and Its Applications, Vol. 1 (3rd ed.), New York, Wiley

Gikhman, I. I., and Skorokhod, A. V. Introduction into the Theory of Stochastic Processes (Russian). Moscow, Nauka, 1977

Gikhman, I. I., Skorokhod, A. V., and Yadrenko, M. I. Theory of Probability and Mathematical Statistics (Russian). Kiev, Vystcha Shkola, 1979

Gnedenko, B. V., and Kolmogorov, A. N. Limit Distributions for Sums of Independent Random variables. Reading, Mass, Addision–Wesley, 1954

Klimov, G. P. Probability Theory and Mathematical Statistics (Russian). Moscow, Moscow State University, 1983

Klimov, G. P., and Kuz'min, A. D. Probability, Processes, Statistics: Problems and Solutions (Russian). Moscow, Moscow State University, 1985

Korolyuk, V. S., ed. Handbook of Probability Theory and Mathematical Statistics (Russian). Kiev, Naukova Dumka, 1978

Kovalenko, I. N., and Filippova, A. A. Probability Theory and Mathematical Statistics (Russian). Moscow, Vyshaya Shkola, 1973

Kovalenko, I. N., Kjuznetsov, N. Yu., and Shurenkov, V. M. Stochastic Processes: Handbook (Russian). Kiev, Naukova Dumka, 1983

Loev, M. Probability Theory (3rd Ed.), New York, D. Van Nostrand Co., Inc.

Lukacs, E. Characteristic Functions. London, Charles Griffin and Co., Ltd.

Maystrov, L. E. Developing of Probability Conception (Russian). Moscow, Nauka, 1980

Meshalkin, L. D. Collection of Problems on Probability Theory (Russian). Moscow, Moscow State University, 1963

Prokhorov, Yu. V., and Rozanov, Yu. A. Probability Theory (Russian), 2nd ed. Moscow, Nauka, 1985

Rozanov, Yu. A. Stochastic Processes (Russian). Moscow, Nauka, 1971

Savel'ev, L. Ya. Combinatorics and Probability (Russian). Novosibirsk, Nauka, 1975

Sevast'yanov, B. A. Course of Probability Theory and Mathematical Statistics (Russian). Moscow, Nauka, 1982

Sevast'yanov, B. A., Chistyakov, V. P., and Zubkov, A. M. Collection of Problems on Probability Theory (Russian). Moscow, Nauka, 1982

Shiryaev, A. N. Probability (Russian). Moscow, Nauka, 1980

Ventsel, A. D. Course of the Theory of Stochastic Processes (Russian). Moscow, Nauka, 1975

Index

Adrain, R., 456, 457
Algebra of sets, 38
Arbuthnot, D., 431
Arbuthnot, J., 438
  published works by, 438
'At random', 19, 20
Axiom(s)
  of addition, extended, 42–43
  of continuity, 42–43
  of Kolmogorov, 40, 41
Axiom(s) of probability theory, 39–43
  incompleteness of, 41

Baldericus, 414
Banach's match box problem, 102
Barbier, J., 441
Batiste, Jean, 427
Bavli, G. M., 472
Bayes' formula, 50–51
  and conditional distribution functions, 333–337
'Bayes Rule', 447
Bayes' Theorem, 447
  classical, 336
Bayes, Thomas, 9, 50, 445, 446
  published works by, 445
Belyaev, Yu. K., 454
Bernoulli, D., 455, 458
  published works by, 455
Bernoulli, Jacob, 1, 40, 63, 85, 199, 295, 435, 436, 437, 444, 445, 448, 449, 450, 458, 461, 462, 475
  published works by, 435, 436, 437, 450, 475
Bernoulli, Johann, 449
Bernoulli, Nicholas, 444, 448, 449, 450, 458, 475, 476, 483
  published works by, 450, 475
Bernoulli distribution, 272
Bernoulli formula, 65, 313, 444, 454
Bernoulli trials, 63, 64, 84, 179, 473
  examples of, 65–68
Bernoulli's Theorem, 85, 196–197
  for repeated independent trials, 89–90
  See also Chebyshev's Theorem
Bernshtein, Sergei N., 2, 38, 47, 468
Bernshtein's Theorem, 218
Bertrand, Joseph L. F., 29, 443
  published works by, 443
Bertrand's paradox, 29–31
Bessel, F. V., 465
  published works by, 465
Binomial distribution law, 65
Birkhoff, George, 373
Birkhoff-Khinchine Ergodic Theorem, 373–378
Birth process, 322–333, 478, 479
Blanc-Lapierre, 373
Blaschke, V., 442
  published works by, 442
Bobrov, A. A., 463, 468
Bobynin, V., 418
Bochner, 241
Bochner-Khinchine Theorem, 243–247
Boltzmann statistics, 21–22
Borel, Emaile, 203, 462
Borel's Theorem, 212–213
Bose-Einstein statistics, 22–23
Bouhiakowsky, V., 2, 441
  published works by, 441
Bouniakowsky-Cauchy inequality, 167, 247
Bounyakovsky. *See* Bouniakowsky
Brahe, Ticho, 454, 455
Breslau, 434
Buffon, George, 438, 439, 440
  published works by, 439
Buffon's needle problem, 31–32

Cantelli, 462, 463
Cantor function, 124
Cardano, Girolamo, 444, 445
  investigations by, 416–419
  published works by, 416, 417, 418, 444
Carlin, 329
Cauchy, A., 457
Cauchy, O., 469, 471
Cauchy distribution, 469
  See also variance
Cauchy Law, 144, 145, 469

Cauchy-Bouniakowsky inequality, 167, 247
Cauchy-Schwarz inequality, 167
Central Limit Theorem, 463–468
Certain events, 12
Characteristic functions, 219–259
 real, 229–230
 definition and properties of, 219–225
 examples of, 223–225
 of multi-dimensional random variables, 248–253
Chebyshev, P. L., 2, 4, 193, 199, 264, 295, 457, 460, 461, 462, 465, 466, 477
 published works by, 461, 465, 477
Chebyshev's inequality, 194–195
Chebyshev's Theorem, 195–196, 466
 special cases of, 196–198
 See also Markov's Theorem
 See also Poisson's Theorem
Classical Limit Theorem, 263–278
Complementary events, 11, 13
Complete group
 of events, 13
 of pairwise mutually exclusive events, 13
Conditional probability, and independence of events, 46
Confidence limits, 396–403
Confidence probabilities, 396–403
Converse Limit Theorem, 238–240
Correlation coefficient, 171
Cotes, R., 455, 458
Cramer, H., 373, 482
Cramer's Theorem, 140
Critical region, 405
 best, 409
Crofton, M., 441, 442, 443
 published works by, 442
Cumulants of k-th order, 186, 222, 223
Czuber, 461, 477
 published works by, 461

D'Alembert, 438
de Finetti, Bruno, 472
de Fournival, Richard, 414, 415, 421
de Mere, Chevalier, 423, 424
 problem of, 61
Death process, 322–333, 478, 479
Decomposibility of an event into mutually exclusive events, 13, 15
Deltail, 442
DeMoivre, A., 1, 46, 437, 438, 445, 446, 448, 449, 451, 452, 458, 480, 483
 published works by, 438, 445, 446, 448, 452
DeMoivre Theorem, 463
DeMoivre-Laplace formula, asymptotic, 74
DeMoivre-Laplace Local Limit Theorem, generalization of, 275–278
DeMoivre-Laplace Theorem, 80, 263
 problems leading to, 85–87
 See also Limit Theorems
DeMoivre-Stirling formula, 451
Density function, 392–393
 of n-dimensional uniform distribution, 130
 of two-dimensional normal distribution, 130
Deterministic events, 8
Deterministic laws, 5
Deterministic phenomena, 8
Diagram, Venn, 11, 12
Difference of events, 11
Diffusion coefficient, 313
d'Imola, Benvenuto, 415
Direct Limit Theorem, 237
Discrete random variable, lattice distribution of, 272
Discrete spectrum, 367
Discrete variable, 121
Distribution
 Bernoulli, 272
 Cauchy, 469
 continuous, 120–125
 discrete, 120–125
 empirical, 210
 exponential, 254
 gamma, 254
 geometrical, 254
 infinitely divisible, 281–308
 Laplace, 187
 lattice, 272
 Maxwell, 134, 142, 187
 normal, 115, 122–124, 130, 133, 383
 Pascal, 186
 Poisson, 115, 254, 272, 395

Polya, 186
Simpson, 137
stable, 470
Student's, 144–145
threshold, 253
χ^2, 140
Distribution function, 114, 129
as a step function, 121
conditional, 125
constant, 121
continuous, 119
multi-dimensional, 125–134
no-dimensional, 125
of a sum, 135–136
unconditional, 163
Distribution parameters, classical procedure for estimation of, 383–394
Distribution span, 273–274
maximal, 274
Döbline, W., 470
Drift coefficient, 313
Dubrovsky, V. M., 482

Einstein, A., 315, 479
Elementary events, 9
Elementary outcome, 9
Elementary systems, of random variables, 295
Ellipses of equal probability, 130
Empirical distribution function, 210
Empty set, 10
Equiprobable (equally likely) events, 9
Equivalent events, 11
Erlang, A. K., 327, 478, 479
Erlang formulas, 329, 331
Errors of observations, 476
Euler, L., 455
Event(s)
certain, 12
collectively independent, 47
complete group of, 13
complete group of pairwise mutually exclusive, 13
complimentary, 11, 13
dnecomposability into mutually exclusive, 13, 15
deterministic, 8
difference of, 11
elementary, 9
equiprobable (equally likely), 9
equivalent, 11
field of, 14
implication of, 11
impossible, 13
mutually exclusive, 9, 13
probabilistic, 7
probability of, 9, 16
product (or intersection) of, 11, 12
simultaneous, 11
stochastic, 7
sum (or union) of, 11, 12
true, 8
uncertain, 8
unique, 7
Expectation. *See* mathematical expectation
Expected value. *See* expectation
Extended axiom of probabilities addition, 41, 42, 43, 126

Feller, W., 302, 308, 329, 350, 464, 467, 468, 474, 481, 482
published works by, 481
Feller's Theorem, 324–325
Fermat, Pierre, 1, 428, 443, 445, 475
contributions to development of probability theory by, 422–427
Fermi-Dirac statistics, 24
Field of events, 14
First Helly's Theorem, 232–234
First Kolmogorov equation, 340–342
Fisher, R., 394
Fokker, 312, 343, 480
Fokker-Planck equation, 314
Franc-Careau game, 440
Functions, positive semi-definite, 241–247

Galilei, Galileo, 454, 458
investigations by, 420–422
published works by, 420, 422
'Gambler ruin problem', 52–54, 447–449
Gauss, Karl, 1, 456, 457, 458, 459, 460, 465, 476, 480
published works by, 456
Gaussian distribution law, 465
Generalised Second Theorem of Helly, 235–237

Glivenko, V. I., 463
Glivenko's Theorem, 210–217
Gnedenko, B. V., 165, 278, 281, 303, 463, 472, 475
published works by, 472
Gosset, W. L., 145
Graunt, John, 432, 433, 434, 435, 436
published works by, 432

Halley, Edmond, 434, 435
published works by, 434
Hardy, G., 473
Hartmann, 474
Hausdorff, F., 473
Helly's Theorems, 231–237
Herglots' Theorem, 245
Huygens, Christian, 1, 52, 413, 427–432, 435, 438, 443, 448, 449, 475
published works by, 428, 435, 438, 448

Impossible events, 13
probability of, 16
Infinitely divisible distributions
canonical representation of, 285–290
characteristic function of, 283–284
fundamental properties of, 282–285
Limit Theorem for, 291–294
theory of, 281–308
Integral Limit Theorem, 75–84
of repeated independent trials, 80–82
Integral of stochastic process, 369
Integral, Stieltjes, 147–153, 371
Integral Theorem, 452
Integral Theorem of DeMoivre-Laplace, 76–80
applications of, 84–90
Intersection of events, 11, 12
Inversion formula for characteristic function, 225–228

Karhunen, 373
Kepler, Johann, 454
Khinchine, A. Ya., 2, 37, 199, 241, 314, 363, 373, 463, 464, 468, 470, 471, 472, 473, 474, 478, 480, 482
published works by, 480
Khinchine's Theorem, on correlation coefficient, 363–368
Kolmogorov, A. N., 2, 38, 43, 165, 199, 207, 278, 281, 303, 314, 339, 350, 359, 373, 394, 454, 461, 462, 472, 474, 478, 480, 481, 482
published works by, 461, 462, 472, 480
Kolmogorov-Feller equations, 348–356
Kolmogorov's axioms, 40, 41
Kolmogorov's equations, 339–348
special case of, 347–348
Kolmogorov's inequality, 203–205
Kolmogorov's Theorem, 205–207, 285–289
See also Borel's Theorem
Krein, M., 260
Kuzmin, R. O., 460

L-class, characterization property of laws of, 471
Lagrange, J., 456
Lambert, I., 456
Lame, G., 441
Laplace, P., 1, 48, 50, 89, 438, 440, 441, 447, 448, 451, 456, 458, 459, 463, 464, 465, 476, 480
published works by, 438, 440, 447, 451, 459, 463–464, 476
Laplace distribution, 187, 307
Laplace-Stieltjes transform, 253–259
Law of Double Logarithm, 473, 474, 475
Law of Large Numbers, 85, 191–217, 461–463
Chebyshev's form of, 194–199
necessary and sufficient condition for, 199–203
Law of Repetitive Logarithm, 473–475
Legendre, Adrien, 456, 457, 458, 465, 480
published works by, 456
Levy, P., 2, 460, 468, 469, 470, 471, 472, 474
published works by, 469
Liapounov, A. M., 264, 266, 460, 466, 467
published works by, 460
Liapounov condition, 272
Liapounov's Theorem, 271–272

Limit distribution, 472
Limit Theorems
appearance of, 449–452
central, 463–468
classical, 263–278
converse, 238–240
direct, 237
for characteristic functions, 237–241
integral, 75–84
See also infinitely divisible distributions
Limit Theorems for sums, 296–300
formulation of problem, 294–295
Lindeberg, 464, 467
condition, 266, 267, 301–302
Lindeberg's Theorem, 266–272, 468
Littlewood, J., 473
Lobachevsky, Nikolai, 1, 137–138
Local DeMoivre-Laplace Theorem, 68–70
See also Local Theorem for polynomial distribution
Local Limit Theorem, 68–75, 272–278
Local Theorem, 452
for polynomial distribution, 70–71
Loew, 373
Lyapounov. *See* Liapounov

MacLaurin expansion, 241
Maistrov, L. P., 454
published works by, 454
Makeham's formula, 57
Marcinkiewicz, 474
Markov, A. A., 2, 105, 111, 193, 199, 264, 266, 295, 457, 466, 467, 480
Markov chains, 105–111, 315, 338
definition of, 105–106
homogeneous, 106
Markov processes, 314, 315, 481
Markov's condition, for pairwise independent random variables, 199
Markov's equation, 107
generalised, 337–339
Markov's Theorem, 198–199, 218
Martikainen, A. I., 475
Mass production, quality control of, 452–454
Mass phenomena. *See also* Law of Large Numbers
Mathematical expectation, 159–166, 219, 222, 428, 475
conditional, 163
theorems on, 172–179
Mathematical statistics, problems of, 379–382
Matrix of transition probabilities, 106–107
Maxwell, J. C., 459
Maxwell distribution, 134, 142, 187
McGregor, 330
Mean, formation of concept of, 475–478
Median life length, 434
Module, 452
Moments, 179–186
absolute, 180
central, 179
first-order, 222
initial, 179
method of, 466
of order k, 179–186
problem of, 185
Montmort, P., 437, 445, 448, 449, 458
published works by, 437, 449
Most probable value, 67
Multi-dimensional random variables, characteristic functions of, 248–253
Multinomial distribution law, 65
Mutually exclusive events, 9, 13

Normal distribution, 115, 122–124, 130, 133, 383
of two-dimensional random variable, 146
See also characteristic functions
See also infinitely divisible distributions
See also mathematical expectation
See also moments
See also variance
Normal law
one-dimensional, 128
two-dimensional, 128
Normalised deviation, 178

Ore, O., 418
Ostrogradsky, M. V., 453
published works by, 453

Pacioli, Luke, 415, 416, 418
published works by, 415
Pascal, Blaise, 1, 428, 443, 444, 445, 475
contribution to development of probability theory by, 422–427
Pascal distribution, 186
Peres, M. T. Lorino, 442
Petrov, V. V., 474
Petty, William, 432, 433, 434, 435, 436
published works by, 433
Planck, M., 312, 343, 480
Poincare, Henry, 461, 465, 477
published works by, 461, 465
Poisson, S., 1, 90, 199, 272, 457, 460, 464, 465, 469, 471, 472, 479
published works by, 460, 461, 464, 469
Poisson distribution, 115, 254, 272, 395
See also infinitely divisible distributions
See also mathematical expectation
Poisson integral, 123
Poisson process, 315–322, 480
See also Birth and Death processes
Poisson scheme, 474
Poisson's law, 92, 229, 321, 354
Poisson's Theorem, 90–95, 197
See also Chebyshev's Theorem
Polya distribution, 186
Probability
axioms of, 39
classical definition of, 816, 435–438
classical definition, difficulties of, 34
classical determination of, 417
conditional, 43–52
confidence, 396–403
formula of total, 48
geometrical, 26–34, 438–443
of transition, 106
space, 43
subjective, 8
unconditional, 44
unknown, statistical estimation of, 34–37
von Mises' definition of, 35–37
Probability density function, 122, 130, 455
Probability function, 14–15
Probability theory
axiomatic construction of, 37–43
basic theorems of, 443–447
history of, 413–484
Probable life length, 434
Problem
Banach's match box, 102
Buffon's needle, 31–32
of Chevalier de Mere, 61
of gambler's ruin, 52–54, 447–449
Process(es)
of Birth, 322–333, 478, 479
of Death, 322–333, 478, 479
stationary, 316
without aftereffect, 316
Product of events, 11, 12
Prokhorov, Yu. V., 462
published works by, 462
Pure Birth, 323
Pure Death, 323

Quetelet, A., 459, 477

Raikov, D. A., 229, 468
'Random error', 458
Random events, intuitive understanding of, 58
Random phenomena, 5–6
Random process. *See* stochastic process
Random variable(s)
discrete, 121
distribution law of, 120
formation of concept of, 454–478
general limit distribution for sum of, 469–473
independent, 132
multi-dimensional, 125, 127
n-dimensional, 125
numerical characteristics of, 159–186
Random variable(s), two-dimensional
normally distributed, 127
See also mathematical expectation
See also variance
Random vector, 125
uniformly distributed, 127
Random walk, 95–99
Reuter, 329
Roberval, 426
Romanovsky, V. I., 454

Sample space, 8–16
Sarymsakov, T. A., 474
Second Helly's Theorem, 234–235
Second Kolmogorov equation, 343–345
Semi-invariant (or cumulants) of k-th order, 186, 222, 223
Silvester, D., 441, 442
Simpson, T., 439, 452, 453, 455, 458
  published works by, 439, 452, 455
Simpson distribution, 137
Simultaneous events, 11
Sirazhdinov, S.H., 454
Slutsky, E., 2, 368, 478, 480
Smoluchowsky, M., 479
Smolukhovsky, M., 315, 321
Spinoza, Benedict, 431
  published works by, 431
Stable distribution, 470
Stationary process, 316
  stochastic integral spectral decomposition of, 369–373
Statistical hypothesis, test of, 403–411
Statistical stability, 6–7
Statistics
  Boltzmann, 21–22
  Bose-Einstein, 22–23
  exhaustive, 394–395
  Fermi-Dirac, 24
Stieltjes integral, 147–153
  improper, 148
  stochastic, 371
Stirling, James, 451
Stirling's formula, 20, 69, 390, 451
Stochastic events, 7
Stochastic process(es)
  as process without aftereffect, 314, 315
  continuous, 339–348
  definition of, 315
  history of, 478–484
  homogeneous, with independent increments, 357–362
  integral of, 369
  purely discontinuous, 348–356
  stationary, 363–368
  theory of, 311–378
Strengthened Law of Large Numbers, 462
Strong Law of Large Numbers, 203–209
'Student's' distribution, 144–145
'Student's' law, 145
Sum of events, 11, 12
'Systematic error', 458

Tartaglia, Niccolo
  investigations by, 416–419
  published works by, 418
Taylor's formula, 344
Theorem of probabilities addition, 15, 48
Theorem of probabilities multiplication, 48
  of independent events, 45
Theorem on Limiting Probabilities, 108–111
Theory of Errors of Measurement, development of, 454–457
Transition matrix, 106–108

Uncertain events, 8
Unconditional distribution function, 163
Unconditional probability, 44
Uniform distribution
  mathematical expectation of, 162
  variance of, 168
Union of events, 11, 12
Unique events, 7
Uniqueness Theorem, 228–230

Value, most probable, 67
Van Schooten, Frantz, 428, 475
Variance, 166–172, 222
  formation of concept of, 475–478
  theorems on, 172–179
Variation sequence, 210
Vasilyev, A. V., 466
Venn diagrams, 11, 12
Vibold, Bishop, 414, 421
Volterra, Vito, 478
von Mises, R., 2, 35–37
  published works by, 37

Wald, A., 482
Wallhouse, 442
Whiteworth, 442, 443
  published works by, 442
Wiener, Norbert, 478, 480

Zygmund, 474

Printed in the United States
by Baker & Taylor Publisher Services